# Biowaste Treatment and Valorization

# Biowaste Treatment and Valorization

Editor

**Carlos Rico de la Hera**

MDPI • Basel • Beijing • Wuhan • Barcelona • Belgrade • Manchester • Tokyo • Cluj • Tianjin

*Editor*
Carlos Rico de la Hera
Universidad de Cantabria
Santander
Spain

*Editorial Office*
MDPI
St. Alban-Anlage 66
4052 Basel, Switzerland

This is a reprint of articles from the Special Issue published online in the open access journal *Applied Sciences* (ISSN 2076-3417) (available at: https://www.mdpi.com/journal/applsci/special_issues/biowaste_treatment_and_valorization).

For citation purposes, cite each article independently as indicated on the article page online and as indicated below:

LastName, A.A.; LastName, B.B.; LastName, C.C. Article Title. *Journal Name* **Year**, *Volume Number*, Page Range.

**ISBN 978-3-0365-6285-8 (Hbk)**
**ISBN 978-3-0365-6286-5 (PDF)**

Cover image courtesy of Carlos Rico de la Hera

# Contents

*applied sciences*

MDPI

*Editorial*

# Special Issue on Biowaste Treatment and Valorization

**Carlos Rico De La Hera**

Department of Water and Environmental Science and Technologies, Universidad de Cantabria, 39005 Santander, Spain; carlos.rico@unican.es

## 1. Introduction

Biowaste has been defined as "Biodegradable waste from gardens and parks, food and kitchen waste from homes, restaurants, collective catering services and retail establishments, and comparable waste from food processing plants" (Directive (2008)/98/EC (EC—European Commission, 2008). Biowaste includes food and kitchen waste from homes and small-size green waste, but also several types of organic wastes collected from different origins. The most significant benefits of the adequate management of biowaste, in addition to avoiding the emission of greenhouse gases, would be the production of good-quality compost or other organic products, as well as biogas that would contribute to improving soil quality and resource efficiency, together with a higher level of energy self-sufficiency. For the management of biodegradable waste diverted from landfills, there appear to be several environmentally friendly options. While the waste-management hierarchy also applies to biowaste management, in specific cases it may be justified to deviate from it, as the environmental balance of the various options available for managing biowaste depends on a number of local factors, such as the collection systems; the composition and quality of the waste; the climatic conditions; and the potential use of various products derived from waste, such as electricity, heat, and gas rich in methane or compost. Therefore, strategies for the management of this type of waste must be determined based on a structured and comprehensive approach, such as the life cycle concept.

**Citation:** Rico De La Hera, C. Special Issue on Biowaste Treatment and Valorization. *Appl. Sci.* **2022**, *12*, 11217. https://doi.org/10.3390/app122111217

Received: 3 November 2022
Accepted: 4 November 2022
Published: 5 November 2022



## 2. Strategies to Deal with Biowaste

In light of the above points, this Special Issue aims to collect the latest research on the relevant topics, the mature, conventional, and innovative technologies used to address the challenging issues, at present, with the treatment and valorization of different flows of biowaste. Fourteen papers have been published in this Special Issue. When revisiting past Special Issues, it can be observed that various topics have been addressed, mainly concerning biotreatment processes, such as anaerobic digestion, composting, and bio-augmentation, but also other areas about physical processes, such as evaporation, membrane separation, or the production of bio-adsorbents and other processes that promote the recovery of valuable products obtained from biowaste, such as proteins or volatile fatty acids.

Three studies address anaerobic digestion and biogas production. Sailer et al. [1] studied the anaerobic digestion of the organic fraction of municipal solid waste and digested sewage sludge at lowered temperatures, indicating that operation at 25 °C could be beneficial regarding energy input (heating costs) and output in terms of methane yield. Montes et al. [2] assessed the possibility of energetic valorization for solid wastes from alcoholic beverage production. They concluded that brewers' spent grains were a suitable feedstock for anaerobic digestion, whereas distilled gin spent botanicals presented problems for its mono-digestion due to the presence of toxic compounds for anaerobic digestion. The energy balance of lean-burn turbocharged engines using biogas as fuel for a wastewater treatment plant enhanced by thermal hydrolysis and co-digestion with cheese whey and microalgae was also reported [3]. Other biological treatments reported in the research included backyard composting [4]. The results of the research show that the quality properties of the composts are greatly influenced by the different techniques and raw materials used, showing that with a plan of basic education for composting, there is potential to encourage farmyard composting.

Some approximations for the recovery of other kinds of resources obtained from biowastes were also presented. Lara-Musule et al. [5] studied the feasibility of using multiscale analysis to diagnose and monitor the key variables in VFA production by the anaerobic treatment of raw-cheese whey. Facing the problem of the world's significant demand for protein in the subsequent few decades, an emerging alternative for the recovery of proteins from microalgae by means of biodegradable solvents was reviewed by Moldes et al. [6]. Protein separation by the use of membranes was also reported by Mazzei et al. [7]. In line with the biorefinery concept, García-Montalvo et al. [8] contributed with a study on the extraction of antioxidants from grape and apple pomace using different solvents of industrial interest, concluding that that ethanol/water mixtures are adequate solvents for the extraction of polyphenols due to their high efficiency and environmentally benign nature.

Other strategies used to enhance the biological processes for biowaste involve bioaugmentation. In their study, Ravi et al. [9] evaluated the efficacy of anaerobic fungi as a hydrolytic pretreatment of lignocellulosic biomass for agricultural biomass. Biyada et al. [10] developed the potential of autochthonous inoculums through bio-augmentation tests to improve the compost quality of textile waste. They observed a reduction in the composting time from 44 to only 12 weeks in comparison to classical composting treatments used for the assayed waste.

Emerging uses for biowaste have been proposed by other authors. For instance, Aguilar-Rosero et al. [11] presented a review on the production of bio-adsorbents obtained from agricultural wastes with the aim of being applied in wastewater treatment plants to remove toxic compounds. Ali et al. [12] studied the impact of biochar applications on the germination behavior and early growth of maize seedlings. They observed the biochar application to be an attractive approach to improve the initial phase of plant growth and provide better crop stand and essential, sustainable high yields.

On the other hand, two studies addressed with the treatment processes used for municipal landfill leachate and the wastes derived from olive oil production. Pandis et al. [13] proposed an easy way to manufacture tubular ceramic electrodes, coated with an oxygen-reduction catalyst, to be used in a single-chamber MFC for the treatment of municipal landfill leachate. Slama et al. [14] reported on the environmental threat of uncontrolled olive-mill wastewater and olive-mill pomace disposal in evaporation ponds.

## 3. Perspectives on the Future of Biowaste Treatment

Biowaste poses a considerable problem to the environment due to the high volume of waste generated worldwide, and its easy degradability when deposited in landfill sites, which in turn represent a great source of methane emissions. It must be acknowledged that landfilling is not a good management option for biowaste. Therefore, the future research should focus on treatment processes aiming to recover the resources contained in the biowaste, such as energy, nutrients, and organic compounds. Also on the use of biowaste to replace tools that are produced, at present, using other non-renewable materials.

**Funding:** This research received no external funding.

**Conflicts of Interest:** The authors declare no conflict of interest.

## References

1. Sailer, G.; Silberhorn, M.; Eichermüller, J.; Poetsch, J.; Pelz, S.; Oechsner, H.; Müller, J. Influence of Digester Temperature on Methane Yield of Organic Fraction of Municipal Solid Waste (OFMSW). *Appl. Sci.* **2021**, *11*, 2907. [CrossRef]
2. Montes, J.A.; Rico, C. Energetic Valorization of Solid Wastes from the Alcoholic Beverage Production Industry: Distilled Gin Spent Botanicals and Brewers' Spent Grains. *Appl. Sci.* **2021**, *11*, 10158. [CrossRef]

3. García-Cascallana, J.; Carrillo-Peña, D.; Morán, A.; Smith, R.; Gómez, X. Energy Balance of Turbocharged Engines Operating in a WWTP with Thermal Hydrolysis. Co-Digestion Provides the Full Plant Energy Demand. *Appl. Sci.* **2021**, *11*, 11103. [CrossRef]
4. Ujj, A.; Percsi, K.; Beres, A.; Aleksza, L.; Diaz, F.R.; Gyuricza, C.; Fogarassy, C. Analysis of Quality of Backyard Compost and Its Potential Utilization as a Circular Bio-Waste Source. *Appl. Sci.* **2021**, *11*, 4392. [CrossRef]
5. Lara-Musule, A.; Alvarez-Sanchez, E.; Trejo-Aguilar, G.; Acosta-Dominguez, L.; Puebla, H.; Hernandez-Martinez, E. Diagnosis and Monitoring of Volatile Fatty Acids Production from Raw Cheese Whey by Multiscale Time-Series Analysis. *Appl. Sci.* **2021**, *11*, 5803. [CrossRef]
6. Moldes, D.; Rojo, E.M.; Bolado, S.; García-Encina, P.A.; Comesaña-Gándara, B. Biodegradable Solvents: A Promising Tool to Recover Proteins from Microalgae. *Appl. Sci.* **2022**, *12*, 2391. [CrossRef]
7. Mazzei, R.; Szymczak, A.M.; Drioli, E.; Al-Fageeh, M.; Aljohi, M.A.; Giorno, L. High Purity of α-Lactalbumin from Binary Protein Mixture by Charged UF Membrane Far from the Isoelectric Point to Limit Fouling. *Appl. Sci.* **2021**, *11*, 9167. [CrossRef]
8. Garcia-Montalvo, J.; Garcia-Martín, A.; Ibañez Bujan, J.; Santos Mazorra, V.E.; Yustos Cuesta, P.; Bolivar, J.M.; Ladero, M. Extraction of Antioxidants from Grape and Apple Pomace: Solvent Selection and Process Kinetics. *Appl. Sci.* **2022**, *12*, 4901. [CrossRef]
9. Ravi, B.; Nkongndem Nkemka, V.; Hao, X.; Yanke, J.; McAllister, T.A.; Lee, H.; Veluchamy, C.; Gilroyed, B.H. Effect of Bioaugmentation with Anaerobic Fungi Isolated from Ruminants on the Hydrolysis of Corn Silage and *Phragmites australis*. *Appl. Sci.* **2021**, *11*, 9123. [CrossRef]
10. Biyada, S.; Imtara, H.; Elkarrach, K.; Laidi, O.; Saleh, A.; Al Kamaly, O.; Merzouki, M. Bio-Augmentation as an Emerging Strategy to Improve the Textile Compost Quality Using Identified Autochthonous Strains. *Appl. Sci.* **2022**, *12*, 3160. [CrossRef]
11. Aguilar-Rosero, J.; Urbina-López, M.E.; Rodríguez-González, B.E.; León-Villegas, S.X.; Luna-Cruz, I.E.; Cárdenas-Chávez, D.L. Development and Characterization of Bioadsorbents Derived from Different Agricultural Wastes for Water Reclamation: A Review. *Appl. Sci.* **2022**, *12*, 2740. [CrossRef]
12. Ali, L.; Xiukang, W.; Naveed, M.; Ashraf, S.; Nadeem, S.M.; Haider, F.U.; Mustafa, A. Impact of Biochar Application on Germination Behavior and Early Growth of Maize Seedlings: Insights from a Growth Room Experiment. *Appl. Sci.* **2021**, *11*, 11666. [CrossRef]
13. Pandis, P.K.; Kamperidis, T.; Bariamis, K.; Vlachos, I.; Argirusis, C.; Stathopoulos, V.N.; Lyberatos, G.; Tremouli, A. Comparative Study of Different Production Methods of Activated Carbon Cathodic Electrodes in Single Chamber MFC Treating Municipal Landfill Leachate. *Appl. Sci.* **2022**, *12*, 2991. [CrossRef]
14. Slama, H.B.; Chenari Bouket, A.; Alenezi, F.N.; Khardani, A.; Luptakova, L.; Vallat, A.; Oszako, T.; Rateb, M.E.; Belbahri, L. Olive Mill and Olive Pomace Evaporation Pond's By-Products: Toxic Level Determination and Role of Indigenous Microbiota in Toxicity Alleviation. *Appl. Sci.* **2021**, *11*, 5131. [CrossRef]

*applied sciences*

MDPI

Article

# Extraction of Antioxidants from Grape and Apple Pomace: Solvent Selection and Process Kinetics

Jorge Garcia-Montalvo †, Alberto Garcia-Martín †, Jon Ibañez Bujan, Victoria E. Santos Mazorra, Pedro Yustos Cuesta, Juan M. Bolivar and Miguel Ladero *

FQPIMA Group, Materials and Chemical Engineering Department, Chemical Science School, Complutense University of Madrid, 28040 Madrid, Spain; jorgar10@ucm.es (J.G.-M.); albega13@ucm.es (A.G.-M.); joniba01@ucm.es (J.I.B.); vesantos@ucm.es (V.E.S.M.); pyustosc@ucm.es (P.Y.C.); juanmbol@ucm.es (J.M.B.)

* Correspondence: mladerog@ucm.es; Tel.: +34-913-944-164

† These authors contributed equally to this work.

**Featured Application: Several promising aqueous mixtures are selected for the fast extraction of functional antioxidant mixtures from apple and grape pomaces.**

**Abstract:** Polyphenols have become a research target due to their antioxidant, anti-inflammatory and antimicrobial activity. Obtention via extraction from natural sources includes the revalorization of food wastes such as grape pomace (GP) or apple pomace (AP). In this work, GP and AP were submitted to a liquid–solid extraction using different solvents of industrial interest. Process kinetics were studied measuring the total phenolic content (TPC) and antioxidant capacity (AC), while the extraction liquor composition was analyzed employing chromatographic methods. Extraction processes using water-solvent mixtures stood out as the better options, with a particular preference for water 30%–ethanol 70% ($v/v$) at 90 °C, a mixture that quickly extracts up to 68.46 mg GAE/gds (Gallic Acid Equivalent per gram dry solid) and 122.67 TEAC/gds (TROLOX equivalent antioxidant capacity per gram dry solid) in case of GP, while ethylene water 10%–ethylene glycol 90% ($v/v$) at 70 °C allows to reach 27.19 mg GAE/gds and 27.45 TEAC/gds, in the case of AP. These extraction processes can be well-described by a second-order kinetic model that includes a solubility-related parameter for the first and fast-washing and two parameters for the slow mass transfer controlled second extraction phase. AP liquors were found to be rich in quercetin with different sugar moieties and GP extracts highlighted flavonols, cinnamic acids, and anthocyanins. Therefore, using identical extraction conditions for AP and GP and a comparative kinetic analysis of TPC and AC results for the first time, we concluded that ethanol/water mixtures are adequate solvents for polyphenols extraction due to their high efficiency and environmentally benign nature.

**Keywords:** phenolics; antioxidants; food wastes; green solvents; extraction kinetics; grape pomace; apple pomace; biorefinery

**Citation:** Garcia-Montalvo, J.; Garcia-Martín, A.; Ibañez Bujan, J.; Santos Mazorra, V.E.; Yustos Cuesta, P.; Bolivar, J.M.; Ladero, M. Extraction of Antioxidants from Grape and Apple Pomace: Solvent Selection and Process Kinetics. *Appl. Sci.* **2022**, *12*, 4901. https://doi.org/10.3390/app12104901

Academic Editors: Carlos Rico de la Hera and Dibyendu Sarkar

Received: 22 February 2022
Accepted: 10 May 2022
Published: 12 May 2022



## 1. Introduction

Nowadays, with the growth of the population, the scarcity of natural resources and the necessity of their reuse is one of the main concerns for society and, therefore, for scientists and engineers [1]. This problem is reflected in the United Nations Sustainable Development Goals from Agenda 2030. It can be applied in the first goal, which targets poverty, the second one aiming at zero hunger, and the third one worried about good health and well-being. In addition, the quantity of agro-industrial waste is increasing considerably in recent years, meaning more than 90 million tons of food waste created in the EU each year. This issue is generating an increasing problem of waste management and means a notable inefficiency in terms of water, energy, and food loss [2]. However, those residues, especially agro-food wastes, possess high industrial potential due to the presence of different polysaccharides (cellulose, hemicellulose, and pectin), polymers (lignin, proteins), antioxidant compounds, essential oils, etc. [3]. In the present work, we paid attention to two types of residues of importance in the agro-food sector: grape (GP) and apple pomaces (AP).

Cider production produces a waste called apple pomace. A mixture of pulp, peels, pips, and stems of the whole apple, which is produced in the peeling, slicing, and pressing operations in the production of cider. AP is rich in sugars and in certain types of phenolics such as quercetin. So, although the residue is not very rich in phenolics, the phenolics from AP have high industrial interest. Furthermore, in 2020, worldwide apple production accounted for 86.4 million of tons of waste. However, compared to other fruit wastes such as citrus peels, around 30% of the initial weight is lost in their processing in the case of apples. However, from the total apple production only 30–40% of them are processed for different purposes, principally for juice obtention prior to cider production. In Spain, the annual production in 2020 was of 522,000 metric tons according to FAO statistics [4]. If the aforementioned estimations are considered, apple waste production amounted to 156,000 tons in Spain in that year, which were used for animal feeding and compost production. Additionally, AP has a valuable composition of other polymeric compounds that are the main constituents: cellulose, hemicellulose, pectin, and lignin [5]. These relatively poor added value applications are a waste in the sense of the potential of this biomass as a source of sugars, polysaccharides, and antioxidant molecules. In any case, the recovery of the aforementioned chemicals requires the development of processes that respond to the needs of economical optimisation [6]. AP is, in particular, a natural source of interesting antioxidant molecules and, more specifically, phenolics. In these fruit residues, some examples of the different families of natural phenolics are found: hydroxybenzoic acids, cinnamic acids, and the group of flavonoids. Despite the variety of this kinds of molecules, apple-derived wastes are enriched in particular classes. Flavanols, for instance, catechin and epicatechin and their polymeric derivatives, as well as other abundant flavonoids belong to the flavonols group, which consists mainly of quercetin and its glucosides [7].

In the case of grape residues, they are created during the production of wine. The winemaking industry is the principal beverage industry worldwide according to FAO, reaching a production of 77.8 Mt [4], especially in Italy, Spain, and France [8]. Nearly half of grape production is destined for the winemaking process, predicted to reach 258 million hectoliters in 2020, according to the Organization of Vine and Wine (OIV) [8]. The vinification process is composed of different stages, so a wide variety of residues are produced like grape pomace (GP), grape stalks (GK), grape seeds (GS), wastewater and wine lees (WL), among others. Traditionally, grape-derived byproducts have been employed as dyes or fertilizers, or in the food industry. However, those residues have a higher potential, as all of those residues are rich in added value products such as oils, phenolic acids, and polymers such as cellulose, hemicellulose, or lignin. GP is a residue very rich in antioxidants as it comes from grapes, a natural source of these compounds. This residue is produced after vinification, a fermentation process where a partial extraction process takes place (around 40% of phenolics are extracted from grapes during vinification [9]). Thus, there is still a notable amount of remaining antioxidants in GP. Anthocyanins, flavonols, flavonoids, phenolic acids, and stilbenes are the most important phenolic compounds found in this residue [3].

Polyphenols are secondary metabolites produced by plants as a mechanism of defense with potential applications in the pharmaceutical, cosmetic, and food industries. The particularity of those molecules is their antioxidant activity that provides beneficial actions in human health, as they are able to reduce oxidative stress by donating hydrogen atoms or electrons, preventing atherosclerosis, cardiovascular diseases, acute hypertension, and diverse types of cancer. Those polyphenols are classified in carotenoids and phenolic compounds. The latter are classified in phenolic acids, flavonoids (that includes anthocyanins, flavanols, flavanones, and isoflavones), stilbenes, lignans, and tannins [10].

Polyphenols are obtained via extraction of different plant materials and plant-derived materials and phases, both solid and liquid. Most common methods are solid–liquid extraction, Soxhlet extraction, liquid–liquid extraction or maceration. However, those methods have some drawbacks such as requiring a large volume of extraction, low efficiency, and environmental problems due to the use of organic solvents. To avoid those disadvantages, unconventional methods have been developed: for instance, microwave-assisted extraction, ultrasound assisted extraction, and enzyme-assisted extraction, among others. In this case, the advantages are lower solvent requirements, higher yields, less extraction time, and better reproducibility. Still, unconventional methods have generally a higher cost and more complex operation conditions [6,11].

The aim of this work was to develop fast and efficient extraction processes for both wastes (AP and GP), leading to rich antioxidant liquors, by screening different solvents and conditions and to understand process efficiency by a comparative kinetic analysis of the total phenolic content (TPC) and antioxidant capacity evolution with extraction time. Furthermore, a compositional study was performed with the liquors obtained in the best conditions.

## 2. Materials and Methods

### 2.1. Chemicals

Diverse solvents were used as extractants: ethanol, acetone, ethylene glycol, propylene glycol, and choline chloride, all from analytical grade. Other reagents employed include Folin–Ciocalteu's phenol reagent (analytical grade, Chemical Lab, Belgium), 2,2-diphenyl 1 picrylhydrazyl radical (DPPH) (Analytical grade, Scharlab, Barcelona, Spain), gallic acid (Analytical grade, Sigma-Aldrich, St Louis, MO, USA), Rutin 97% (Acros, Waltham, MA, USA), and sodium carbonate anhydrous (Panreac, Barcelona, Spain). Additionally, formic acid (Scharlab, Barcelona, Spain) and Acetonitrile gradient 240 nM, Far UV (Scharlab, Barcelona, Spain) were employed as mobile phases.

### 2.2. Residues

Apple pomace (AP) residue from Reineta and Fuji varieties was provided by Covillasa S.A, Zaragoza, Spain. The AP was frozen and stored at −20 °C and thawed for the extraction process.

Grape pomace residue from *Vitis vinifera* was kindly provided by Pago de Carraovejas (Valladolid, Spain) coming from the 2019 harvest. At the time of reception, the grape samples were frozen at −20 °C and were subsequently freeze-dried using freeze-dryer Virtis Benchtop Pro Dried (Omnitronic, Perth, Australia). Then, samples were crushed using a blender, Moulinex hv2 (Moulinex, Paris, France), and grape seeds were separated. The remaining grape skin and flesh (GP) was ground and sieved, selecting the <1 mm fraction for the extraction experiments, which was subsequently stored at −20 °C.

Before each extraction process, the moisture of the residues was measured employing a moisture analyzer, Kern MLB 50-3N (Kern, Frankfurt am Main, Germany).

### 2.3. DES Preparation

Choline chloride: EG, a classical deep eutectic solvent (DES) was prepared in a relation molar 1:4 according to Ozturk et al. [12]. Briefly, the appropriate mass of choline chloride (hydrogen acceptor) was weighed and left at 80 °C for 16 h to dry the solid. The ethylene glycol (hydrogen donor) necessary was weighed and heated to 80 °C and the choline chloride was added and mixed until complete dissolution. Finally, the solution was maintained at 80 °C overnight and cooled down afterwards. When the temperature decreased to 30 °C, a 10% (*w*/*w*) of water was added to stabilize the solution.

### 2.4. Extraction Process

In this work, a batch solid–liquid extraction method was employed using different solvents and temperatures. These experiments were carried out in 250 mL round bottom flasks with continuous stirring at 500 r.p.m. and a solid/solvent ratio of 1/40 in the case of GP and 1/20 in the case of AP. The kinetics of the process were followed for one hour. After extraction, the samples were centrifuged at 13,000× *g* for 5 min and the supernatants were stored at −20 °C in darkness until analysis. All experiments were performed in duplicate.

### 2.5. Determination of Total Phenolic Content (TPC)

TPC was measured in triplicate according to the Folin–Ciocalteu method as described by Ribeiro et al. [13]. Briefly, 30 µL of the sample's appropriate dilution was mixed with 1500 µL of MilliQ water and 150 µL of the Folin–Ciocalteu's reagent. Then, 450 µL of 15% sodium carbonate and 870 µL of MilliQ water were added. The mixture was incubated in the dark at room temperature for 2 h. Afterwards, the absorbance was measured at 765 nm using an UV-vis spectrophotometer V600 Jasco (Tokyo, Japan). Results were expressed as gallic acid equivalents per gram of dry solid (GAE/gds) using a calibration curve between 0 and 500 mg/L of gallic acid. Controls of the different solvents were prepared for each experiment.

### 2.6. Antioxidant Capacity Determination

Antioxidant capacity was determined employing the DPPH method as described by Ozturk et al. [12] (three times each sample). In this assay, 0.2 mL of each sample was mixed with 3.8 mL of 0.1 mM DPPH diluted in 96% (*v*/*v*) ethanol and shaken vigorously. The homogenized mixture was incubated in the dark for 1 h at room temperature. Later, the absorbance was measured at 517 nm using an UV-vis spectrophotometer V600 Jasco (Japan). Solvent controls were prepared for each extraction. Results were expressed in total equivalent antioxidant capacity per gram of dry solid. Standard curves were prepared between 0 and 500 mg/L employing TROLOX diluted in ethanol 70% (*v*/*v*) as standard, and the antioxidant activity was expressed as TROLOX equivalent antioxidant capacity per gram dry solid (TEAC/gds).

### 2.7. Phenolic Compositional Analysis

The chromatographic analysis (HPLC) was performed employing a reversed phase C18 column, 250 mm × 4.6 mm × 0.5 µm (Fortis, Liverpool, UK) using a Jasco series 2000 HPLC modular system, employing a diode array detector (MD-2010). As a mobile phase, a mixture of formic acid 1.5 % (*v*/*v*) (phase A) and acetonitrile (phase B) was used while employing a gradient to improve peak separation. The gradient used in the case of AP samples was as follows: From 0 min to 10 min, 98% of A and 2% of B; from 10 min until 55 min, the gradient decreased to 60% of phase A and 40% of phase B; and from 55 min to 65 min there was a gradient decrease until 40% of phase A and 60% of phase B was applied. Finally, from 65 min to 80 min, the mobile phase was composed of 98% of phase A and 2% of phase B to stabilize the column for the next sample.

In the case of GP, the gradient applied was as follows: From 0 min to 5 min, the mobile phase composition was 98% of A and 2% of B; from 5 min to 60 min, there was a gradient decrease until 60% of A and 40% of B was applied; and between 60 min and 70 min, the gradient employed was 20% of A and 80% of B. Again, to stabilize the column, from 70 min to 85 min, a 98% A and 2% B was employed as mobile phase.

For determining the phenolic composition of the most relevant extraction liquors, a LC/MS-MS analysis was performed using a LC-ESI-QTOF (Impact, Bruker, Billerica, MA, USA) and measuring m/z between 100 and 1200, under the chromatographic conditions described for the previous HPLC method.

### 2.8. Kinetics of Solid–Liquid Extraction

The extraction process kinetics were fitted to an empirical adsorption/desorption model based on the Langmuir adsorption concept as developed by Islam et al. [14]. This second-order model contains three parameters: $q_0$ represents the concentration of adsorbate at zero time, thus showing the effect of waste washing (a very fast dynamic phenomenon due to the dissolution of compounds on or very near to the surface of the waste particles); $q_1$ indicates the concentration of adsorbate liberated in the slow phase, which can be controlled by surface and/or mass transfer phenomena in the pores; and $k_d$, which is the kinetic constant of desorption. Finally, $q_e$ represents the concentration of adsorbate in equilibrium in the liquid–solid suspension.

$$q_l = q_0 + q_1 \frac{b \cdot t}{1 + b_1 \cdot t} \quad \text{being } b = k_d \cdot q_1 \text{ and } q_e = q_0 + q_1 \tag{1}$$

OriginLab 2019® (OriginLab Corporation, Northampton, MA, USA) was the software employed to perform fitting to this hyperbolic model. Nonlinear fitting was performed employing the Levenberg–Marquardt algorithm [15].

All the results are shown as mean value and standard deviation (SD), being that this last parameter was evaluated by Student's *t*-test at 95% confidence. The R-squared parameter was used to evaluate the goodness-of-fit in all cases. A *p*-value below 0.05 was considered significant [16].

## 3. Results and Discussion

### *3.1. Kinetic Study of Extraction Processes*

The first step was to perform solid–liquid extractions of the selected residues with different solvents of industrial interest: ethanol, acetone, ethylene glycol, propylene glycol, water, and water:solvent mixtures, as well as a known deep eutectic solvent (choline chloride-ethylene glycol -ChCl:EG- with a molar ratio of 1:4). After the extraction process, the TPC and antioxidant capacity of the samples were measured. The results from the AP extraction are shown in Figures 1–6 while the extraction results from the GP extraction are displayed in Figures 7–10. In these figures, experimental data is shown as points while the fitting of the empirical second-order kinetic model is displayed as lines.

When referring to results for the AP extraction using ethanol 70% (*v/v*) extraction at several temperatures (Figure 1), final TPC values increase with temperature from 10.57 ± 1.21 mg GAE/gds at the final extraction point (60 min) at 25 °C to 13.13 ± 1.43 mg GAE/g gds at 70 °C, a value that is very similar to the one obtained at 90 °C. The effect of the temperature is also perceived in the initial extraction rate, thus a fast washing of the residue in the first minute leads to 1.8 mg GAE/gds at 25 °C, 3.0 at 50 °C, 4.3 at 70 °C, and 5.7 mg GAE/gds at 90 °C. Considering the analytical method here performed, sampling at particular time values followed by a subsequent spectrophotometric analysis, the dissolution of readily available material on or near the surface of waste particles was so fast that it cannot be followed, thus leading to an apparent value at zero time, as reflected by $q_0$ in the kinetic model or, more visibly, by a perceived TPC value at zero time in Figure 1 (and for this parameter and the antioxidant activity in all other figures). In this case of AP extraction with hydroalcoholic solutions at several temperatures, the increment in TPC at zero time indicates a sharp increase in phenolic solubility with temperature. In fact, this increment is slightly exponential—TPC = 1.18·exp(0.0177·T[°C]); $R^2$ = 0.995, a phenomenon that is also observed for (+)-catechin solubility in water and water–ethanol mixtures in similar temperature intervals [17]. After this fast solubilization, a progressive extraction occurs, reaching values that are maximal at 70 °C. The subsequent reduction in final TPC values observed at higher temperature could be ascribed to pore collapse by dehydration or a similar phenomenon that hinders mass transfer of phenolics out of the porous structure, as an increase in temperature always favors the desorption of the compounds out of the pore inner surface. Alternatively, phenolics degradation due to a relative high temperature usually happens, and affects TPC measurement, as indicated by Sólyom et al. [18]. According to these authors, in the case of grape marc or pomace, polyphenol oxidase can reduce TPC values during the first hour of extraction at temperatures of 80 °C or higher. This deactivation is evident for raw grape marc but does not happen in the case of filtered extracts, suggesting that the phenolic composition is key to the behavior of phenolics under oxidizing conditions at mild–high temperatures.

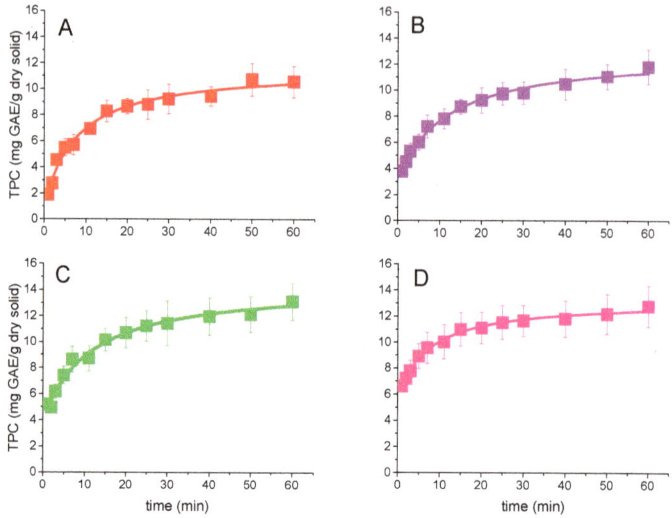


**Figure 1.** Results of total phenolic content (TPC) during extraction processes using AP and ethanol/water 70/30 (% $v/v$) at different temperatures: (**A**) 25 °C, (**B**) 50 °C, (**C**) 70 °C, and (**D**) 90 °C. Data were obtained by the Folin–Ciocalteu method. Kinetic analysis corresponding to the second-order kinetic model is displayed as lines.

A solvent comparison is shown in Figure 2 where the effect of water in the solvent mixture was studied. Water extraction presents a final yield for TPC in the case of AP, 11.56 ± 0.61 mg GAE/gds, which is similar to those obtained with a hydroalcoholic mixture containing 70% ethanol and 30% $v/v$ water. In the case of acetone and acetone 80%/water 20% AP extraction, the results are also similar (10.42 ± 1.08 and 10.90 ± 1.09 mg GAE/gds, respectively), so, in the extraction of AP, water content seems not relevant for this process. This could be attributed to the fact that as AP is not a dry solid (AP here used had a dry solid content of 33.5 ± 0.9%), the residue confers a local content of water to the mixture that is relevant to equalize, at pore scale, the amount of water irrespective of the percentage of water in the solvent, thus making an extraction without water not feasible. Ethanol 96% extraction is similar to acetone: the final solvent will be a mixture of water contained in the residue and ethanol, so the final yield (10.33 ± 1.29 mg GAE/gds) was similar to the one obtained in the ethanol 70% extraction.

The effect of water can be observed in the slower initial extraction rate. This can be attributed to the effect of the high polarity of water with yields at 5 min in the range of 4 mg GAE/gds in comparison to 5, 6, and 7 for aqueous mixtures of ethanol 70% and acetone 80%, and pure acetone, respectively. The difference between solvents may also be due, in part, to the viscosity at 25 °C for the solvents under comparison: 0.91 cP (water), 1.07 cP (ethanol), and 0.31 cP (acetone). One should keep in mind that, according to the Einstein–Stokes equation, and most empirical equations for the calculation of molecular diffusivities in liquids' diffusivity, and, thus, according to the first Fick law on diffusional mass transfer, mass flow rates are inversely proportional to viscosity. Acetone has, most evidently, the lowest viscosity. In the case of an almost pure ethanol (96%) at 70 °C, viscosity equals 0.5231 cP and cannot explain the low extraction rate (6 mg GAE/gds) achieved in the first 5 min, a fact that is further supported by a relatively low final yield slightly over 10 mg GAE/gds. Thus, the water content in hydroalcoholic mixtures seems critical to achieve high extraction yields, improving solvation power of the mixture in comparison to pure ethanol, while medium to high temperatures will result (Figure 1) in fast to very fast extraction processes. Possibly, in view of the results of acetone:water and ethanol:water mixtures, this observation can be extended to acetone:water mixtures applied at medium-high temperature and pressure-values.

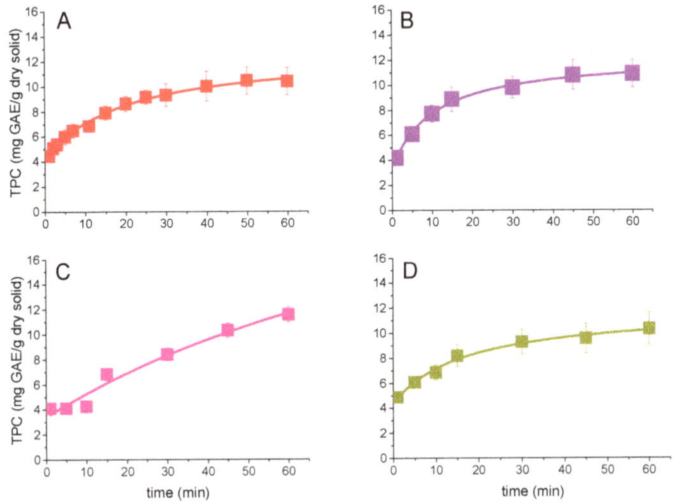


**Figure 2.** Kinetic analysis of the extraction processes of AP employing (**A**) acetone at 25 °C, (**B**) acetone 80% at 25 °C, (**C**) water at 25 °C, and (**D**) ethanol 96% ($v/v$) at 70 °C obtained by the Folin–Ciocalteu method. Fitting lines correspond to the second-order kinetic model shown in Equation (1).

Figure 3 displays the results for TPC extraction when using glycol solutions with a small amount (10%) of water. From the perspective of TPC final yields, these glycol/water mixtures seem to be optimal solvents to extract phenolics in the case of AP at 70 °C, as TPC values reach 25 mg GAE/gds in 60 min and 13 mg GAE/gds in 15 min (equal to the TPC value for ethanol 70% aqueous solution in 60 min). Moreover, the apparent TPC value at zero time (the result of washing or dissolution of readily available phenolics on or near the waste particles' surface) is 6 mg GAE/gds for the ethyleneglycol solution and almost 5 GAE/gds for the propyleneglycol solution (comparing to 4.3 GAE/gds for the ethanol 70% watery solution). These numbers are a further confirmation of the fast nature of the dissolution or washing process: as the stirring speed was set to a high value (500 r.p.m.) for a low volume of liquid–solid suspension, no external mass transfer hindrances are expected, thus favoring the solubilization phenomena taking place in the first minute of contact. Here, mass transfer is immediate and solvent viscosity plays no role, being that its solvation power is the most important factor to achieve high TPC values.

The kinetic trends and model curves reflecting the antioxidant capacity extracted from AP are displayed in Figures 4–6. Considering this functional parameter, the extraction driven by an aqueous solution of ethanol at 70% ($v/v$) shows differences depending on the temperature value. We can see that the maximum antioxidant activity, according to a DPPH test, is obtained at 70 °C and 90 °C, with results of 24.05 ± 1.94 and 24.02 ± 1.89 mg TEAC/gds at 60 min, respectively. It is interesting to observe that these results at both temperatures are identical from a statistical perspective, while a slightly higher initial extraction rate is observed at 90 °C, indicating that this temperature, even in the absence of light, can be deleterious of the antioxidant activity of the extract (as suggested before by TPC analysis) or that an asymptotic value for antioxidant activity is being reached. While the average slopes of tangents to the curve along the extraction period (slow phase) indicate an increasing extraction rate with temperature, the value of TPC at zero time increases slowly from 25 to 70 °C: 3 mg TEAC/gds (25 °C), 4.5 mg TEAC/gds (50 °C), 5.5 mg TEAC/gds (70 °C), and 7 mg TEAC/gds (90 °C), with a sudden increment from

70 to 90 °C coherent with a certain physical modification of the structure near the boiling point observed for the solvent at atmospheric pressure.

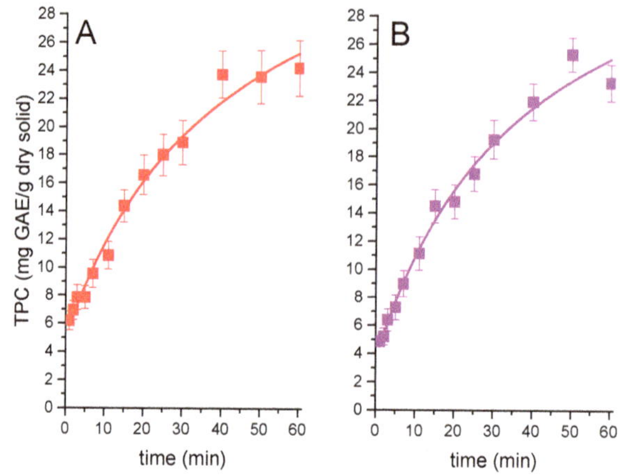


**Figure 3.** Kinetic analysis of the extraction process of AP employing (**A**) ethylene glycol 90% at 70 °C and (**B**) propylene glycol 90% at 70 °C obtained by the Folin–Ciocalteu method. The lines show the fitting of the second-order kinetic model shown in Equation (1).

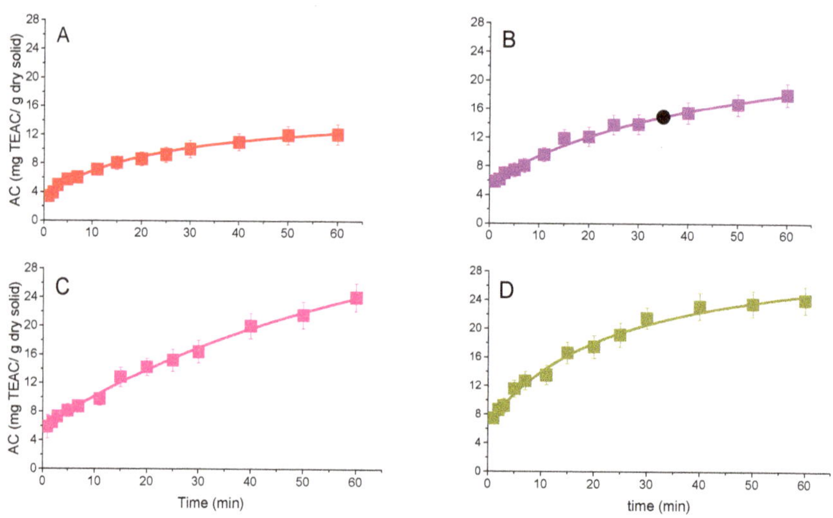


**Figure 4.** Kinetic analysis of the antioxidant activity as measured by the DPPH method during the extraction processes of AP employing a 70%/30% *v/v* ethanol/water mixture at several temperature values: (**A**) 25 °C, (**B**) 50 °C, (**C**) 70 °C, and (**D**) 90 °C. The lines show the fitting of a one-term second-order kinetic model (Equation (1)).

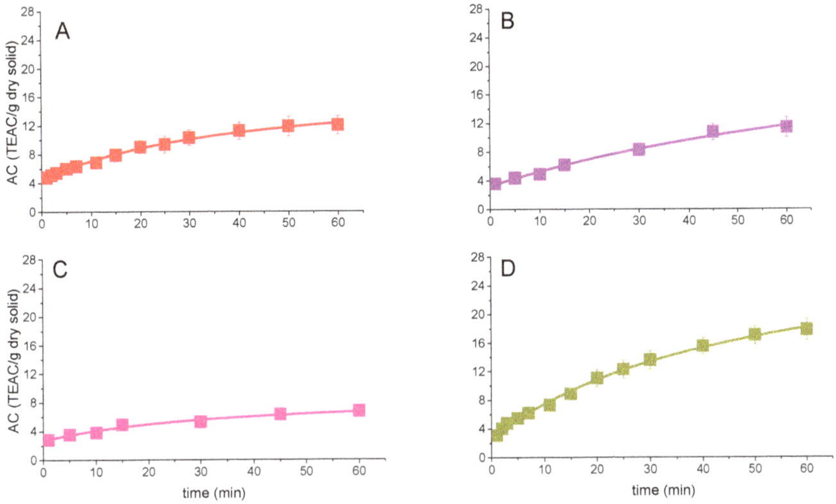


**Figure 5.** Kinetic analysis of the extraction processes of AP employing (**A**) acetone at 25 °C, (**B**) acetone/water 80%/20% *v/v* at 25 °C, (**C**) water at 25 °C, and (**D**) ethanol 96% at 70 °C: antioxidant activity as obtained by DPPH method. Kinetic model fitting is presented in lines.

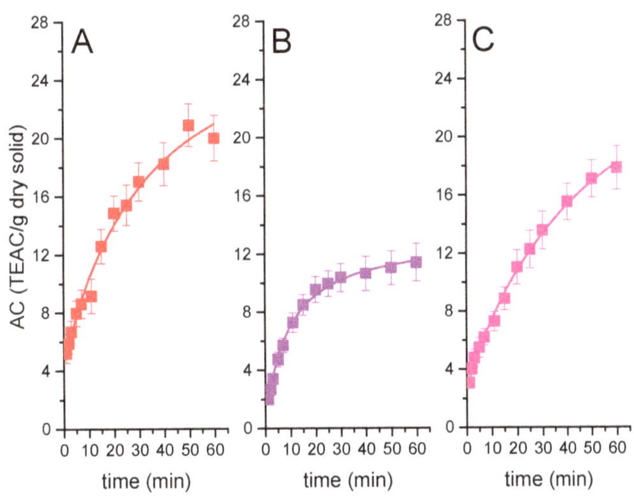


**Figure 6.** Kinetic analysis of the extraction processes of AP with glycols and a glycol-based DES at 70 °C. (**A**) ethylene glycol 90%, (**B**) propylene glycol 90%, and (**C**) ChCl:Gly 90%. The kinetic model fitting is presented in lines.

In the case of propylene glycol and ethylene glycol, despite being the best extractives in terms of TPC, the antioxidant extraction capacity is lower, being higher in the case of EG (19.29 ± 1.58 TEAC/gds in 60 min), while the nontoxic PG scarcely extracts 11.41 ± 1.30 TEAC/gds in the same time. A deep eutectic solvent (DES) composed of CHCl and EG is as good as EG (17.83 ± 1.49 TEAC/gds in 60 min). This may be caused by the nature of the phenolics extracted with these solvents in comparison to those dissolved in mixtures of water/ethanol, which show an almost-double antioxidant-specific activity in terms of AC/TPC (1.5 for ethanol 70% at 70 °C versus 0.77 for ethyleneglycol and 0.46

for propyleneglycol), showing that, on top of being greener and less expensive, ethanol–water mixtures are more specific for phenolics with a high antioxidant activity when valorizing AP.

Acetone and a mixture of acetone (80% *v/v*) and water, when compared to ethanol 70% in water at 25 °C (12.08 ± 1.39 TEAC/gds in 60 min) are slightly better extractants (12.06 ± 1.28 TEAC/gds and 11.45 ± 1.39 in the same time, respectively), although water is the worst solvent in this case, with an antioxidant capacity of the aqueous extract at 60 min of 6.75 ± 0.58 TEAC/gds. However, AC values at zero time are very interesting for acetone (4.5 mg TEAC/gds and 3.5 mg TEAC/gds) compared to the benchmark ethanol 70% solution (3 mg TEAC/gds). In general, these results show again the superiority of ethanol:water 70/30 *v/v* mixture for the extraction of antioxidant power from AP, although pure acetone allows for a very fast washing, suggesting a good solvation power or a certain role of viscosity—very low for the acetone at 25 °C—in this first phase. Moreover, the ratio AC/TPC is 1.8 for ethanol 96% and 1.5 for ethanol 70%, showing a more specific extraction for the hydroalcoholic solvent richer in ethanol.

When applying these solvents and conditions to the extraction of GP components (Figures 7–10), the better extraction process in terms of TPC and antioxidant capacity yields is performed again with the mixture ethanol/water 70% (*v/v*), increasing yield values as the operating temperature increases. Therefore, as expected, temperature affects the solubility of phenolics and the rate of the extraction process from the matrix, in this latter case due to the increase in the mass transfer rate and/or solvation power (viscosity reduction, increment in solubility). The GP extraction yields ten times higher TPC than when extracting AP. Moreover, the process is more affected by temperature: it increases from 44.05 ± 1.78 mg GAE/gds at 25 °C to 68.46 ± 1.03 mg GAE/gds at 90 °C (a 55.4% increment in comparison to a 24.2% increment for AP with the same temperature variation). This fact could suggest that AP porosity structure is less affected by this variable, so the effective diffusivity that depends on it changes only slightly.

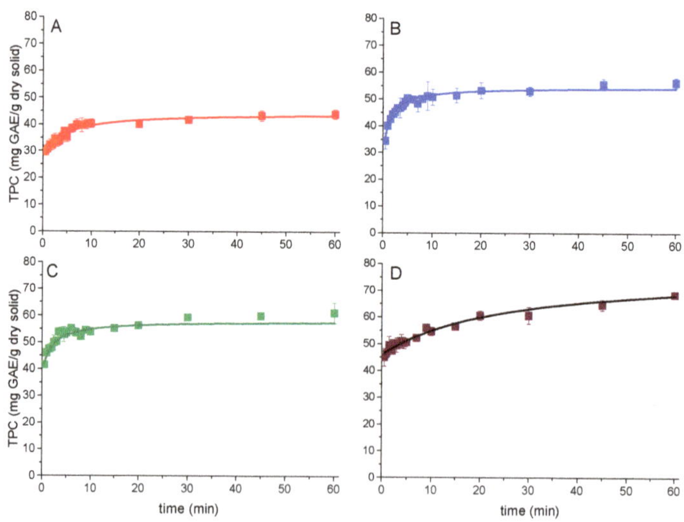


**Figure 7.** Kinetic analysis of the extraction process from grape pomace (GP) with a hydroalcoholic solution (ethanol/water 70%/30% *v/v*) at several temperatures: (**A**) 25 °C, (**B**) 50 °C, (**C**) 70 °C, and (**D**) 90 °C. Data, in points, reflects the total phenolic content (TPC) obtained by the Folin–Ciocalteu method. The lines indicate the fitting of the second-order kinetic model presented in Equation (1).

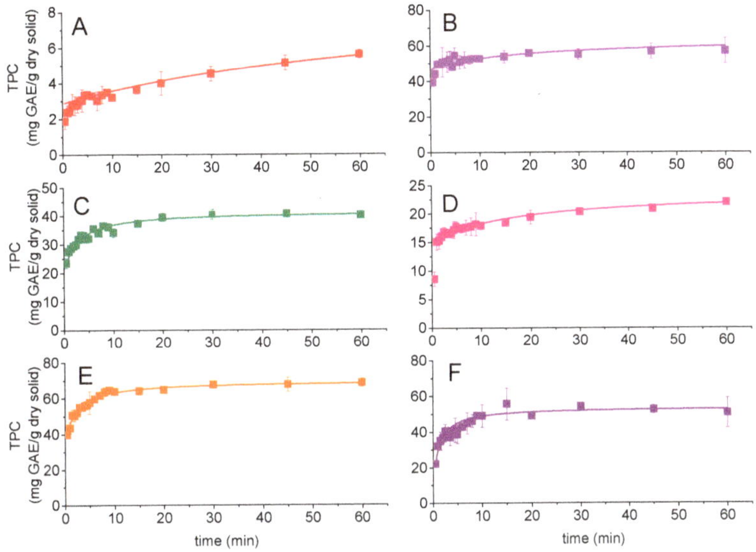


**Figure 8.** Total phenolic content results, and their kinetic analysis, obtained during the extraction processes applied to GP using several solvents: (**A**) acetone at 25 °C, (**B**) acetone 80%: water 20% at 25 °C, (**C**) ethanol 96% at 70 °C, (**D**) water at 25 °C, (**E**) ethylene glycol at 70 °C, and (**F**) propylene glycol at 70 °C. The lines show the second-order kinetic model (Equation (1)) fitting to data.

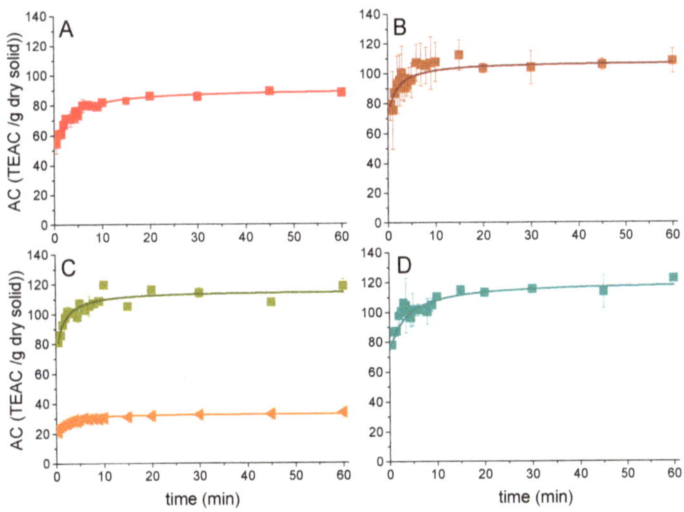


**Figure 9.** Antioxidant activity by DPPH method in points (data) and fit of the second-order kinetic model in lines. Results from grape pomace (GP) with a hydroalcoholic solution (ethanol/water 70%/30% *v/v*) at several temperatures: (**A**) 25 °C, (**B**) 50 °C, (**C**) 70 °C (green line and points: ethanol 70%; orange line and points: ethanol 96%), and (**D**) 90 °C.

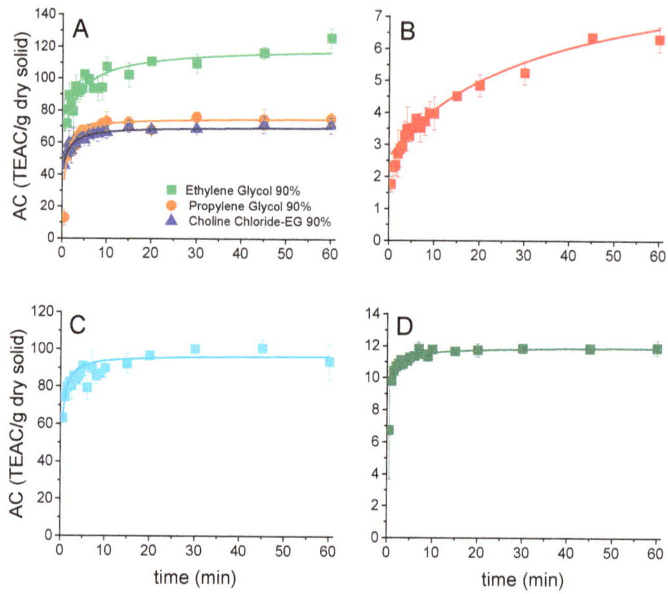


**Figure 10.** Antioxidant activity (DPPH) results, and their kinetic analysis, obtained during the extraction processes applied to GP using several solvents: (**A**) glycols at 70 °C; (**B**) acetone at 25 °C, (**C**) acetone 80%: water 20% at 25 °C, and (**D**) water at 25 °C. The lines show the second-order kinetic model (Equation (1)) fitting to data.

Again, there is a notable difference between the TPC and antioxidant capacity (AC) results: a 30% water content leads to a slightly better extraction of phenolics, but, for GP extractions, their AC is more than double that of the phenolics extracted with ethanol containing 4% water at 70 °C. Curiously, the AC/TPC ratio is worse for ethanol 96% (0.5) than for ethanol 70% (1.83, the best one for GP extraction), contrary to what is observed for AP extraction. As in the case of TPC, when comparing with AP extracts, these GP extracts have a much higher AC value in the best conditions (5 times higher).

In the case of the other extraction solvents, the ethylene glycol is highlighted as the better one, with high AC and TPC values, and a high AC/TPC ratio (1.7), followed by acetone 80% (*v/v*) extraction (AC/TPC ratio = 1.5), with the worst solvents being pure acetone and water. In the case of propylenglycol and the tested DES, the results are intermediate with both having similar AC results.

Contrary to the extraction of AP, GP extractions show high differences between solvents and solvents/water mixture due to the fact that the GP residue employed is a dry solid, so, as the residues do not add water, the extraction is only produced by the solvent. An addition of water to solvents is a good option as the mixtures produce higher extraction yields. Acetone 80%, ethyleneglycol 90%, and ethanol 70% have a high extractive capacity, emphasizing the importance of adding a notable percentage of water to these solvents. Possibly, the increase of extraction yield is due to the optimal hydrophilicity–hydrophobicity balance of the solvent mixture, more adequate to extract GP phenolics.

### 3.2. Kinetic Modeling of Extraction Process

All extraction results concerning TPC and AC have been used to fit a second-order kinetic model with two phases: an almost immediate washing or dissolution process followed by a slow extraction step, as indicated by Equation (1). After the adjustment of the mentioned kinetic equation to the relevant experimental data employing a Levenberg–Marquardt nonlinear regression algorithm, the kinetic constants and correlation values were calculated. The results have been compiled in Tables 1–4. The values of $R^2$ were superior to 0.9 in all cases, while the F values varied from 1757 to 10,854, leading to a *p*-value much lower than 0.01 at 95% confidence. Therefore, the mathematical model here proposed fits to the experimental data in each run. This can be further observed in Figures 1–10 when comparing the coincidence between data points and fitting curves.

**Table 1.** Kinetic parameters from the extraction process of AP measuring total phenolic content (TPC) by Folin–Ciocalteu's method; $q_0$, $q_1$, and $q_e$ are measured in mg GAE/gds, while $k_d$ units are gds/(mg GAE·min).

| Extraction | $q_0$ | $q_1$ | $q_e$ | $k_d$ | $r^2$ |
|---|---|---|---|---|---|
| Ethanol 70% (*v/v*) 25 °C | 1.3 ± 0.2 | 9.6 ± 0.4 | 10.9 ± 0.6 | $0.02 \pm 3 \times 10^{-3}$ | 0.99 |
| Ethanol 70% (*v/v*) 50 °C | 3.0 ± 0.1 | 9.6 ± 0.2 | 12.6 ± 0.2 | $0.01 \pm 1 \times 10^{-3}$ | 0.99 |
| Ethanol 70% (*v/v*) 70 °C | 3.9 ± 0.4 | 10.6 ± 0.7 | 14.5 ± 1.1 | $9 \times 10^{-3} \pm 3 \times 10^{-3}$ | 0.97 |
| Ethanol 70% (*v/v*) 90 °C | 5.7 ± 0.1 | 7.5 ± 0.2 | 13.2 ± 0.3 | $0.02 \pm 2 \times 10^{-3}$ | 0.99 |
| Ethanol 96% (*v/v*) 70 °C | 4.5 ± 0.2 | 9.0 ± 0.9 | 13.5 ± 1.1 | $5 \times 10^{-3} \pm 2 \times 10^{-3}$ | 0.98 |
| Acetone 100% (*v/v*) 25 °C | 4.3 ± 0.1 | 8.8 ± 0.6 | 13.1 ± 0.7 | $5 \times 10^{-3} \pm 1 \times 10^{-3}$ | 0.99 |
| Acetone 80% (*v/v*) 25 °C | 3.4 ± 0.2 | 9.0 ± 0.3 | 12.4 ± 0.5 | $0.01 \pm 2 \times 10^{-3}$ | 0.99 |
| Water 25 °C | 4.0 ± 0.2 | 19.3 ± 3.8 | 23.3 ± 4.0 | $6 \times 10^{-4} \pm 3 \times 10^{-4}$ | 0.99 |
| Ethylene glycol 90% (*v/v*) 70 °C | 5.4 ± 0.4 | 39.3 ± 6.5 | 44.7 ± 6.9 | $5 \times 10^{-4} \pm 2 \times 10^{-4}$ | 0.98 |
| Propylene glycol 90% (*v/v*) 70 °C | 3.9 ± 0.3 | 36.3 ± 4.0 | 40.2 ± 4.3 | $7 \times 10^{-4} \pm 2 \times 10^{-4}$ | 0.99 |

**Table 2.** Kinetic parameters from the extraction process of AP measuring antioxidant capacity by DPPH method; $q_0$, $q_1$, and $q_e$ are measured in TEAC/gds, while $k_d$ units are gds/(TEAC·min).

| Extraction | $q_0$ | $q_1$ | $q_e$ | $k_d$ | $r^2$ |
|---|---|---|---|---|---|
| Ethanol 70% (*v/v*) 25 °C | 3.0 ± 0.2 | 11.8 ± 0.7 | 14.8 ± 0.6 | $5 \times 10^{-3} \pm 1 \times 10^{-3}$ | 0.99 |
| Ethanol 70% (*v/v*) 50 °C | 5.2 ± 0.2 | 20.5 ± 1.7 | 25.7 ± 0.2 | $1 \times 10^{-3} \pm 3 \times 10^{-4}$ | 0.99 |
| Ethanol 70% (*v/v*) 70 °C | 5.7 ± 0.2 | 25.5 ± 2.9 | 31.2 ± 1.1 | $7 \times 10^{-4} \pm 9 \times 10^{-5}$ | 0.99 |
| Ethanol 70% (*v/v*) 90 °C | 6.5 ± 0.3 | 25.1 ± 1.4 | 31.6 ± 0.3 | $2 \times 10^{-3} \pm 4 \times 10^{-4}$ | 0.99 |
| Ethanol 96% (*v/v*) 70 °C | 3.7 ± 0.2 | 5.6 ± 0.9 | 9.3 ± 1.1 | $4 \times 10^{-3} \pm 1 \times 10^{-3}$ | 0.96 |
| Acetone 100% (*v/v*) 25 °C | 4.5 ± 0.1 | 13.7 ± 1.1 | 18.2 ± 0.7 | $2 \times 10^{-3} \pm 4 \times 10^{-4}$ | 0.99 |
| Acetone 80% (*v/v*) 25 °C | 3.4 ± 0.2 | 30.5 ± 3.2 | 33.7 ± 0.5 | $2 \times 10^{-4} \pm 2 \times 10^{-4}$ | 0.99 |
| Water 25 °C | 2.7 ± 0.3 | 6.4 ± 1.3 | 8.6 ± 4.0 | $4 \times 10^{-3} \pm 3 \times 10^{-3}$ | 0.96 |
| Ethylene glycol 90% (*v/v*) 70 °C | 4.5 ± 0.3 | 26.5 ± 2.7 | 31.0 ± 6.9 | $1 \times 10^{-3} \pm 3 \times 10^{-4}$ | 0.99 |
| Propylene glycol 90% (*v/v*) 70 °C | 0.8 ± 0.2 | 13.0 ± 0.3 | 13.8 ± 6.9 | $7 \times 10^{-3} \pm 7 \times 10^{-4}$ | 0.97 |
| ChCl: EG (1:4) 90% (*v/v*) 70 °C | 2.8 ± 0.2 | 29.6 ± 3.0 | 31.4 ± 4.3 | $6 \times 10^{-4} \pm 2 \times 10^{-4}$ | 0.99 |

**Table 3.** Kinetic parameters from the extraction process of GP measuring total phenolic content (TPC) by Folin–Ciocalteu's method; $q_0$, $q_1$, and $q_e$ are measured in mg GAE/gds, while $k_d$ units are gds/(mg GAE·min).

| Extraction | $q_0$ | $q_1$ | $q_e$ | $k_d$ | $r^2$ |
|---|---|---|---|---|---|
| Ethanol 70% (*v/v*) 25 °C | 27.4 ± 1.1 | 17.0 ± 1.0 | 34.4 ± 2.1 | $0.02 \pm 3 \times 10^{-3}$ | 0.95 |
| Ethanol 70% (*v/v*) 50 °C | 28.9 ± 3.2 | 25.7 ± 2.7 | 54.6 ± 5.9 | $0.03 \pm 5 \times 10^{-3}$ | 0.98 |
| Ethanol 70% (*v/v*) 70 °C | 37.7 ± 3.2 | 20.3 ± 1.5 | 58.0 ± 4.7 | $0.03 \pm 4 \times 10^{-3}$ | 0.98 |
| Ethanol 70% (*v/v*) 90 °C | 45.8 ± 0.8 | 31.3 ± 2.1 | 76.1 ± 2.9 | $0.01 \pm 4 \times 10^{-3}$ | 0.98 |
| Ethanol 96% (*v/v*) 70 °C | 19.4 ± 10.5 | 22.0 ± 9.0 | 41.4 ± 19.5 | $0.02 \pm 7 \times 10^{-3}$ | 0.96 |
| Acetone 100% (*v/v*) 25 °C | 2.9 ± 0.2 | 6.1 ± 3.0 | 9.0 ± 3.2 | $2 \times 10^{-3} \pm 1 \times 10^{-3}$ | 0.84 |
| Acetone 80% (*v/v*) 25 °C | 46.8 ± 1.2 | 16.8 ± 4.8 | 63.6 ± 6.0 | $3 \times 10^{-3} \pm 2 \times 10^{-3}$ | 0.82 |
| Water 25 °C | 15.0 ± 0.5 | 8.8 ± 0.7 | 23.8 ± 1.2 | $7 \times 10^{-3} \pm 4 \times 10^{-3}$ | 0.95 |
| Ethylene glycol 90% (*v/v*) 70 °C | 34.4 ± 3.7 | 35.1 ± 3.3 | 69.5 ± 7.0 | $0.01 \pm 2 \times 10^{-3}$ | 0.96 |
| Propylene glycol 90% (*v/v*) 70 °C | 9.3 ± 2.8 | 44.5 ± 2.1 | 53.8 ± 4.9 | $0.02 \pm 4 \times 10^{-3}$ | 0.99 |

**Table 4.** Kinetic parameters from the extraction process of GP measuring antioxidant capacity by DPPH method; $q_0$, $q_1$, and $q_e$ are measured in TEAC/gds, while $k_d$ units are gds/(TEAC·min).

| Extraction | $q_0$ | $q_1$ | $q_e$ | $k_d$ | $r^2$ |
|---|---|---|---|---|---|
| Ethanol 70% (*v/v*) 25 °C | $50 \pm 0.5$ | $41 \pm 0.5$ | $91 \pm 1.0$ | $9 \times 10^{-3} \pm 5 \times 10^{-4}$ | 0.99 |
| Ethanol 70% (*v/v*) 50 °C | $70 \pm 8.2$ | $38 \pm 7.7$ | $108 \pm 16$ | $0.015 \pm 5 \times 10^{-3}$ | 0.82 |
| Ethanol 70% (*v/v*) 70 °C | $69 \pm 6.8$ | $47 \pm 6.0$ | $116 \pm 13$ | $0.015 \pm 4 \times 10^{-3}$ | 0.95 |
| Ethanol 70% (*v/v*) 90 °C | $73 \pm 1.8$ | $48 \pm 1.4$ | $121 \pm 3.2$ | $6 \times 10^{-3} \pm 1 \times 10^{-3}$ | 0.99 |
| Ethanol 96% (*v/v*) 70 °C | $19 \pm 3.2$ | $15 \pm 3.2$ | $34 \pm 6.4$ | $0.05 \pm 6 \times 10^{-3}$ | 0.98 |
| Acetone 100% (*v/v*) 25 °C | $2.6 \pm 0.2$ | $6.2 \pm 0.7$ | $8.8 \pm 0.9$ | $5 \times 10^{-3} \pm 2 \times 10^{-3}$ | 0.96 |
| Acetone 80% (*v/v*) 25 °C | $28 \pm 8.8$ | $69 \pm 21$ | $98 \pm 30$ | $0.030 \pm 0.011$ | 0.95 |
| Water 25 °C | $8.8 \pm 0.4$ | $3.2 \pm 0.4$ | $12 \pm 0.8$ | $0.23 \pm 0.023$ | 0.98 |
| Ethylene glycol 90% (*v/v*) 70 °C | $60 \pm 4.4$ | $60 \pm 3.4$ | $120 \pm 7.8$ | $4 \times 10^{-3} \pm 2 \times 10^{-3}$ | 0.95 |
| Propylene glycol 90% (*v/v*) 70 °C | - | $75 \pm 21$ | $75 \pm 21$ | $0.025 \pm 9 \times 10^{-3}$ | 0.90 |
| ChCl: EG (1:4) 90% (*v/v*) 70 °C | $41 \pm 4.2$ | $30 \pm 3.9$ | $71 \pm 8.1$ | $0.024 \pm 4 \times 10^{-3}$ | 0.94 |

When fitting Equation (1) to the TPC results from the AP extraction (Table 1), the first observation is that $q_0$, the initial TPC value, dramatically increases with temperature, suggesting that the solubility of the compounds that are dissolved during this initial phase increases exponentially with this variable. After calculating an apparent extraction rate ($r_0$) in this phase by dividing $q_0$ by the first sampling time, a finer study, representing $\ln(r_0)$ versus the inverse value of absolute temperature (1/T)—Arrhenius equation—an exponential trend is again observed (as for $q_0$ with T), but, in this case, the apparent activation energy ($E_a$) can be calculated, being equal to 19.92 kJ/mol, a value that should be considered essentially empirical, as it shows the exponential increase of solubility with temperature for this particular case: apple pomace and ethanol 70%: water 30% solution as solvent. This is also evident when fitting this same second-order equation to AC results; in this case, the $E_a$ is 10.68 kg/mol, so temperature affects the extraction of phenolic compounds more than the extraction of their antioxidant activity during the washing phase. Thus, the nature of the compounds being dissolved could change with the temperature, suggesting that most active compounds in AP are more polar (for $q_0$, time values are so low that no thermal deactivation exists). For $q_1$ and for $q_e$, there are maximum TPC values at 70 °C ($10.6 \pm 0.7$ and $14.5 \pm 1.1$ mg GAE/gds, respectively) with a decrease between this temperature and 90 °C. This trend can be ascribed to a reduction in redox potential or to a certain collapse of the inner porous structure as the temperature approaches the boiling point of the extraction liquid. Again, there is no trend with temperature for the kinetic constant $k_d$ for TPC analysis, possibly indicating that there is a trade-off between accelerating effects due to increased diffusivity and phenolics desorption rate from pore surface and decelerating effects possibly due to pore structure collapse. Considering the experimental error, this idea can be extended to the analysis of AC $k_d$ values.

If we compare the results for different solvents at room temperature to extract active components from AP, we can observe that acetone 80% *v/v* is the best extractant for a fast extraction of total phenolics ($q_0$ value similar to that of ethanol 70% or 96% at 70 °C); curiously, the addition of water to either ethanol or acetone reduces $q_0$, suggesting a lower solubility of phenolics extracted during the washing phase when water is added. Values for $q_1$ are similar for all acetone and ethanol mixtures, irrespective of water concentration, suggesting that the total amount of phenolics extracted is essentially constant for these solvents. In the case of acetone, the addition of water helps in the extraction slow phase, increasing two-fold the $k_d$ value: as this increase cannot be due to solvent viscosity (it is lower in the case of pure acetone), it should be due to the solvation power of the solvent for the phenolics extracted in this slow phase; this factor is key to explaining a higher solvent–solute interaction and, thus, the ability of the solvent to desorb the solute from the pore surface. The case of water is interesting: the relatively high value of $q_0$ indicates a mild solvation power for phenolics during washing (therefore, these phenolics should be notably polar in nature, for example, due to their glycosylation), while a high value for $q_1$ shows a good solvation power for those phenolics remaining in the porous structure at the beginning, compounds that are slowly extracted in the second extraction period, which can be expected when using a relatively viscous solvent (in comparison to the others here tested). The low value for $k_d$, similar to that of pure acetone, can be explained, again, by the high viscosity of water, especially at 25 °C. Glycols are even better solvents from the TPC perspective in terms of $q_0$ and $q_e$, and their high viscosity, as in the case of water, can be the reason for the low values of $k_d$.

Considering the results for TPC and AC values, as ethyleneglycol is toxic and due to economic and logistic reasons, for food and pharma applications, the ethanol 70%–water 30% mixture working at relatively high temperature (70 °C) is the best solvent to obtain active extracts from apple pomace, considering that a higher temperature, though it can be better for a fast extraction in the washing stage, in the end is deleterious for the antioxidant capacity of the extract. On average, the antioxidant mixtures obtained in these conditions, in terms of AC/TPC value (2.15), have a higher antioxidant potential than the one obtained with ethyleneglycol at 70 °C (AC/TPC = 0.95). As for acetone 80%, AC/TPC equals to 1.17, while ethanol 70% AC/TPC at 25 °C leads to a 1.36 value, which are fair ratios obtained with green solvent mixtures at low temperature, thus low thermal energy inputs during the extraction process.

The results from GP extraction indicate much higher final values for TPC (almost 10 times more) and AC (about 5 times higher), which reflects on $q_0$, $q_1$, $q_e$, and $k_d$ values, with the values of this latter parameter being much dependent on the solvent. When ethanol 70% is used as solvent at several temperatures, $q_0$ and $q_1$ values show a mild exponential growth for TPC ($E_a$ = 7.57 and 6.52 kJ/mol for $q_0$ and $q_1$, respectively) and a mild hyperbolic increase towards an asymptotic value around 70 ($q_0$) and 48 ($q_1$) mg TEAC/gds for the AC analysis (so a maximum extraction qe value of 118 mg TEAC/gds can be reached); this diversity indicates thermal degradation of AC power as the temperature increases. It is interesting to observe that $k_d$ reaches a maximum between 50 and 70 °C both for TPC and AC values, suggesting, again, thermal degradation at high temperature.

A more detailed analysis using the Arrhenius equation for the 25–50 and the 70–90 °C intervals for both TPC and AC $k_d$ results indicate $E_a$ values of 21.1 and 16.0 kJ/mol for the low temperature interval (TPC and AC, respectively), indicating an activation of the slow extraction step, while $E_a$ values for TPC and AC in the high temperature interval are −35.8 and −47.4 kJ/mol, indicating a sharp deactivation of the extraction rate (possibly, because extraction and thermal degradation happens at the same time, being the last phenomenon dominant in the high temperature interval). Thus, when aiming for high TPC and AC values (high $q_0$ and $q_e$), it is of interest to use relatively mild temperature values, designing process units and methods to avoid thermal degradation during phenolics extraction [19,20]. Moreover, if we compare $k_d$ trends with temperature for AP and GP phenolic extracts, it can be seen that AP extracts are more stable towards temperature.

Ethanol 96%, ethylene glycol 90%, and acetone 80% show a very fast TPC release in the first and the second extraction phases ($q_0$ and $q_1$), emphasizing once again the important role of low to medium water content to reach a good hydrophilic/hydrophobic balance and high solvation power [20]. Low or high values for $k_d$ for the solvents under consideration cannot be explained only on the basis of viscosity as, for example, acetone extraction in the second phase is very slow, while those with propylene glycol with 10% water is much faster. Thus, an explanation should be sought in a structure modification: solvents can play a role in the shrinkage or swelling of the porous structure of waste particles and acetone is a well-known dehydration agent, so shrinkage can be expected when using this solvent. Particle shrinkage results in the reduction of average pore diameter and an increase in mass transfer hindrances, thus explaining low $k_d$ values. At 70 °C, ethyleneglycol 90% extracts phenolics and AC with a similar performance to ethanol 70% in the long term. However, for a fast and effective extraction with an inexpensive, highly available, and nontoxic mixture, ethanol 70% is the obvious choice.

To extract phenolic compounds from AP or GP, several methods have been used (leaching with reflux using a Soxhlet apparatus, direct contact leaching with agitation or by arranging the material to be extracted in a fixed bed, macerating or homogenizing the solid in the solvent, electric pulses, pressure pulses, ultrasonic cavitation, microwaves, and pressurized liquid extraction), even combining various technologies and solvents (MeOH, EtOH, acetone, water, ionic liquids, DEPs, and supercritical fluids) [6,21]. For apple residues, TPC values vary between 3 and 200 mg GAE/gds, with most values between 3 and 7 mg GAE/gds [6], while these values have been reported to be between 6.2 and 196 mg GAE/gds for GP, with most values in the 30–80 mg GAE/gds range [3]. In this work, these values are around 13–26 mg GAE/gds for AP, and near 120 mg GAE/gds for GP.

Most classic extraction procedures seem to take between 60 min and 24 h to achieve high TPC values in the case of AP, while operation time reduces to 5–20 min for intensified processing (ultrasounds, microwaves, pressurized liquid extraction) [6]. The same can be said for GP, with optimal time values around 15 min for intensified methods and in the range 60–120 min for classic solid–liquid extraction [3,22]. These values are confirmed here, observing, in addition, that a very fast extraction, mostly boosted by ethanol–water mixtures at medium-high temperatures, results in a high percentage of total polyphenols being extracted in the first 2–3 min.

In recent years, some kinetic modeling studies showed that second-order kinetic models can better explain extraction from AP [23] and GP [24], showing in this latter case that biphasic models explain better the lixiviation kinetics. As indicated in this work, activation energy values for relevant kinetic constants show what type of phenomenon is controlling the extraction process, either mass-transfer (low $E_a$) or chemical phenomena (desorption followed by solvation on the solid pore inner surface, also known as solubilization or thermal deactivation) (high $E_a$). While second-order apparent behavior in mass-transfer driven processes can be explained by a sum of first-order phenomena [24], real second-order behavior can be expected when slow chemical phenomena such as site-specific desorption take place on the pore surface [14].

### 3.3. Compositional Analysis of Extraction Liquors

After selecting the optimal extraction method that uses ethanol 70% as the extractant and 90 °C as the operating temperature, an analysis of the AP and GP extracts' composition was determined by HPLC and HPLC-MS, obtaining the results compiled in Tables 5 and 6.

**Table 5.** List of principal compounds obtained from the apple pomace extraction with ethanol 70% (*v/v*) and 90 °C and their mass spectrometric data.

| Compound | $t_r$ (min) | MW (g/mol) | UV Spectrum Pattern ($\lambda^*_{máx}$) |
|---|---|---|---|
| Quercetin-Arabinose-Arabinose | 32.67 | 565.3 | 254 |
| Rutin | 41.68 | 610.5 | 254, 354 |
| Isoquercitrin | 42.30 | 464.4 | 254, 354 |
| Quercetin-Arabinose 1 * | 43.97 | 434.3 | 254, 354 |
| Quercetin-Arabinose 2 * | 45.04 | 434.3 | 254, 354 |
| Quercetin-Fucose or Rhamnose | 45.46 | 448.4 | 254, 354 |

* The compound was glycosylated one time at two different positions, but the position could not be determined.

**Table 6.** List of principal compounds obtained from the grape skin extraction with ethanol 70% (*v/v*) and 90 °C and their mass spectrometric data.

| Compound | MW (g/mol) |
| --- | --- |
| Tetramer 2,3-dihydroxycinnamic acid | 684 |
| Catechin | 290 |
| Di-Catechin + Gallic acid | 730 |
| Malvidin-O-Glucoside | 493 |
| Dihydro 2,3-dihydroxycinnamic acid + quinic acid | 356 |
| Sinapic acid | 224 |
| 2,3-dihydroxybenzoic acid + quinic acid | 328 |
| Dimer (2,3-dihydroxybenzoic acid + quinic acid) | 638 |
| Kaempferol-O-xiloside | 418 |
| Myricetin-O-rhamnoside | 464 |
| Procyanidin dimer | 578 |
| Delphinidin-O-glucoside | 465 |
| Isorhamnetin + acetylglucuronic acid | 534 |
| Petunidin + glucuronic acid | 493 |
| Malvidin acetyl glucoside | 535 |

In the case of the AP extracts, despite having six different important peaks, there are only two different compounds: isoquercetin and quercetin glycosylated at different positions and with different sugar moieties such as rutin, a quercetin linked to rutinose or linked to arabinose. This composition is similar to others obtained with similar residues [13,14].

In the case of the GP, the extraction liquors are more complex with a high number of different molecules. The most abundant antioxidants detected are shown in Table 6. In the GP extracts, some of the compounds seem to be larger than the limit of detection employed, so we can only analyze the fragments of those compounds detected such as the tetra 2,3-dihydroxycinnamic acid or Di 2,3-dihydroxybenzoic acid + quinic acid.

As Table 6 reflects, GP extracts have more variability, presenting phenolic acids such as hydroxycinnamic acid derivatives, sinapic acid, and 2,3-dihydroxycinnamic acid with quinic acid and flavonols such as catechin and its derivatives, kaempferol, myricetin, isorhamnetin and derivatives, and anthocyanins such as procyanidin, delphinidin, petunidin, and malvidin. This composition is similar to those obtained in other works [3,25–27].

## 4. Conclusions

Apple and grape pomaces were valorized successfully to obtain extracts rich in antioxidant compounds with simple, efficient, and fast extractions, highlighting the extractions employing ethanol/water mixtures as solvents at mild-high temperature values. However, more solvents could be useful like ethyleneglycol or acetone/water mixtures, especially in the case of GP.

We can conclude that GP is the better raw material for the obtention of liquors with high antioxidant capacity, reaching a maximum of 131 TEAC/g dry solid in the extraction employing ethanol 70% (*v/v*) and 90 °C, that matches with the better extraction solvents in terms of TPC reaching a value of 73 mg GAE/g dry solid. Ethyleneglycol is a promising solvent too, but its toxicity reduces the range of applications, especially in the food and health sectors. Nevertheless, AP can be used to obtain more pure liquors as its composition is less complex, despite its TPC and antioxidant capacity being notably lower.

An in-depth, though empirical, kinetic analysis has been applied to all time-course TPC and AC data, achieving high regression coefficients and high to very high F-values, showing an almost perfect match of the kinetic model to all data. This kinetic model is identical for AP and GP: a second-order kinetic model that reflects an immediate washing or dissolution (first extraction phase) in the parameter $q_0$ and a second extraction phase controlled by mass transport that is much slower and is characterized by the two parameters $q_1$ and $k_d$.

Viscosity, solid structure modification by solvents, and the solvation power of such solvents seem to explain the differences between solvents while the Arrhenius equation suggests different controlling phenomena at diverse temperature intervals for TPC and AC with both wastes, including internal mass transfer and phenolic thermal degradation.

Finally, a fine characterization of liquors in terms of phenolic composition was achieved thanks to HPLC and HPLC-MS analysis.

**Author Contributions:** Conceptualization, V.E.S.M., P.Y.C., J.M.B. and M.L.; data curation, J.G.-M. and A.G.-M.; formal analysis, A.G.-M., P.Y.C. and M.L.; funding acquisition, V.E.S.M. and M.L.; investigation, J.G.-M., A.G.-M. and J.I.B.; methodology, A.G.-M., P.Y.C., J.M.B. and M.L.; project administration, V.E.S.M. and M.L.; resources, V.E.S.M. and J.M.B.; software, J.G.-M., J.I.B. and M.L.; supervision, V.E.S.M. and M.L.; validation, A.G.-M. and M.L.; writing—original draft, J.G.-M. and M.L.; writing—review and editing, V.E.S.M., P.Y.C., J.M.B. and M.L. All authors have read and agreed to the published version of the manuscript.

**Funding:** This work was kindly supported by the Spanish Science and Innovation Ministry though three research projects: PCI2018-093114, CTQ2017-84963-C2-1-R, and PID2020-114365RB-C21, funding that is gratefully acknowledged.

**Institutional Review Board Statement:** Not applicable.

**Informed Consent Statement:** Not applicable.

**Data Availability Statement:** All relevant data are included in the manuscript. Other data are available from the authors upon reasonable request.

**Acknowledgments:** We are very grateful to Eva Navascues López-Cordón (Pago de Carraovejas SL, Valladolid, Spain) and Eva Romeo Longares (COVILLASA, Zaragoza, Spain) for the gift for grape and apple pomaces, respectively, and their kind explanations of the processes where these wastes are created. The assistance and support in the pretreatment of grape pomace residues of Cynthia Hopson and the research group "Development of Processes and Products of Low Environmental Impact" (UCM) is gratefully acknowledged.

**Conflicts of Interest:** The authors declare no conflict of interest.

## References

1. Fierascu, R.C.; Fierascu, I.; Avramescu, S.M.; Sieniawska, E. Recovery of natural antioxidants from agro-industrial side streams through advanced extraction techniques. *Molecules* **2019**, *24*, 4212. [CrossRef] [PubMed]
2. UN General Assembly Transforming Our World: The 2030 Agenda for Sustainable Development. Available online: https://www.refworld.org/docid/57b6e3e44.html (accessed on 15 February 2022).
3. Tapia Quirós, P.; Reig, M.; Vecino, X.; Cortina, L.; Saurina, J.; Granados, M. Recovery of polyphenols from agri-food by-products: The olive oil and winery industries cases. *Foods* **2022**, *11*, 362. [CrossRef] [PubMed]
4. Food and Agriculture Organization of the United Nations. FAOSTAT Data Crops. Available online: https://www.fao.org/faostat/en/#home (accessed on 15 February 2022).
5. Cano, M.E.; García-Martin, A.; Comendador Morales, P.; Wojtusik, M.; Santos, V.E.; Kovensky, J.; Ladero, M. Production of oligosaccharides from agrofood wastes. *Fermentation* **2020**, *6*, 31. [CrossRef]
6. Da Silva, L.C.; Viganó, J.; de Souza Mesquita, L.M.; Dias, A.L.B.; de Souza, M.C.; Sanches, V.L.; Chaves, J.O.; Pizani, R.S.; Contieri, L.S.; Rostagno, M.A. Recent advances and trends in extraction techniques to recover polyphenols compounds from apple by-products. *Food Chem. X* **2021**, *12*, 100133. [CrossRef]
7. Barreira, J.C.; Arraibi, A.A.; Ferreira, I.C. Bioactive and functional compounds in apple pomace from juice and cider manufacturing: Potential use in dermal formulations. *Trends Food Sci. Technol.* **2019**, *90*, 76–87. [CrossRef]
8. Ahmad, B.; Yadav, V.; Yadav, A.; Rahman, M.U.; Yuan, W.Z.; Li, Z.; Wang, X. Integrated biorefinery approach to valorize winery waste: A review from waste to energy perspectives. *Sci. Total Environ.* **2020**, *719*, 137315. [CrossRef]
9. Miller, K.V.; Noguera, R.; Beaver, J.; Medina-Plaza, C.; Oberholster, A.; Block, D.E. A mechanistic model for the extraction of phenolics from grapes during red wine fermentation. *Molecules* **2019**, *24*, 1275. [CrossRef]
10. Durazzo, A.; Lucarini, M.; Souto, E.B.; Cicala, C.; Caiazzo, E.; Izzo, A.A.; Novellino, E.; Santini, A. Polyphenols: A concise overview on the chemistry, occurrence, and human health. *Phytother. Res.* **2019**, *33*, 2221–2243. [CrossRef]
11. Alara, O.R.; Abdurahman, N.H.; Ukaegbu, C.I. Extraction of phenolic compounds: A review. *Curr. Res. Food Sci.* **2021**, *4*, 200–214. [CrossRef]
12. Ozturk, B.; Parkinson, C.; Gonzalez-Miquel, M. Extraction of polyphenolic antioxidants from orange peel waste using deep eutectic solvents. *Sep. Purif. Technol.* **2018**, *206*, 1–13. [CrossRef]

13. Ribeiro, L.F.; Ribani, R.H.; Francisco, T.M.G.; Soares, A.A.; Pontarolo, R.; Haminiuk, C.W.I. Profile of bioactive compounds from grape pomace (*Vitis vinifera* and *Vitis labrusca*) by spectrophotometric, chromatographic and spectral analyses. *J. Chromatogr. B Anal. Technol. Biomed. Life Sci.* **2015**, *1007*, 72–80. [CrossRef] [PubMed]
14. Islam, M.A.; Chowdhury, M.A.; Mozumder, M.S.I.; Uddin, M.T. Langmuir adsorption kinetics in liquid media: Interface reaction model. *ACS Omega* **2021**, *6*, 14481–14492. [CrossRef] [PubMed]
15. Martin-Dominguez, V.; Aleman Cabrera, P.I.; Eidt, L.; Pruesse, U.; Kuenz, A.; Ladero, M.; Santos, V.E. Production of fumaric acid by *Rhizopus arrhizus* NRRL 1526: A simple production medium and the kinetic modelling of the bioprocess. *Fermentation* **2022**, *8*, 64. [CrossRef]
16. Suárez, B.; Álvarez, Á.L.; García, Y.D.; del Barrio, G.; Lobo, A.P.; Parra, F. Phenolic profiles, antioxidant activity and in vitro antiviral properties of apple pomace. *Food Chem.* **2010**, *120*, 339–342. [CrossRef]
17. Cuevas-Valenzuela, J.; González-Rojas, Á.; Wisniak, J.; Apelblat, A.; Pérez-Correa, J.R. Solubility of (+)-catechin in water and water-ethanol mixtures within the temperature range 277.6–331.2 K: Fundamental data to design polyphenol extraction processes. *Fluid Phase Equilibria* **2014**, *382*, 279–285. [CrossRef]
18. Sólyom, K.; Solá, R.; Cocero, M.J.; Mato, R.B. Thermal degradation of grape marc polyphenols. *Food Chem.* **2014**, *159*, 361–366. [CrossRef]
19. Manara, P.; Zabaniotou, A.; Vanderghem, C.; Richel, A. Lignin extraction from Mediterranean agro-wastes: Impact of pretreatment conditions on lignin chemical structure and thermal degradation behavior. *Catal. Today* **2014**, *223*, 25–34. [CrossRef]
20. Reis, S.F.; Rai, D.K.; Abu-Ghannam, N. Water at room temperature as a solvent for the extraction of apple pomace phenolic compounds. *Food Chem.* **2012**, *135*, 1991–1998. [CrossRef]
21. Waldbauer, K.; McKinnon, R.; Kopp, B. Apple pomace as potential source of natural active compounds. *Planta Med.* **2017**, *83*, 994–1010. [CrossRef]
22. Razgonova, M.; Zakharenko, A.; Pikula, K.; Manakov, Y.; Ercisli, S.; Derbush, I.; Kislin, E.; Seryodkin, I.; Sabitov, A.; Kalenik, T.; et al. Lc-ms/ms screening of phenolic compounds in wild and cultivated grapes vitis amurensis rupr. *Molecules* **2021**, *26*, 3650. [CrossRef]
23. Hobbi, P.; Okoro, O.V.; Delporte, C.; Alimoradi, H.; Podstawczyk, D.; Nie, L.; Bernaerts, K.V.; Shavandi, A. Kinetic modelling of the solid–liquid extraction process of polyphenolic compounds from apple pomace: Influence of solvent composition and temperature. *Bioresour. Bioprocess.* **2021**, *8*, 114. [CrossRef]
24. Natolino, A.; Da Porto, C. Kinetic models for conventional and ultrasound assistant extraction of polyphenols from defatted fresh and distilled grape marc and its main components skins and seeds. *Chem. Eng. Res. Des.* **2020**, *156*, 1–12. [CrossRef]
25. Flamini, R. Recent applications of mass spectrometry in the study of grape and wine polyphenols. *ISRN Spectrosc.* **2013**, *2013*, 813563. [CrossRef]
26. De Villiers, A.; Vanhoenacker, G.; Majek, P.; Sandra, P. Determination of anthocyanins in wine by direct injection liquid chromatography-diode array detection-mass spectrometry and classification of wines using discriminat analysis. *J. Chromatogr. A* **2004**, *1054*, 195–204. [CrossRef]
27. Li, J.; Zhang, S.; Zhang, M.; Sun, B. Novel approach for extraction of grape skin antioxidants by accelerated solvent extraction: Box–Behnken design optimization. *J. Food Sci. Technol.* **2019**, *56*, 4879–4890. [CrossRef]

*applied sciences*

MDPI

*Article*

# Bio-Augmentation as an Emerging Strategy to Improve the Textile Compost Quality Using Identified Autochthonous Strains

**Saloua Biyada [1,*], Hamada Imtara [2,*], Karima Elkarrach [1], Omar Laidi [1], Asmaa Saleh [3,*], Omkulthom Al Kamaly [3] and Mohammed Merzouki [1]**

1 Laboratory of Biotechnology, Environment, Agrifood and Health, Faculty of Sciences Dhar El Mahraz, Sidi Mohamed Ben Abdellah University, BP: 1796, Atlas, Fez 30 000, Morocco; karima.elkarrach@usmba.ac.ma (K.E.); omar.laidi@usmba.ac.ma (O.L.); mohammed.merzouki@usmb.ma.ac (M.M.)

2 Faculty of Arts and Sciences, Arab American University Palestine, P.O. Box 240, Jenin 44862, Palestine

3 Department of Pharmaceutical Sciences, College of Pharmacy, Princess Nourah Bint Abdulrahman University, P.O. Box 84428, Riyadh 11671, Saudi Arabia; omalkmali@pnu.edu.sa

* Correspondence: salouabiyada@gmail.com (S.B.); hamada.tarayrah@gmail.com (H.I.); asali@pnu.edu.sa (A.S.)

**Abstract:** The present investigation is devoted, for the first time, to the potential of autochthonous inoculums through bio-augmentation tests to improve the compost quality and to decrease the composting time during composting of textile waste. For this reason, three strains were isolated from a mixture of textile waste, green waste, paper, and cardboard waste, and therefore identified as *Streptomyces cellulosae*, *Achromobacter xylosoxidans*, and *Serratia liquefaciens*, employed using bio-augmentation test. The organic matter decaying was assessed according to three different inoculums doses, separately and in consortium (4%, 6%, and 8%), to describe the effect of bio-augmentation process on the organic matter decaying. Indeed, these three strains and their consortium have shown a strong potential of organic matter degradation, equally the bacterial consortium showed a total organic carbon degradation of 20.3%, total Kjeldahl nitrogen of 1.52%, and a Carbon/Nitrogen ratio of 13.36. Compost maturity has been completed after only 12 weeks of treatment instead of 44 weeks using the classical treatment by composting. Ultimately, according to these results, bio-augmentation could be an emerging and promising strategy to accelerate the composting process of solid waste, especially in the case of industrial waste. Equally, it could be an effective tool to avoid the accumulation of industrial waste disposal in public landfills and/or nature while allowing their treatment.

**Keywords:** bio-augmentation; composting time; organic matter; textile solid waste; composting



**Citation:** Biyada, S.; Imtara, H.; Elkarrach, K.; Laidi, O.; Saleh, A.; Al Kamaly, O.; Merzouki, M. Bio-Augmentation as an Emerging Strategy to Improve the Textile Compost Quality Using Identified Autochthonous Strains. *Appl. Sci.* **2022**, *12*, 3160. https://doi.org/10.3390/app12063160

Academic Editor: Carlos Rico de la Hera

Received: 31 January 2022
Accepted: 16 March 2022
Published: 20 March 2022



## 1. Introduction

The textile industry depicts a substantial economic sector worldwide, and it is considered one of the largest industrial sectors from the aspect of solid waste generation [1]. Despite their economic value, this industry is ranked among the most polluting industries, which generates huge amounts of solid waste. Its toxicity is based on the fact that these wastes contain various pollutants such as degradable organics, dyes, salts, sulfuric acid, toxicants, as well as refractory organics [2]. These pollutants depict a significant environmental and health issue, especially in the case of dyes such as azo dyes, anthraquinones, phthalocyanines, etc., which are toxic, persistent in nature, and affect the biotic species of the environment. Equally, they have a genotoxic and carcinogenic effect on the health public [1], in the whole world, particularly Fez city in Morocco. This problem is further aggravated in Fez city, because these wastes are disposed of directly into the environment without any treatment. Several chemical and physical methods are used in color removal, such as precipitation, flocculation, membrane filtration, adsorption, and wet oxidation.

These methods are quite expensive, could generate the production of toxic by-products, and have posed operational problems [2]. For this reason, it is mandatory to promote the stabilization of these wastes to avoid their innumerous environmental problems. On the contrary, textile wastes are a source of organic matter, and these wastes could be bio-converted into a nutrient source to the soil if they are treated properly [3,4].

For many years, composting has been considered as the most efficient technology for the stabilization of the solid waste generated [5]. Although composting has several advantages regarding processing cost and product quality, these do not preclude that there are equally some disadvantages, especially the presence of recalcitrant molecules, which are not easily dissipated by microorganisms, thus allowing a slowdown in biodegradation during composting was observed. According to a previous study, textile waste had only 31% of organic matter that may be degradable by microorganisms [6], so it is necessary to increase this percentage by adding other organic waste such as green waste. In this sense, the implementation of this process to improve some biological and physical–chemical parameters is widely recommended.

Bioaugmentation is defined as a technique for improvement of the degradative capacity of contaminated areas by the introduction of specific competent strains or consortia of microorganisms.

Otherwise, bio-augmentation is recognized as a cost-effective technology for improving the biodegradability of organic matter, even with the presence of recalcitrant molecules [7], by the introduction of specific competent strains or consortia of microorganisms [8]. This technology involves the inoculation of potential microorganisms to improve organic matter degradation, in which these microorganisms may be autochthonous or allochthonous, as well as potentially being a pure culture and/or a consortium [9–12]. The selection of appropriate strains for bio-augmentation should take into account the following characteristics of microorganisms: high degradation potential, rapid growth, and ease of culture [11]. Abed et al. [13] have revealed that bio-augmentation allows shortening the period required for the degradation of compounds relatively recalcitrant to degradation. Several studies are devoted to the effect of bio-augmentation on composting performance using commercial strains [5,8], but few of them are interested in the use of autochthonous strains during composting. For this reason, a deep understanding of the effect of autochthonous strains on composting is recommended.

No published work was reported for the use of the bio-augmentation technique during textile waste composting as an additional technology to improve the composting performance and the final compost quality, or even for the isolation of autochthonous strains and their re-inoculation during textile waste composting. The main aim of the present work was to isolate and identify bacterial strains from a mixture of textile waste, green waste, and waste paper and cardboard intended for treatment by composting and re-inoculating them during the bio-augmentation test. More specifically, this study focuses on the use of different doses of autochthonous inoculum to show their effect on the composting process. The proposed method could be an effective tool to treat and recover solid textile waste in a short time and with surprising results from the aspect of organic matter decaying, therefore reducing the number of landfills built and their harmful impact on the environment.

## 2. Materials and Methods

### 2.1. Characterization of Feedstock

Preparation of the mixture was followed according to the procedure explained previously [14]. The composition of the feedstock was summarized in Table 1.

**Table 1.** Physical–chemical characterization of the feedstock.

| Physical–Chemical Parameters | Textile Waste | Green Waste | Paper and Cardboard Waste | Norm NF U44-051/A2 |
|---|---|---|---|---|
| Moisture% | 51.28 ±1.03 | 61.49 ± 1.41 | 11.28 ±1.07 | 40–60 |
| pH | 7.4 ± 0.15 | 6.6 ± 0.53 | 7.2 ± 0.35 | 6.5–8.5 |
| Total Organic Carbon (TOC%) | 31.63 ± 1.48 | 45.67 ± 1.37 | 59.35 ± 0.90 | >20 |
| Total Nitrogen (TN%) | 0.57 ± 0.04 | 1.23 ± 0.04 | 1.05 ± 0.05 | - |
| C/N ratio | 55.1 | 37.2 | 56.7 | 20–40 |

Values designate mean ± standard deviation based on 3 samples.

In a previous study [15], a mixture symbolized as Mix C of the three wastes was set up for composting using silo technique according to the following proportion ratio of 40%:30%:30%, respectively, for textile waste, green waste, and paper and cardboard waste. A total of 40 kg was used during composting at room temperature and turned at least three times per week for 44 weeks. Samples were collected according to the four cardinal positions (north, east, south, and west).

### 2.2. Isolation of Bacterial Strains from Compost Mixture

Isolation of the bacterial strains was carried out using a medium based on the mix of the initial substrate, prepared as follows: 35 g of initial substrate (textile waste, green waste, and paper and cardboard waste) in one liter of distilled water, macerated overnight, and subsequently the filtrate was supplemented with 15 g of agar. Isolates of bacteria morphologically distinct were selected and purified by successive preculture until the pure strains were obtained [16]. Pure cultures were stored on nutrient agar slants at 4 °C as working stock cultures.

### 2.3. Bio-Augmentation Test

In previously sterilized pots (with an effective size of 0.17 × 0.17 m), 1 kg of sterile waste from Mix C (textile waste, green waste and paper and cardboard waste) was inoculated on the surface using different inoculums doses. All the pots were mixed daily for aeration throughout 12 weeks of treatment at room temperature. Three inoculums' doses were studied, which were 4%, 6%, and 8% ($v/w$), as well as a consortium of the three isolates with the same concentrations. A negative test (T) was prepared under the same conditions and used without inoculum to reveal the bio-augmentation efficiency of our isolates.

### 2.4. Experimental Analysis

Physical–chemical parameters of the mixtures, such as total organic content (TOC%), was carried out by methods as described by [9]. The total Kjeldahl nitrogen (TKN%) content was determined using the standard method of nitrogen Kjeldahl analysis according to [17]. The C/N ratio was calculated from the percentage of total organic carbon (TOC%) and the total Kjeldahl nitrogen (TKN%). Measurement of the temperature was carried out according to the protocol described by the French Association for Standardization [18]. Microbial analysis was assessed using a standard serial dilution procedure. All assays were estimated in triplicate (in weeks 0, 2, 4, 6, 8, 10, and 12), and were then placed into polythene bags and stored at 4 °C until further analysis was conducted.

### 2.5. Molecular Identification of Isolated Strains

DNA extraction was carried out according to kit protocol using the Genomic DNA purification Kit of Thermo Fisher. Then, 16S rRNAs gene targeting was amplified by polymerase chain reaction (PCR) using the universal bacterial primers: FD1 (5′ AGA GTT TGA TCC TGG CTC AG 3′) and RP2 (5′ ACG GCT ACC TTG TTA CGA CTT 3′) and a thermal cycler (Verity ABI). Ultimately, the amplified fragments were sequenced using ABI 3130XL capillary sequencer. Obtained sequences were aligned using the BLASTn software.

Additionally, the neighbor-joining tree was constructed using neighbor-joining (NJ) method MEGA_11-0.2 software. Molecular identification was performed at the laboratory of the National Center of Scientific and Technical Research (CNRST) at the Functional Genomics Platform in Rabat city, Morocco.

### 2.6. Statistical Analysis

The data are shown as the mean value ± standard deviation (SD) from three independent experiments. ANOVA two-way tests were performed to determine significant differences between the different strains, consortium, and organic matter degradation; the differences were additionally performed using Tukey's tests. A $p$-value of <0.05 was considered to be statistically significant. These tests were performed using Graph-Pad prism.

## 3. Results and Discussion

### 3.1. Morphological Appearance of the Isolated Strains

Three morphologically distinct isolates with opaque forms were isolated and purified. The isolates are symbolized as S1, S2, and S3.

### 3.2. Effect of the Bio-Augmentation on the Composting Process

#### 3.2.1. Evolution of Bacterial Growth

Bacterial growth evolution is depicted in Figure 1. From the beginning, the three isolated strains (S1, S2, and S3) and their consortium experienced considerable growth. All of the isolates grew swiftly with a maximum value at a time corresponding to the 8th week (Figure 1a–d). Nevertheless, the bacterial growth was significantly higher with the consortium relative to the pure culture ($p$-value < 0.05). Equally, the bacterial growth was considerably higher using these selected strains, with maximum growth reached in 8 weeks compared to the basic composting without inoculation, with a maximum of growth in 28 weeks [6], which could be assigned to the antibiosis phenomenon between the different microbial strains existing during the classical composting, and consequently prevent the growth of the strains which allows organic matter degradation, including those isolated. No growth of bacteria was noticed with a negative test (T) without bacteria, thus proving that organic matter decaying has been accomplished by the strains isolated. Moreover, the increase in microbial activity using these strains could be explained on the one hand by the fact that these isolates have already been adapted to the compost environment and have the ability to resist toxic molecules of textile waste. On the other hand, it could be explained by the abundance of organic matter with nutrients such as carbon and nitrogen, which are indispensable for microorganism's growth and therefore stimulates their proliferation. The decrease in microbial growth towards the end of the test could be linked to the depletion of medium [19].

#### 3.2.2. Evaluation of Temperature, Total Organic Carbon, (TOC) Total Kjeldahl Nitrogen (TKN), and C/N Ratio

The effects of inoculum doses on some physical–chemical parameters (Temperature, TOC, TKN, and C/N ratio) are depicted during this investigation. Measurement of temperature using different inoculums doses is presented in Figure 2. A noteworthy increase in temperature was noticed with the three strains and their consortium, depending on different inoculum concentrations (4%, 6%, and 8%). The highest temperature was recorded with strain S3 whatever the inoculum dose, and for strain S3 (42 °C) and the consortium (54.17 °C) at the inoculum dose of 8% (Figure 1). The increase in temperature could be mainly ascribed according to several authors to microbial metabolisms resulting from the decomposition of organic matter, particularly carbohydrates, thus releasing energy during their degradation [6,15,20]. Biyada et al. [21] have shown the presence of carbohydrates in composted waste like cellulose through FTIR analysis, thus confirming that this increase of the temperature could be explained by the break of these carbohydrates by the isolated strains used.

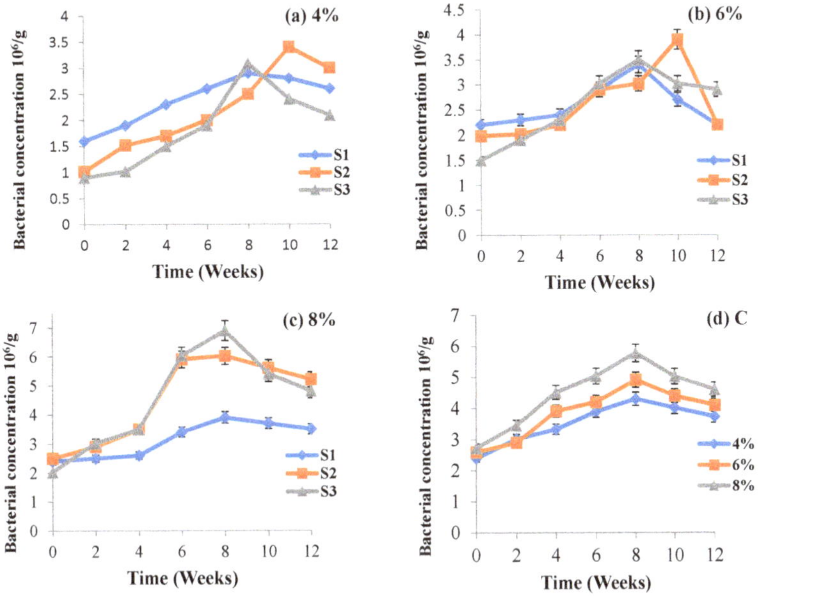


**Figure 1.** Evolution of bacterial growth of the three strains isolated and their consortium as function of the time at different inoculum doses (4% (**a**), 6% (**b**), and 8% (**c**)). S1: Strain 1; S2: Strain 2; S3: Strain 3; (**d**) C: Consortium.

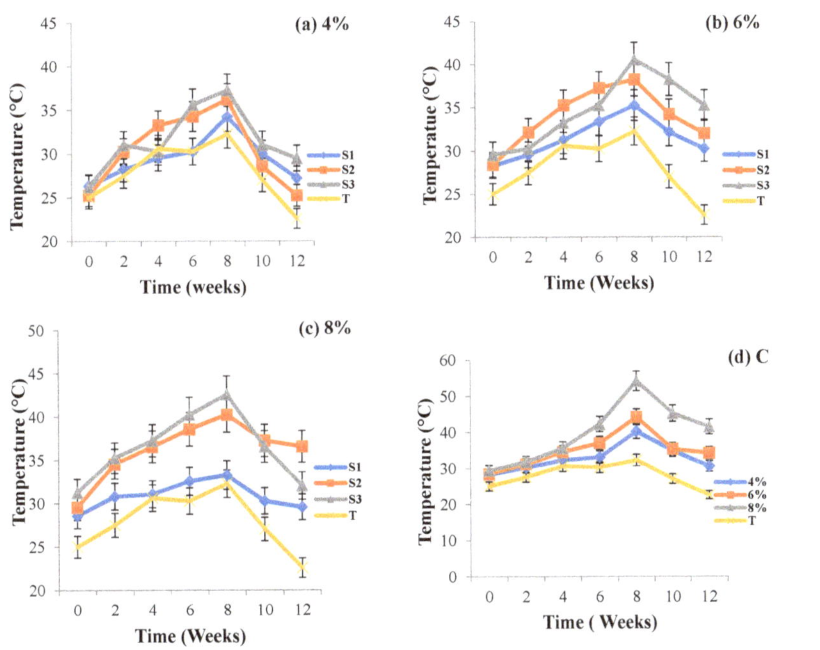


**Figure 2.** Evolution of temperature (°C) of the three strains isolated and their consortium as function of the time at different inoculum doses (4% (**a**), 6% (**b**), and 8% (**c**)), T; negative test. S1: Strain 1; S2: Strain 2; S3: Strain 3; (**d**) C: Consortium.

Table 2 depicts the total organic carbon (TOC%) evolution using isolated strains and their consortium. In fact, a significant decrease in TOC was recorded ($p < 0.05$) towards the end for the three strains and their consortium depending on different inoculums doses (4%, 6%, and 8%). The decrease in TOC was together with the increase in bacterial growth, which confirms the capacity of these strains to degrade carbonaceous compounds during their growth [20,22]. This was confirmed in a previous study using infrared spectroscopy analysis, thus showing that the wastes used in this study are rich in carbohydrates and polysaccharide compounds, which is considered as a source of carbon and energy for the bacterial growth as mentioned above [21]. Through this investigation, a decrease attributed to these compounds (carbohydrates and polysaccharide, which are rich in carbon) depending on the composting time was recorded, which could be assigned to the degradation and/or the assimilation of these compounds by the microorganisms of the compost, especially as TOC in negative test remains stable during the composting process (Table 2). Indeed, this decrease of TOC could explain the increase of the temperature and confirm our previous hypothesis.

Similarly, for TKN from the beginning, a significant increase was recorded ($p < 0.05$) (Table 3), which could be attributed to the organic matter decaying (proteins and peptides) to ammoniacal compounds, which are assimilated forms by bacterial isolates and their consortium [23,24]. Together with a decrease of TOC and the increase of TKN, a significant decrease in the C/N ratio was noticed, but it still remains above the maturity standard (15–20), with values of 20.00, 19.87, and 19.30 for 4%, 6%, and 8%, respectively, in the case of pure strains, and below the maturity standard in the case of bacterial consortium 18.69, 15.70, and 13.36 for 4%, 6%, and 8%, respectively (Table 4). Nevertheless, whether for TOC, TKN, or even C/N ratios, the content remained stable over time within the negative Test (T). This could confirm that the organic matter degradation was carried out by the three strains used. Additionally, the decrease in C/N ratios was higher using the consortium compared to that using pure culture.

Using these strains, whether on pure culture or their consortium, the results obtained are significantly higher and swiftly compared to the results obtained during classic composting without selected inoculums [6].

In this respect, Tukey's multiple comparison test, revealed that the use of the inoculum affected positively the quality of the final compost produced. Equally, by increasing the inoculums concentration (from 4%, 6%, to 8%), the TOC content and C/N ratios were decreased; at the same time, the temperature and TKN were increased. For the consortium of the three strains (S1, S2, and S3), there is a significant effect ($p$-value $< 0.05$) between the consortium concentration and the degradation of organic matter.

### 3.3. Molecular Identification of Isolated Strains

The FASTA formats were analyzed using the database of BLASTn software. The identification was based on the score, the percentage of identity, and Query covers, which must be greater than 96%, thus identifying these isolates as: *Streptomyces cellulosae*, *Achromobacter xylosoxidans*, and *Serratia liquefaciens* for S1, S2, and S3, respectively (similarity more than 98%). Briefly, a sequence from these strains was submitted to GenBank with their accession numbers KC429648, JX050258, and CP033893, respectively, for *Streptomyces cellulosae*, *Achromobacter xylosoxidans*, and *Serratia liquefaciens*. Based on the 16S RNA gene sequence of the strains isolated, three neighbor-joining trees were formed during this investigation and some other phylogenetically related taxa (Figure 3). These phylogenetic trees indicated that S1, S2, and S3 were closely related to *Streptomyces cellulosae*, *Achromobacter xylosoxidans*, and *Serratia liquefaciens*, respectively, thus revealing that the strains isolated were identified as *Streptomyces cellulosae*, *Achromobacter xylosoxidans*, and *Serratia liquefaciens*, respectively.

**Table 2.** Evolution of TOC degradation of the three isolated strains (S1: Strain 1; S2: Strain 2; S3: Strain 3) and their consortium as function of the time at different inoculum doses (4%, 6%, and 8%).

| TOC% | [4%] | | | | [6%] | | | | [8%] | | | | Negative Test |
|---|---|---|---|---|---|---|---|---|---|---|---|---|---|
| | Week (0) | Weeks (4) | Weeks (8) | Weeks (12) | Week (0) | Weeks (4) | Weeks (8) | Weeks (12) | Week (0) | Weeks (4) | Weeks (8) | Weeks (12) | |
| S1 | 32.90 ± 0.44 | 31.00 ± 0.09 | 30.50 ± 0.96 | 26.4 ± 0.30 | 34.00 ± 0.20 | 31.20 ± 0.26 | 25.20 ± 1.02 | 24.20 ± 0.95 | 33.70 ± 0.02 | 29.10 ± 0.56 | 26.70 ± 0.52 | 24.60 ± 0.30 | 32.90 ± 0.10 |
| S2 | 32.74 ± 0.13 | 30.27 ± 0.36 | 26.67 ± 0.30 | 25.04 ± 1.01 | 31.23 ± 0.10 | 29.37 ± 0.95 | 26.25 ± 0.26 | 24.15 ± 0.12 | 31.60 ± 0.10 | 27.55 ± 1.02 | 24.30 ± 0.10 | 23.45 ± 0.20 | 32.90 ± 0.12 |
| S3 | 33.01 ± 0.23 | 30.72 ± 0.40 | 29.35 ± 0.50 | 28.11 ± 1.05 | 31.46 ± 0.50 | 29.49 ± 0.86 | 26.16 ± 0.15 | 24.46 ± 0.50 | 31.50 ± 0.23 | 26.50 ± 0.20 | 23.50 ± 0.20 | 22.20 ± 0.15 | 32.90 ± 0.20 |
| Consortium | 32.10 ± 0.01 | 30.10 ± 0.04 | 28.50 ± 0.30 | 24.30 ± 0.12 | 31.70 ± 0.40 | 29.40 ± 0.20 | 26.50 ± 0.13 | 22.30 ± 0.30 | 30.30 ± 0.15 | 26.90 ± 0.15 | 23.60 ± 0.01 | 20.30 ± 0.12 | 32.90 ± 0.60 |

Values designate mean ± standard deviation based on 3 samples.

**Table 3.** Evolution of TKN degradation of the three strains isolated (Strain 1; S2: Strain 2; S3: Strain 3) and their consortium as function of the time at different inoculum doses (4%, 6%, and 8%).

| TKN% | [4%] | | | | [6%] | | | | [8%] | | | | Negative Test |
|---|---|---|---|---|---|---|---|---|---|---|---|---|---|
| | Week (0) | Weeks (4) | Weeks (8) | Weeks (12) | Week (0) | Weeks (4) | Weeks (8) | Weeks (12) | Week (0) | Weeks (4) | Weeks (8) | Weeks (12) | |
| S1 | 0.89 ± 0.02 | 0.98 ± 0.01 | 1.03 ± 0.03 | 1.13 ± 0.01 | 0.75 ± 0.03 | 0.90 ± 0.01 | 1.01 ± 0.03 | 1.10 ± 0.02 | 0.76 ± 0.01 | 1.00 ± 0.01 | 1.09 ± 0.02 | 1.23 ± 0.02 | 0.50 ± 0.01 |
| S2 | 0.50 ± 0.01 | 0.86 ± 0.02 | 1.02 ± 0.12 | 1.09 ± 0.03 | 0.59 ± 0.04 | 1.01 ± 0.02 | 1.09 ± 0.02 | 1.12 ± 0.02 | 0.65 ± 0.01 | 1.09 ± 0.03 | 1.13 ± 0.03 | 1.18 ± 0.01 | 0.50 ± 0.02 |
| S3 | 0.7 ± 00.10 | 0.84 ± 0.01 | 1.00 ± 0.01 | 1.20 ± 0.01 | 0.75 ± 0.05 | 1.01 ± 0.01 | 1.12 ± 0.05 | 1.15 ± 0.02 | 0.69 ± 0.01 | 1.09 ± 0.02 | 1.14 ± 0.01 | 1.15 ± 0.02 | 0.50 ± 0.01 |
| Consortium | 0.68 ± 0.03 | 0.77 ± 0.04 | 0.98 ± 0.01 | 1.30 ± 0.02 | 0.70 ± 0.01 | 0.92 ± 0.02 | 1.09 ± 0.04 | 1.42 ± 0.01 | 0.69 ± 0.02 | 1.03 ± 0.04 | 1.19 ± 0.02 | 1.52 ± 0.01 | 0.50 ± 0.02 |

Values designate mean ± standard deviation based on 3 samples.

**Table 4.** Evolution of C/N ratio of the three strains isolated (S1: Strain 1; S2: Strain 2; S3: Strain 3) and their consortium as function of the time at different inoculum doses (4%, 6%, and 8%).

| C/N | [4%] | | | | [6%] | | | | [8%] | | | | Negative Test |
|---|---|---|---|---|---|---|---|---|---|---|---|---|---|
| | Week (0) | Weeks (4) | Weeks (8) | Weeks (12) | Week (0) | Weeks (4) | Weeks (8) | Weeks (12) | Week (0) | Weeks (4) | Weeks (8) | Weeks (12) | |
| S1 | 36.97 | 31.63 | 29.61 | 23.36 | 45.33 | 34.67 | 24.95 | 22.00 | 44.34 | 29.10 | 24.50 | 20.00 | 65.80 |
| S2 | 65.48 | 35.20 | 26.15 | 22.97 | 52.93 | 29.08 | 24.08 | 21.56 | 48.62 | 25.23 | 21.50 | 19.87 | 65.80 |
| S3 | 47.16 | 36.57 | 29.35 | 23.43 | 41.95 | 29.20 | 23.36 | 21.27 | 45.65 | 24.31 | 20.61 | 19.30 | 65.80 |
| Consortium | 47.21 | 39.06 | 29.08 | 18.69 | 45.29 | 31.96 | 24.31 | 15.70 | 43.91 | 26.12 | 19.83 | 13.36 | 65.80 |

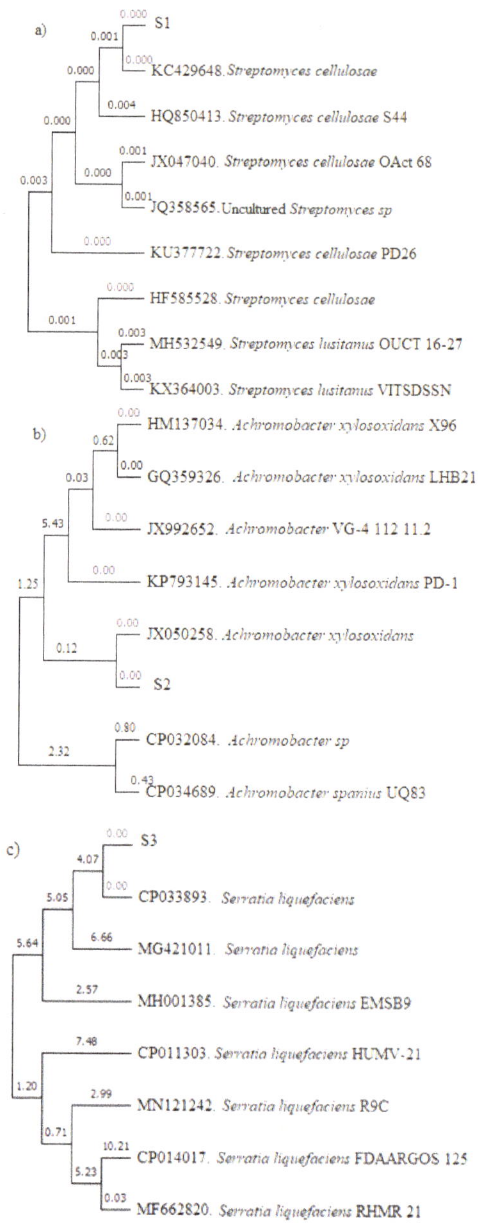


**Figure 3.** Neighbor-joining phylogenetic tree based on the 16S rRNA gene sequence of the isolates and other reference sequences ((**a**) S1, (**b**) S2, and (**c**) S3) (Numbers in figures show branch lengths).

Several authors reveal the effects of bio augmentation on the composting process using a wide range of microorganisms as well as different types of solid waste [5,7,11]. Regarding the literature, it was demonstrated that species belonging to *Streptomyces*, *Achromobacter*, and *Serratia* are able to use a wide range of organic matter, particularly in ligno-cellulosic compounds, as a fountainhead of carbon and energy and consequently use it for their growth [25]. Several authors revealed that *Streptomycetes cellulosae* was studied among potential microorganisms in nitrogen biodegradation systems using their enzymatic complex [26,27], which is in concordance with the results obtained by *Streptomycetes cellulosae* mentioned overhead regarding TKN. Additionally, Li, et al. [28] identified species belonging to *Achromobacter* genera as cellulolytic strains with high ability of polysaccharides compounds decaying, especially those in paper waste, which is owing to its enzymatic system of mainly carbohydrate hydrolases, endoglucanase, and endoxylanase [29]. These findings could explain the results obtained by *A. xylosoxidans* regarding the decrease in TOC, and therefore, C/N ratio. The ability of *Achromobacter* to degrade the lignocellulosic compounds in solid waste is identified by several authors that worked on household waste discharged in landfills as well as in paper waste [28,30]. Indeed, Haq et al. [31] identified *S. liquefaciens* as a ligninolytic strain, which is due to their ability to degrade lignin from solid waste and based on its detoxification and reduction of pollutants activity, as well as the azo dyes containing in paper waste, which is the case for the present study. In fact, the abundance of textile waste, paper and cardboard waste, and green waste in ligno-cellulosic compounds and azo dyes, which is confirmed beforehand [14,21], allows to induce lignin peroxidase enzymes (LiP) in *S. liquefaciens*. Biyada et al. [14] depicted using FTIR and XRD analysis, which the intensities of pics attributed to lignin, carbohydrates, hemicelluloses, and/or cellulosic in compost samples appeared, thus proving the decaying of polysaccharides, aliphatic, and carbohydrates from the compost samples, which could be a good indication of compost stability and maturity. These findings could explain the high effectiveness of *S. liquefaciens* in the degradation of the organic matter represented by TOC, TKN, and C/N compared to the other strains tested, equally to their effect on the time of composting, which is significantly much less than that during classical composting. It can be noted that the use of *S. liquefaciens* allows the dyes present in textile waste to decay.

Nevertheless, when a consortium of isolates was used, the organic matter degradation was considerably more intense, relative, and swift compared to the use of these isolates separately, which could be due to their ability to coexist with each other. The results recorded during this investigation could be explained by the fact that different metabolism pathways of the bacteria within this consortium are involved to degrade organic matter [10]. Several studies indicate that there might be a synergistic cooperation between bio-augmented bacterial species into consortia, which has strengthening their ability to accelerate the conversion of organic matter, especially recalcitrant substances [10,20]. These findings are consistent with the results obtained in this investigation. Other studies deduced that recalcitrant compounds, such as lingo-cellulosic, are degraded by microbial consortia in which each strain has specialized roles: some of them attack the complex substrate, others provide essential nutrients [32,33]. Li et al. and Ariffin et al. [28,34] have demonstrated that *S. liquefaciens* and *A. xylosoxidans* could fulfill the role of a degrading agent for lignocellulosic compounds owing to their enzyme system. Indeed, [35] investigated that *Streptomycetes cellulosae* have the ability to use cellulose and other lignin byproducts as a source of carbon and energy.

Ultimately, as demonstrated during this study, the bioaugmentation could be a useful tool to improve the performance of composting process and the quality of final compost either with industrials wastes, which is the case in the present study or even with other kind of wastes, which has been confirmed previously by several authors [5,7,11]. For this reason, the choice of microorganisms is crucial, and should be based on the nature of waste used, particularly their composition.

## 4. Conclusions

The degradation of the organic matter during composting was reinforced by bioaugmentation tests with three well isolated strains: *Streptomyces cellulosae*, *Achromobacter xylosoxidans*, *Serratia liquefaciens*, and their consortium. *Serratia liquefaciens* recorded an intense and prompt degradation of organic matter compared to other strains and to the classical composting without selected inoculum. In addition, with an inoculum dose of 8%, a value of total organic carbon degradation of 22.2%, total Kjeldahl nitrogen of 1.15%, and a C/N ratio of 19.30, the degradation was considerably more intense compared to 4%

and 6% doses, either by testing the strains separately or in consortium with a value of total organic carbon degradation of 20.3%, total Kjeldahl nitrogen of 1.52%, and a C/N ratio of 13.36, which is statistically confirmed. The present study revealed for the first time that the bio-augmentation with isolated strains can be a useful tool to reduce organic matter degradation time of textile waste during composting from 44 weeks to only 12 weeks, thus proving the effectiveness of the bio-augmentation on the composting process and the quality of the compost from textile waste. This investigation proves that the coupling of bio-augmentation with composting could be an effective tool to reduce the cost of building landfills, and equally to adopt a sustainable alternative to protect the environment from the side effects of waste disposal in landfills and/or directly in the nature.

**Author Contributions:** Conceptualization, S.B., M.M. and H.I.; methodology, S.B., K.E. and O.L.; data curation, H.I. and S.B.; Software, S.B.; Validation, M.M; writing—original draft preparation, S.B.; writing—review and editing, H.I., A.S., O.A.K. and M.M.; supervision, M.M. All authors have read and agreed to the published version of the manuscript.

**Funding:** This study was funded by the Princess Nourah bint Abdulrahman University Researchers Supporting Project number (PNURSP2022R141), Princess Nourah bint Abdulrahman University, Riyadh, Saudi Arabia.

**Institutional Review Board Statement:** Not applicable.

**Informed Consent Statement:** Not applicable.

**Data Availability Statement:** Not applicable.

**Acknowledgments:** The authors express their gratitude to Princess Nourah bint Abdulrahman University Researchers Supporting Project number (PNURSP2022R141),Princess Nourah bint Abdulrahman University, Riyadh, Saudi Arabia. Also, the authors would like to acknowledge the National Center for Scientific Research in Rabat (CNRST), for the crucial facilities and support provided to carry out the molecular identification of our strains.

**Conflicts of Interest:** The authors declare no conflict of interest.

**Sample Availability:** Samples are available from the authors upon reasonable request.

## References

1. Behera, M.; Nayak, J.; Banerjee, S.; Chakrabortty, S. A review on the treatment of textile industry waste effluents towards the development of efficient mitigation strategy: An integrated system design approach. *J. Environ. Chem. Eng.* **2021**, *9*, 105277. [CrossRef]
2. Talouizte, H.; Merzouki, M.; Benlemlih, M. Treatment of Real Textile Wastwater Using SBR Technology: Effect of Sludge Age and Operational Parameters. *J. Biotechnol. Lett.* **2017**, *4*, 79–83.
3. Biyada, S.; Merzouki, M.; Dėmčėnko, T.; Vasiliauskienė, D.; Ivanec-Goranina, R.; Urbonavičius, J.; Marčiulaitienė, E.; Vasarevičius, S.; Benlemlih, M. Microbial community dynamics in the mesophilic and thermophilic phases of textile waste composting identified through next—Generation sequencing. *Sci. Rep.* **2021**, *11*, 23624. [CrossRef] [PubMed]
4. Kashif, M.; Kashif, A.; Fuwad, A.; Choi, Y. Current advances in treatment technologies for removal of emerging contaminants from water – A critical review. *Coord. Chem. Rev.* **2021**, *442*, 213993. [CrossRef]
5. Antônio, L.; Costa, D.M.; Sarolli, M.; De Mendonça, S.; Gazzola, W. Bioaugmentation as a Strategy to Improve the Compost Quality in the Composting Process of Agro-Industrial Wastes. *Environ. Technol. Innov.* **2021**, *22*, 101478. [CrossRef]
6. Biyada, S.; Merzouki, M.; Imtara, H.; Benlemlih, M. Assessment of the Maturity of Textile Waste Compost and Their Capacity of Fertilization. *Eur. J. Sci. Res.* **2019**, *154*, 399–412.
7. Wu, M.; Dick, W.A.; Li, W.; Wang, X.; Yang, Q.; Wang, T.; Xu, L.; Zhang, M.; Chen, L. Bioaugmentation and Biostimulation of Hydrocarbon Degradation and the Microbial Community in a Petroleum-Contaminated Soil. *Int. Biodeterior. Biodegrad.* **2016**, *107*, 158–164. [CrossRef]
8. Villaverde, J.; Láiz, L.; Lara-Moreno, A.; González-Pimentel, J.L.; Morillo, E. Bioaugmentation of PAH-Contaminated Soils With Novel Specific Degrader Strains Isolated From a Contaminated Industrial Site. Effect of Hydroxypropyl- β -Cyclodextrin as PAH Bioavailability Enhancer. *Front. Microbiol.* **2019**, *10*, 2588. [CrossRef]
9. Abdulsalam, S.; Omale, A.B. Comparison of Biostimulation and Bioaugmentation Techniques for the Remediation of Used Motor Oil Contaminated Soil. *Braz. Arch. Biol. Technol.* **2009**, *52*, 747–754. [CrossRef]
10. Mrozik, A.; Piotrowska-seget, Z. Bioaugmentation as a Strategy for Cleaning up of Soils Contaminated with Aromatic Compounds. *Microbiol. Res.* **2010**, *165*, 363–375. [CrossRef]

11. Cycoń, M.; Mrozik, A.; Piotrowska-Seget, Z. Bioaugmentation as a Strategy for the Remediation of Pesticide-Polluted Soil: A Review. *Chemosphere* **2016**, *172*, 52–71. [CrossRef] [PubMed]
12. Elkarrach, K.; Merzouki, M.; Biyada, S.; Benlemlih, M. Bioaugmentation Process for the Treatment of Tannery Effluents in Fez, Morocco: An Eco-Friendly Treatment Using Novel Chromate Bacteria. *J. Water Process Eng.* **2020**, *38*, 101589. [CrossRef]
13. Abed, R.M.M.; Al-sabahi, J.; Al-maqrashi, F.; Al-habsi, A. Characterization of Hydrocarbon-Degrading Bacteria Isolated from Oil-Contaminated Sediments in the Sultanate of Oman and Evaluation of Bioaugmentation and Biostimulation Approaches in Microcosm Experiments. *Int. Biodeterior. Biodegrad.* **2014**, *89*, 58–66. [CrossRef]
14. Biyada, S.; Merzouki, M.; Elkarrach, K.; Benlemlih, M. Spectroscopic Characterization of Organic Matter Transformation during Composting of Textile Solid Waste Using UV—Visible Spectroscopy, Infrared Spectroscopy and X-Ray Diffraction (XRD). *Microchem. J.* **2020**, *159*, 105314. [CrossRef]
15. Biyada, S.; Merzouki, M.; Taisija, D.; Vasiliauskienė, D.; Urbonavičius, J.; Marčiulaitienė, E.; Vasarevicius, S.; Benlemlih, M. Evolution of Microbial Composition and Enzymatic Activities during the Composting of Textile Waste. *Appl. Sci.* **2020**, *10*, 3758. [CrossRef]
16. El Fels, L. *Suivi Physico-Chiique, Microbiologique et Ecotoxicologique du Compostage de Boues de Step Mélangées à des Déchets de Palmier: Validation de Nouveaux Indices de Maturation*; Université de Toulouse: Toulouse, France, 2014.
17. Association Française de Normal. *Afnor Norme T90-1110, Essai Des Eaux: Dosage de l'azote Total Kjeldahl*; Association Française de Normal: Paris, France, 1975.
18. Association Française de Normal. *Afnor Amendements Du Sol et Support de Culture-Préparation Des Échantillons Pour Les Essais Physiques et Chimiques, Détermination de La Teneur En Matière Sèche, Du Taux d'humidité et de La Masse Volumique Compactée En Laboratoire*; Association Française de Normal: Paris, France, 2000.
19. Barje, F.; El Fels, L.; El Hajjouji, H.; Amir, S.; Winterton, P.; Hafidi, M. Molecular Behaviour of Humic Acid-like Substances during Co-Composting of Olive Mill Waste and the Organic Part of Municipal Solid Waste. *Int. Biodeterior. Biodegrad.* **2012**, *74*, 17–23. [CrossRef]
20. Wu, H.; Zhao, Y.; Long, Y.; Zhu, Y.; Wang, H.; Lu, W. Evaluation of the Biological Stability of Waste during Landfill Stabilization by Thermogravimetric Analysis and Fourier Transform Infrared Spectroscopy. *Bioresour. Technol.* **2011**, *102*, 9403–9408. [CrossRef]
21. Biyada, S.; Merzouki, M.; Imtara, H.; Alajmi, M.F.; Elkarrach, K.; Mechchate, H.; Conte, R.; Benlemlih, M. Advanced Characterization of Organic Matter Decaying during Composting of Industrial Waste Using Spectral Methods. *Processes* **2021**, *9*, 1364. [CrossRef]
22. Amir, S.; Hafidi, M.; Merlina, G.; Hamdi, H.; Amir, S.; Hafidi, M.; Merlina, G.; Hamdi, H.; Elemental, J.R. Elemental Analysis, FTIR and 13C-NMR of Humic Acids from Sewage Sludge Composting To Cite This Version: HAL Id: Hal-00886235 Original Article. *Agronomie* **2004**, *24*, 13–18. [CrossRef]
23. Soobhany, N.; Mohee, R.; Kumar, V. Comparative Assessment of Heavy Metals Content during the Composting and Vermicomposting of Municipal Solid Waste Employing Eudrilus Eugeniae. *Waste Manag.* **2015**, *39*, 130–145. [CrossRef]
24. Ilani, T.; Herrmann, I.; Karnieli, A.; Arye, G. Characterization of the Biosolids Composting Process by Hyperspectral Analysis. *Waste Manag.* **2016**, *48*, 106–114. [CrossRef] [PubMed]
25. Lynd, L.R.; Weimer, P.J.; Zyl, W.H.; Van Pretorius, I.S. Microbial Cellulose Utilization: Fundamentals and Biotechnology. *Microbiol. Mol. Biol. Rev.* **2002**, *66*, 506–577. [CrossRef] [PubMed]
26. Guo, Y.; Zhou, X.; Li, Y.; Li, K. Heterotrophic Nitrification and Aerobic Denitrification by a Novel Halomonas Campisalis. *Biotechnol. Lett.* **2013**, *35*, 2045–2049. [CrossRef]
27. Chen, Q.; Ni, J.; Ma, T.; Liu, T.; Zheng, M. Bioaugmentation Treatment of Municipal Wastewater with Heterotrophic-Aerobic Nitrogen Removal Bacteria in a Pilot-Scale SBR. *Bioresour. Technol.* **2015**, *183*, 25–32. [CrossRef]
28. Li, Y.; Zhang, L.; Xian, H.; Zhang, X. Newly Isolated Cellulose-Degrading Bacterium Achromobacter Xylosoxidans L2 Has Deinking Potential. *Bioresources* **2019**, *14*, 2256–2268.
29. Biyada, S.; Merzouki, M.; Demcenko, T.; Vasiliauskien, D.; Marčiulaitienė, E.; Vasarevicius, S.; Urbonavičius, J. The Effect of Feedstock Concentration on the Microbial Community Dynamics During Textile Waste Composting. *Front. Ecol. Evol.* **2022**, *10*, 813488. [CrossRef]
30. Zhang, C.; Zeng, G.; Yuan, L.; Yu, J.; Li, J. Aerobic degradation of bisphenol A by Achromobacter xylosoxidans 287 strain B-16 isolated from compost leachate of municipal solid waste. *Chemosphere* **2007**, *68*, 181–190. [CrossRef]
31. Haq, I.; Kumar, S.; Kumari, V.; Kumar, S.; Raj, A. Evaluation of Bioremediation Potentiality of Ligninolytic Serratia Liquefaciens for Detoxification of Pulp and Paper Mill Effluent. *J. Hazard. Mater.* **2016**, *305*, 190–199. [CrossRef]
32. Goux, S.; Shapir, N.; Fantroussi, S. In soils inoculated with atrazine degraders. *Water Air. Soil Pollut.* **2002**, *3*, 131–142. [CrossRef]
33. Ghazali, F.M.; Noor, R.; Abdul, Z.; Salleh, A.B.; Basri, M. Biodegradation of Hydrocarbons in Soil by Microbial Consortium. *Int. Biodeterior. Biodegrad.* **2004**, *54*, 61–67. [CrossRef]
34. Ariffin, H.; Hassan, M.A.; Shah, U.K.; Abdullah, N.; Ghazali, F.M.; Shirai, Y. Production of Bacterial Endoglucanase from Pretreated Oil Palm Empty Fruit Bunch by Bacillus Pumilus EB3. *J. Biosci. Bioeng.* **2008**, *106*, 231–236. [CrossRef] [PubMed]
35. Sharma, P.; Melkania, U. Effect of Bioaugmentation on Hydrogen Production from Organic Fraction of Municipal Solid Waste. *Int. J. Hydrogen Energy* **2018**, *15*, 7290–7298. [CrossRef]

*applied sciences*

MDPI

*Article*

# Comparative Study of Different Production Methods of Activated Carbon Cathodic Electrodes in Single Chamber MFC Treating Municipal Landfill Leachate

**Pavlos K. Pandis [1,2,*], Theofilos Kamperidis [1], Konstantinos Bariamis [1], Ilias Vlachos [1], Christos Argirusis [1], Vassilis N. Stathopoulos [2], Gerasimos Lyberatos [1,3] and Asimina Tremouli [1,*]**

1 School of Chemical Engineering, Zografou Campus, National Technical University of Athens, 9 Heroon Polytechneiou St., 15780 Athens, Greece; kamp.theo@gmail.com (T.K.); kostas_bariamis@hotmail.com (K.B.); iliasvlachosgr@gmail.com (I.V.); amca@chemeng.ntua.gr (C.A.); lyberatos@chemeng.ntua.gr (G.L.)

2 Laboratory of Chemistry and Materials Technology, Department of Agricultural Development, Agrofood and Management of Natural Resources, Psachna Campus, National and Kapodistrian University of Athens, 34400 Evia, Greece; vasta@uoa.gr

3 Institute of Chemical Engineering Sciences (ICE-HT), Stadiou St., Platani, 26504 Patras, Greece

* Correspondence: ppandis@chemeng.ntua.gr (P.K.P.); atremouli@chemeng.ntua.gr (A.T.)

**Featured Application: Facile production methods of ceramic supported activated carbon cathodic electrodes in single chamber MFC, towards treatment of municipal landfill leachate with simultaneous energy production.**

**Abstract:** The treatment of real waste extracts with simultaneous energy production is currently under research. One method of addressing this dual task is using biochemical reactors named microbial fuel cells (MFCs). MFCs consist of a bioanode and a cathode where the oxygen reduction reaction (ORR) occurs. Cathodes are currently under optimization regarding the nature of their support, their catalytic efficiency and their configurations. In this work, we present facile preparation methods for the production of activated carbon ceramic-supported cathodic electrodes produced with three different techniques (wash-coat, brush-coat, and ultrasound-assisted deposition/infiltration). The produced cathodic electrodes were tested in a single-chamber MFC, filled with the concentrated liquid residue, after the reverse osmosis (RO-CLR) treatment of leachate from a municipal waste landfill, in order to exploit their electrochemical potential for simultaneous waste treatment and energy production. The electrode produced utilizing 20 kHz ultrasounds proved to be more effective in terms of energy harvesting (10.7 mW/g·L of leachate) and wastewater treatment (COD removal 85%). Internal resistances of the ultrasound-produced electrodes are lower, as compared to the other two methods, opening new exploitation pathways in the use of ultrasound as a means in producing electrodes for microbial fuel cells.

**Keywords:** municipal leachate treatment; microbial fuel cell; single chamber; waste treatment; ceramic cathodes; ultrasound-assisted deposition/infiltration; wash coat; brush coat; energy production; facile preparation

**Citation:** Pandis, P.K.; Kamperidis, T.; Bariamis, K.; Vlachos, I.; Argirusis, C.; Stathopoulos, V.N.; Lyberatos, G.; Tremouli, A. Comparative Study of Different Production Methods of Activated Carbon Cathodic Electrodes in Single Chamber MFC Treating Municipal Landfill Leachate. *Appl. Sci.* **2022**, *12*, 2991. https://doi.org/10.3390/app12062991

Academic Editor: Carlos Rico de la Hera

Received: 31 December 2021
Accepted: 11 March 2022
Published: 15 March 2022



## 1. Introduction

Municipal landfill leachate has been a major problem in the municipal waste economy for many years. The accumulation of rainwater along with the humidity of the environment in the waste collection tanks contributes significantly to the problem of waste management [1]. In its unprocessed form, the leachate is reported to be a turbid and hazardous material. The municipal landfill leachate requires treatment so it can be processed in subsequent stages of urban wastewater treatment or discharged immediately with low pollutant values [2]. The management of leachates needs to be seriously addressed before choosing the treatments for lowering its contaminant (inorganic and organic) concentrations. A first step is always established with biological processes mainly for nitrogen removal. Moving bed biofilm reactors (MBBR) and activated sludge loop reactors are the types of reactors commonly employed [3–5]. In order to further remove the biological content, filtration with activated carbon or/and ozonization is used [2,6]. Moreover, due to the large holding time in control tanks the organic content of the leachate is increased. To address this issue, reverse osmosis is commonly used in order to eliminate as many pollutants as possible [7,8]. The reverse osmosis product, in most cases, is returned to the landfill, leading to a stronger leachate. Additionally, the reverse osmosis technology is more expensive than biological processes.

Regarding the electrochemically active bacteria, a time period of acclimation is required to form the biofilm [1,9]. In order for the microorganisms to deposit on the anodic electrode high specific surface is required [10]. Additionally, the anodic electrode should have high conductivity [11] and biocompatibility [12]. Common electrodes used in the MFC anode are plain or coated graphite/carbon-based materials such as graphite granules [13], felt [14], paper [15] and brush [16]. On the other hand, for the cathode electrode depending on the electron acceptor, a presence of catalyst may be necessary. An example of a cathodic reaction requiring catalysis is the oxygen reduction; although spontaneous, it is a slow reaction, so in order to maximize the rate, a catalyst such as platinum is coated on the cathodic surface [17]. Furthermore, instead of expensive catalysts such as platinum, microbia have been used to catalyze the oxygen reduction in MFC [18]. Other electron acceptors such as silver or copper do not require a catalyst and plain carbon or graphite-based materials are suitable.

Between the anode and the cathode, a separator is present in the MFCs. Two basic configurations have been developed, one requiring two chambers and a separator (dual chamber MFC) and a second one requiring one chamber and a separator with the cathode electrode exposed to air (single chamber MFCs). In the case of single chamber MFCs, the electron acceptor is the oxygen, and its abundance in the air is utilized. The separators used in dual chamber (H–type) MFCs are usually membranes (such as proton-exchange membrane), which are expensive materials [19,20]. The advantage of single chamber MFCs is that the expensive separator can be switched with a cheaper one, acting at the same time as the structural material for the cathode electrode [21]. In recent years, the research has focused on the use of ceramic materials as separators and cathode electrodes in MFCs [22,23]. The use of these materials lowers the internal resistance of the MFC and allows improved reduction in the oxygen, as they are resistant to fouling [24] and offer mechanical, thermal and chemical stability [25]. In order for the ceramic materials to be used as a structural component of the electrode, a catalyst needs to be deposited on the surface exposed to air. Various techniques have been employed to accomplish this. Assuming the oxygen reduction catalyst is activated carbon, the following ways of deposition are the most notable. A two-step process was employed to create the activated carbon catalyst, with a phytic acid-doped polyaniline coating and subsequent high-temperature pyrolysis, by [26]. The catalyst was then placed on a stainless steel mesh with polyfluortetraethylene (PTFE) using a rolling method technique [26]. Another approach of utilizing activated carbon was presented by [27], placing an activated carbon layer on a gravel bed with a copper plate for the electrical connection. This set-up was used as the air-exposed cathode electrode of a single chamber MFC [27]. Activated carbon mixed with carbon black placed on a stainless steel mesh was used by [28]. The activated carbon mix was prepared by adding carbon black, activated carbon, PTFE and DI water using ultrasonication and afterwards spreading the paste on stainless steel mesh [28].

The aim of this work is to propose a facile way to manufacture tubular ceramic electrodes, coated with the oxygen reduction catalyst, which are then to be used in a single chamber MFC. The various manufacturing processes employed presented promising results and are facile to reproduce. The selected catalyst (activated carbon) is deposited on the inner surface of the tubes by wash-coat technique, brush-coat technique and ultrasound-assisted deposition/infiltration (sono-coat). The cathode electrodes are tested in a single chamber MFC, operating in batch mode and having as feedstock a leachate originating from a landfill unit in Athens, Greece. The performance of the different electrodes is compared and the ability of the MFC to treat the leachate is examined, in all three cases.

## 2. Materials and Methods

### 2.1. MFC Set up and Operation

A custom-made single-chamber MFC was constructed for this work, made of Ertalon® with an effective volume of 250 cm$^3$ (Figure 1). As the anodic electrode, a commercially available stainless steel sponge (Nimax, Greece, item code 0857) was used, offering a high surface area for the development of the electrochemically active biofilm. The preparation of the cathode electrodes is presented separately. The MFC operated in batch mode and was acclimated with a synthetic glucose feed with anaerobic sludge inoculums (described in detail elsewhere [28,29]. The anaerobic sludge addition in the feed was carried out for the first three operation cycles; after that point, no more sludge was added in the feed. The anaerobic sludge was obtained from the Likovrisi (Athens, Greece) sewage treatment plant. The acclimation period of the MFC was completed once repeatable current output and COD removal were observed.

a) b)

**Figure 1.** (**a**) Single chamber MFC with stainless steel sponge as anode and mullite tube as cathode (**b**) MFC configuration for testing of constructed electrodes.

After the acclimation period of the MFC, the synthetic glucose feed was switched with the landfill leachate. To produce the leachate used as the MFC feedstock, the following process occurred. The liquid generated from a waste decomposition in a major Greek landfill unit in Athens, Greece was collected and treated with reverse osmosis (RO-CLR), resulting in a concentrated liquid residue. The characteristics of the liquid residue and the anaerobic sludge are presented in Table 1.

**Table 1.** Characteristics of processed leachate and anaerobic sludge used in the MFC.

| Substance | Organic Load (g COD/L) | pH | Conductivity (mS/cm) |
|---|---|---|---|
| Leachate | 28 | 7.8 | 70 |
| Anaerobic sludge | 22 | 7.3 | 5.2 |

### 2.2. Cathodic Electrode Preparation

For the preparation of the ceramic cathodes, mullite tubes (Bonis S.A., Athens, Greece) (Figure 2a,b) with an outer diameter of 25 mm, a wall thickness of 2.5 mm and a length of

50 mm were used. Mullite tubes were used as the separators and as a structural support for the catalysts' deposition. An acrylic-based binder (HSF54 paint, YSHIELD GmbH&Co. KG, Ruhstorf, Germany) and powdered activated carbon (Sigma-Aldrich, CAS 7440-44-0) were used as precursors for the production of the cathodic electrode catalyst. A stainless-steel grid was placed along the inner surface of the ceramic tube as the current collector before the application of the above techniques (Figure 2c).

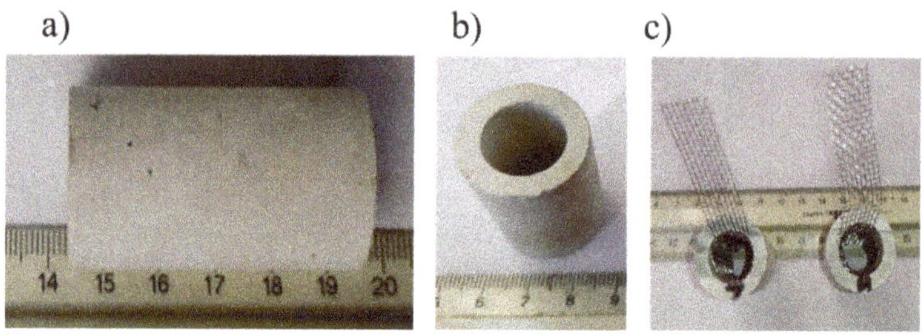

**Figure 2.** (**a**) Horizontal image and (**b**) vertical image of mullite tubes prior to the electrode preparation, (**c**) mullite tubes internally coated with catalytic paste and stainless steel mesh.

To prepare the catalytic paste, 10 g of HSF54 and 5 g of Activated Carbon (AC) were mixed with 45 mL of a solvent consisting of 50% ethanol and 50% isopropanol. The same amounts were maintained across all the electrodes. The produced slurry was placed for 10 min in an EMMI 30 HC (EMMAG, Germany) sonication bath for homogenization. Sedimentation experiments were carried out in order to estimate the stability of the slurry. It was found that the slurry remained stable for at least 30 min, meaning that the subsequent procedures have to be carried out within this time frame. Afterwards, three different techniques were applied in order to produce ceramic-supported cathode electrodes. More specifically:

Wash-coat technique (WC): during this technique, the slurry was poured into the inner surface of the tube and left to dry physically (up to 24 h). Four repetitions at 2 min intervals were performed for the deposition of the slurry.

Brush-coat technique (BC): The slurry was brushed onto the inner surface of the tube with a paintbrush (size 4, Germany).

Sono-coat technique (SC): The slurry was poured up to the surface into the inner volume of the ceramic tube with a rubber cap at its bottom to prevent spilling (Figure 3a). A tip, producing ultrasound at the frequency of 20 kHz (Vibra cell 750, SONICS, USA), was placed 1 cm inside the tube. It was operated at 19% amplitude for 10 min (Figure 3b,c). After that time an ultrasound-assisted ceramic supported electrode was obtained, as depicted in Figure 3d.

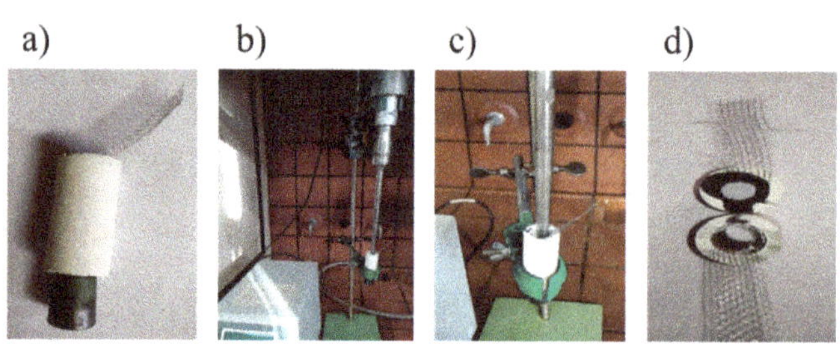

**Figure 3.** (**a**) Mullite tube with rubber cap, (**b**) and (**c**) setup of ultrasound-assisted deposition, (**d**) tube before and after ultrasound-assisted of HSF54 and AC deposition.

### 2.3. Analytical Methods and Calculations

Chemical oxygen demand (COD), pH and conductivity characteristics of the leachate were measured prior and after each batch operation cycle for each produced cathode electrode. The measurements of COD were carried out according to standard methods [30]. The pH and the conductivity values were measured using a digital pH-meter (WinLab Data Line pH-meter, No: 210224Windaus Labortechnik, Germany) and a conductivity meter (CMD83, Radiometer Copenhagen, Analytical instruments Division, Hach, USA).

Electrochemical characterization of the whole cell with the three different cathode electrodes has been performed through a Potentiostat/Galvanostat (BIOLOGIC SP150, France) with a three-electrode setup, using a saturated Ag/AgCl reference electrode. Open-current voltage (OCV) was monitored, polarization curves from OCV to zero potential were obtained at a rate of 0.3 mV/s and electrochemical impedance spectroscopy (EIS) spectra were obtained from 200 kHz to 10 mHz with a sinusoidal amplitude of 10 mV. The power output of the cell was expressed in mW considering a leachate volume of 250 $cm^3$. The mathematical model of the fitting analysis that was used was the same as reported in our previous references [31–33].

## 3. Results and Discussion

### 3.1. Characterization of Ceramic Supports (Mullite Tubes)

The SEM image (cross-section) of the mullite tube is depicted in Figure 4. The open porosity of the tubes was found to be at 20% according to the Archimedes method [34,35]. The tubes were washed and dried with deionized water prior to any use and then weighted in order to calculate the deposition mass of the catalyst (HSF54 + AC).

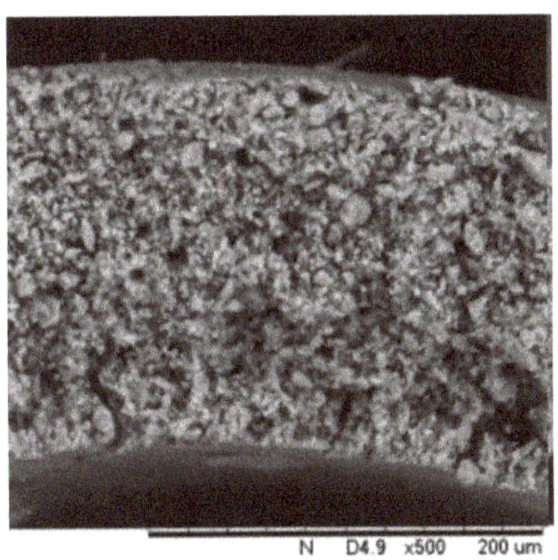


**Figure 4.** Cross-section SEM image of mullite tube with open porosity.

### 3.2. Ceramic-Supported Electrodes Production

All three preparation methods provided stable catalyst loadings (in the form of coatings) into the inner surface of the mullite tubes. The produced electrodes are depicted in Figure 5. In both pictures in Figure 5 the blank tube is shown at the left with the electrodes prepared with the three facile methods (WC, BC, SC) at the right side. The production of all coatings was uniform, and the physical properties of the produced electrodes are presented in Table 2. The values of Table 2 are the mean values of three samples.

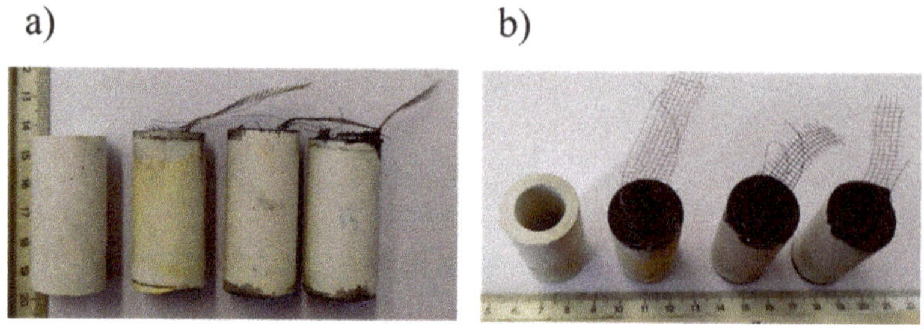


**Figure 5.** (**a**) Horizontal images and (**b**) vertical images of four mullite tubes with stainless steel mesh (**left** to **right**: blank, WC, BC, and SC).

**Table 2.** Physical properties of the three different produced cathode electrodes.

| Electrode | Coat Thickness (mm) | Catalyst Mass (g) | Conditions |
|---|---|---|---|
| WC | 1.5 | 1.5 | 4 wash coats |
| BC | 1.1 | 0.9 | 3 brush coats |
| SC | 0.8 | 0.7 | 10 min under ultrasound |

### 3.3. *Batch Mode Operation Using Leachate as Substrate*

The MFC operated in batch mode using the ceramic electrodes for the three different preparation techniques (WC, BC and SC). Polarization experiments determined the maximum power output values for the three different cases. Figure 6 presents the dependence of the MFC voltage, V, and the produced power, P, on the current passing through the MFC, when the MFC was set up with the different electrodes.

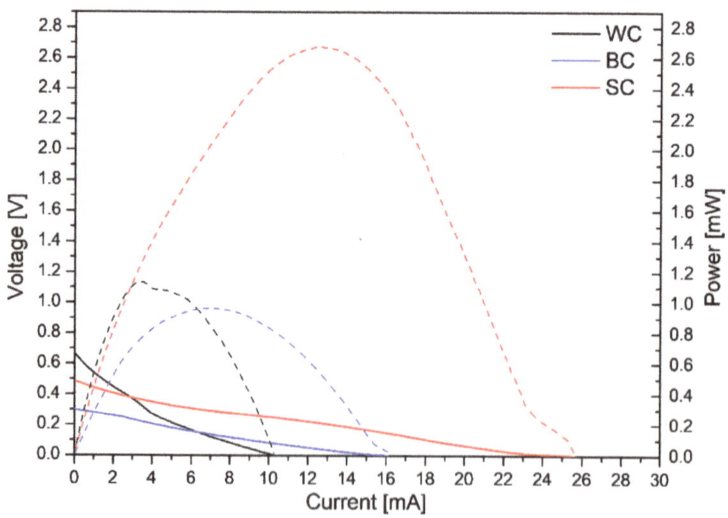


**Figure 6.** Polarization curves (V–I continuous lines) and power curves (P–I dashed lines) of the leachate-fed MFC for the three cathode electrode cases (WC ▬, BC ▬ and SC ▬).

As can be seen from Figure 6, the maximum power output (2.67 mW) was obtained with the ceramic electrode produced with the SC technique. Additionally, the maximum power output for the MFC operation with the WC and BC electrodes was in the same range (WC: 1.12 and BC: 0.96 mW). Moreover, the slope of the V–I curves (Figure 6, continuous lines) indicated that ohmic losses dominated in the MFC, across the three electrode cases. Table 3 presents the open circuit values (OCV) and the maximum power output per catalysts mass and per volume of leachate, for the cycles' operation with the different electrodes.

**Table 3.** OCV and maximum power output values obtained from the leachate-fed MFC with the three different electrode cases examined.

| Electrode | OCV (V) | Power Output (mW) | Power Output ($mW \cdot g^{-1}$) | Power Output ($mW \cdot g^{-1} \cdot L^{-1}$) |
|---|---|---|---|---|
| WC | 0.66 | 1.12 | 1.5 | 6.00 |
| BC | 0.31 | 0.96 | 0.9 | 3.84 |
| SC | 0.48 | 2.67 | 0.7 | 10.68 |

Figure 7 depicts the variations in the conductivity and the COD removal efficiency values at the end of each operation cycle in comparison with the initial values of the leachate: As can be seen from Figure 7, high COD removal efficiencies were achieved in all cases (COD removal > 70%), while the leachate conductivity decreased after its treatment. It is worth mentioning that the conductivity decreased by almost 50% when the wastewater was treated using the electrodes prepared with the SC technique. The conductivity decrease may be attributed to the decomposition of the substances present in the landfill leachate. The electrochemically active biofilm oxidizes the available substrate (leachate), decomposing the substances that contribute to the high conductivity, thus resulting in the decrease in the conductivity. A similar decrease in the anolyte's conductivity has been observed elsewhere [36], although comparison between the two cases is difficult, due to the different acclimation, MFC set-up and feedstock conditions.

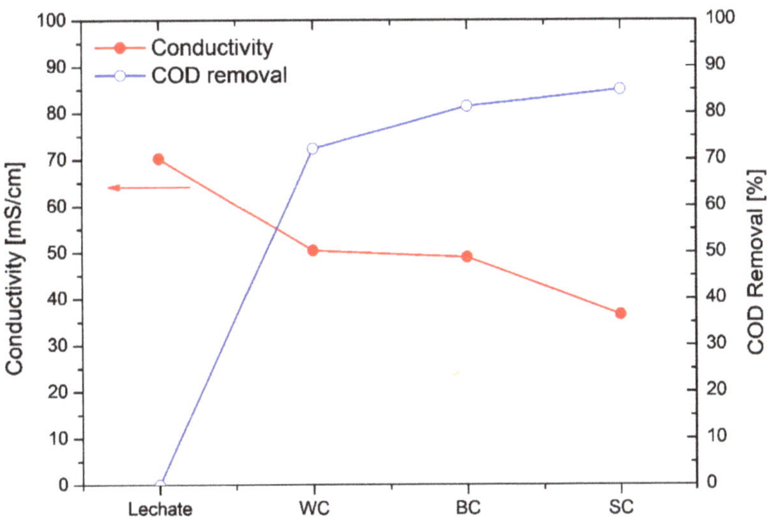


**Figure 7.** Conductivity measurements (**left**) and COD removal values (**right**) for the leachate and the three leachate-fed MFC configurations with different cathode electrodes.

### *3.4. Electrochemical Impendance Spectroscopy Experiments (EIS)*

In order to perform a detailed electrochemical characterization of the cell with the different electrodes, EIS experiments were carried out. From the electrochemical point of view, as reported elsewhere [37–39], the internal resistances of the cells with different electrodes can be calculated. In Figure 8, the Nyquist diagrams show significant information for the electrochemical processes that occur during the simultaneous wastewater treatment and electricity production.

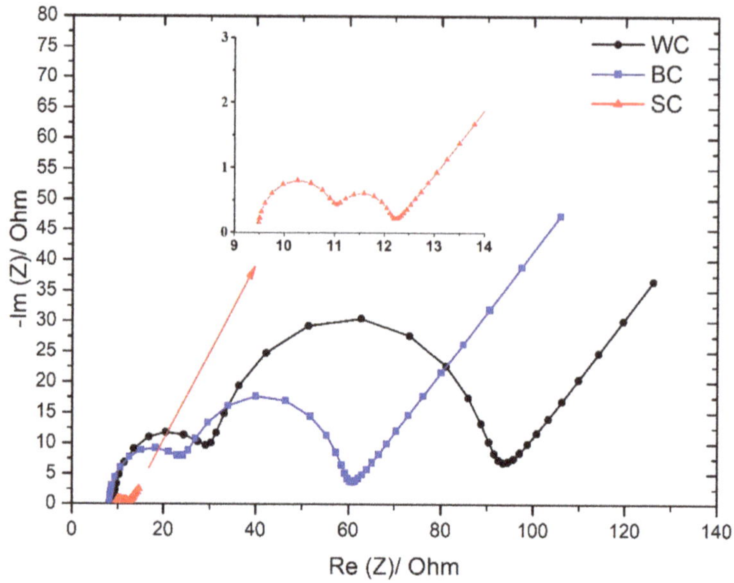


**Figure 8.** Nyquist diagrams of leachate-fed MFC for three different cathode configurations (WC ▬, BC ▬ and SC ▬).

All three setups depict two distinguished arcs followed by a Warburg element, as seen in other microbial fuel cells [40]. The fitted parameters are explained and reported in earlier publications, and the measurements here held through three electrode setups, as in most cases of fuel cells [32,33,41]. Table 4 presents the values of the solution resistance ($R_s$), the biofilm resistance ($R_{BF}$), the charge transfer resistance ($R_{CT}$), biofilm capacitance ($C_{BF}$) and the charge transfer capacitance ($C_{CT}$), which were calculated from the fitting model. Moreover, the internal resistance ($R_{in}$) of each case is presented.

As can be seen from Table 4, the $R_{in}$ of SC electrodes is approximately 40% lower when compared with the $R_{in}$ of the WC electrodes, and 28% lower from the $R_{in}$ of the cell which operated with the BC electrodes. However, from the Nyquist diagrams, it is observed that this difference is attributed to the lower values of biofilm and charge resistances and capacitances of SC electrodes when compared with the respective values of WC and BC electrodes (Table 4). The solution resistance ($R_S$~9.3 Ω) is practically the same for all three operation batch modes. This practically denotes that the nature of the electrolyte/leachate does not affect by any means the three setups. On the other hand, the values of $R_{BF}$, $R_{CT}$, $C_{BF}$ and the $C_{CT}$, in the case of SC electrodes, are significantly lower when compared with the respective values of the BC and WC electrodes. This result indicates that the rate of the leachate oxidation is faster in the case of SC electrodes and also that a better biofilm has been developed in the case of SC electrodes [42,43]. On the contrary, no significant differences on the internal resistances are observed between WC and BC techniques. The above results are in accordance with the polarization experiments, which indicated that the maximum power output was produced with the SC electrodes where the maximum COD removal efficiency also occurred.

**Table 4.** Fitting parameters from EIS measurements for the leachate-fed MFC with the three different electrode cases examined.

| Fitted Parameters | WC | BC | SC |
|---|---|---|---|
| RS (Ω) | 9.3 | 9.2 | 9.4 |
| RBF (Ω) | 8.8 | 8.6 | 1.6 |
| RCT (Ω) | 21.5 | 15.1 | 1.03 |
| CBF (F) | $2.6 \times 10^{-7}$ | $2.5 \times 10^{-7}$ | $0.15 \times 10^{-7}$ |
| CCT (F) | $9.1 \times 10^{-3}$ | $5.0 \times 10^{-3}$ | $0.7 \times 10^{-3}$ |
| $R_{INT}$ (Ω) * | 39.6 | 32.91 | 23.71 |

* calculated.

## 4. Conclusions

Three different preparation methods of producing tubular ceramic-supported electrodes capable of the oxygen reduction reaction (ORR) were evaluated under the view of maximizing the wastewater treatment and the power production of the MFC technology. High COD-removal efficiencies were achieved in wash-coat (WC), brush-coat (BC) and ultrasound-assisted (SC) deposition techniques, although the higher value was obtained using the electrodes produced from the SC technique (COD removal of 72%, 82% and 85% for the WC, BC and SC, respectively). Moreover, amongst WC, BC and SC deposition technique, the latter boosted the MFC performance in terms of the energy production ($P_{max}$: 1.12 mW (WC), 0.96 mW (BC), 2.67 mW (SC). The electrochemical experiments verified that the operation of the cell with the ultrasound-assisted deposition technique operated with lower internal resistance ($R_{in}$ 23.71 Ω) and specifically lower charge transfer and biofilm resistance when compared with the two other techniques ($R_{in}$: 39.6 Ω (WC), 32.91 Ω (BC)). These results indicated that the deposition technique of the cathodic catalyst has a major effect on the overall cell performance affecting not only the oxygen reduction reaction rate but also the biofilm development and the rate of the substrate oxidation. Further study is needed in order to determine the mechanisms that contribute to this performance boost when using the ultrasound-assisted deposition technique.

**Author Contributions:** Conceptualization, P.K.P. and A.T.; methodology, P.K.P., K.B., I.V. and A.T.; writing—original draft preparation, P.K.P.; writing—review and editing, A.T. and T.K.; supervision, P.K.P., C.A., V.N.S., A.T. and G.L.; resources, V.N.S., C.A. and G.L.; project administration, G.L. and A.T.; All authors have read and agreed to the published version of the manuscript.

**Funding:** This project has received funding from the Hellenic Foundation for Research and Innovation (HFRI) and the General Secretariat for Research and Technology (GSRT), under grant agreement No. 862.

**Institutional Review Board Statement:** Not applicable.

**Informed Consent Statement:** Not applicable.

**Data Availability Statement:** The datasets generated and analyzed during the current study are available from the corresponding author on reasonable request.

**Acknowledgments:** This project has received funding from the Hellenic Foundation for Research and Innovation (HFRI) and the General Secretariat for Research and Technology (GSRT) under grant agreement No. 862.

**Conflicts of Interest:** The authors declare no conflict of interest.

## References

1. Renou, S.; Givaudan, J.G.; Poulain, S.; Dirassouyan, F.; Moulin, P. Landfill leachate treatment: Review and opportunity. *J. Hazard. Mater.* **2008**, *150*, 468–493. [CrossRef] [PubMed]
2. Tenodi, S.; Krčmar, D.; Agbaba, J.; Zrnić, K.; Radenović, M.; Ubavin, D.; Dalmacija, B. Assessment of the environmental impact of sanitary and unsanitary parts of a municipal solid waste landfill. *J. Environ. Manag.* **2020**, *258*, 110019. [CrossRef] [PubMed]

3. Hasar, H.; Ipek, U.; Kinaci, C. Joint treatment of landfill leachate with municipal wastewater by submerged membrane bioreactor. *Water Sci. Technol.* **2009**, *60*, 3121–3127. [CrossRef] [PubMed]
4. Roy, D.; Drogui, P.; Tyagi, R.D.; Landry, D.; Rahni, M. Mbr treatment of leachates originating from waste management facilities: A reference study of the design parameters for efficient treatment. *J. Environ. Manag.* **2020**, *259*, 110057. [CrossRef] [PubMed]
5. Gkotsis, P.; Zouboulis, A.; Mitrakas, M. Using additives for fouling control in a lab-scale mbr; comparing the anti-fouling potential of coagulants, pac and bio-film carriers. *Membranes* **2020**, *10*, 42. [CrossRef] [PubMed]
6. Hube, S.; Eskafi, M.; Hrafnkelsdóttir, K.F.; Bjarnadóttir, B.; Bjarnadóttir, M.Á.; Axelsdóttir, S.; Wu, B. Direct membrane filtration for wastewater treatment and resource recovery: A review. *Sci. Total Environ.* **2020**, *710*, 136375. [CrossRef]
7. Chianese, A.; Ranauro, R.; Verdone, N. Treatment of landfill leachate by reverse osmosis. *Water Res.* **1999**, *33*, 647–652. [CrossRef]
8. Ren, X.; Song, K.; Xiao, Y.; Zong, S.; Liu, D. Effective treatment of spacer tube reverse osmosis membrane concentrated leachate from an incineration power plant using coagulation coupled with electrochemical treatment processes. *Chemosphere* **2020**, *244*, 125479. [CrossRef]
9. Santoro, C.; Flores-Cadengo, C.; Soavi, F.; Kodali, M.; Merino-Jimenez, I.; Gajda, I.; Greenman, J.; Ieropoulos, I.; Atanassov, P. Ceramic microbial fuel cells stack: Power generation in standard and supercapacitive mode. *Sci. Rep.* **2018**, *8*, 3281. [CrossRef]
10. Tremouli, A.; Greenman, J.; Ieropoulos, I. Investigation of ceramic MFC stacks for urine energy extraction. *Bioelectrochemistry* **2018**, *123*, 19–25. [CrossRef]
11. Li, W.W.; Sheng, G.P.; Liu, X.W.; Yu, H.Q. Recent advances in the separators for microbial fuel cells. *Bioresour. Technol.* **2011**, *102*, 244–252. [CrossRef] [PubMed]
12. Zhang, P.; Yang, C.; Xu, Y.; Li, H.; Shi, W.; Xie, X.; Lu, M.; Huang, L.; Huang, W. Accelerating the startup of microbial fuel cells by facile microbial acclimation. *Bioresour. Technol. Rep.* **2019**, *8*, 100347. [CrossRef]
13. Sakai, K.; Iwamura, S.; Sumida, R.; Ogino, I.; Mukai, S.R. Carbon paper with a high surface area prepared from carbon nanofibers obtained through the liquid pulse injection technique. *ACS Omega* **2018**, *3*, 691–697. [CrossRef]
14. Li, S.; Cheng, C.; Thomas, A. Carbon-Based Microbial-Fuel-Cell Electrodes: From Conductive Supports to Active Catalysts. *Adv. Mater.* **2017**, *29*, 1–30. [CrossRef] [PubMed]
15. Yaqoob, A.A.; Ibrahim, M.N.M.; Rafatullah, M.; Chua, Y.S.; Ahmad, A.; Umar, K. Recent advances in anodes for microbial fuel cells: An overview. *Materials* **2020**, *13*, 2078. [CrossRef] [PubMed]
16. Tremouli, A.; Kamperidis, T.; Pandis, P.K.; Argirusis, C.; Lyberatos, G. Exploitation of Digestate from Thermophilic and Mesophilic Anaerobic Digesters Fed with Fermentable Food Waste Using the MFC Technology. *Waste Biomass Valorization* **2021**, *12*, 5361–5370. [CrossRef]
17. Wang, P.; Lai, B.; Li, H.; Du, Z. Deposition of Fe on graphite felt by thermal decomposition of Fe(CO)5 for effective cathodic preparation of microbial fuel cells. *Bioresour. Technol.* **2013**, *134*, 30–35. [CrossRef]
18. Guan, Y.F.; Zhang, F.; Huang, B.C.; Yu, H.Q. Enhancing electricity generation of microbial fuel cell for wastewater treatment using nitrogen-doped carbon dots-supported carbon paper anode. *J. Clean. Prod.* **2019**, *229*, 412–419. [CrossRef]
19. Sayed, E.T.; Abdelkareem, M.A.; Alawadhi, H.; Elsaid, K.; Wilberforce, T.; Olabi, A.G. Graphitic carbon nitride/carbon brush composite as a novel anode for yeast-based microbial fuel cells. *Energy* **2021**, *221*, 119849. [CrossRef]
20. Das, I.; Noori, M.T.; Bhowmick, G.D.; Ghangrekar, M.M. Application of low-cost transition metal based Co0.5Zn0.5Fe2O4 as oxygen reduction reaction catalyst for improving performance of microbial fuel cell. *MRS Adv.* **2018**, *3*, 3149–3154. [CrossRef]
21. Rabaey, K.; Read, S.T.; Clauwaert, P.; Freguia, S.; Bond, P.L.; Blackall, L.L.; Keller, J. Cathodic oxygen reduction catalyzed by bacteria in microbial fuel cells. *ISME J.* **2008**, *2*, 519–527. [CrossRef] [PubMed]
22. Daud, S.M.; Kim, B.H.; Ghasemi, M.; Daud, W.R.W. Separators used in microbial electrochemical technologies: Current status and future prospects. *Bioresour. Technol.* **2015**, *195*, 170–179. [CrossRef] [PubMed]
23. Winfield, J.; Gajda, I.; Greenman, J.; Ieropoulos, I. A review into the use of ceramics in microbial fuel cells. *Bioresour. Technol.* **2016**, *215*, 296–303. [CrossRef] [PubMed]
24. Jimenez, I.M.; Greenman, J.; Ieropoulos, I. Electricity and catholyte production from ceramic MFCs treating urine. *Int. J. Hydrogen Energy* **2017**, *42*, 1791–1799. [CrossRef] [PubMed]
25. Chatterjee, P.; Ghangrekar, M.M. Preparation of a fouling-resistant sustainable cathode for a single-chambered microbial fuel cell. *Water Sci. Technol.* **2014**, *69*, 634–639. [CrossRef] [PubMed]
26. Lv, K.; Zhang, H.; Chen, S. Nitrogen and phosphorus co-doped carbon modified activated carbon as an efficient oxygen reduction catalyst for microbial fuel cells. *RSC Adv.* **2018**, *8*, 848–855. [CrossRef]
27. González, M.L.J.; Hernández Benítez, C.; Juarez, Z.A.; Zamudio Pérez, E.; Ramírez Coutiño, V.Á.; Robles, I.; Godínez, L.A.; Rodríguez-Valadez, F.J. Study of the effect of activated carbon cathode configuration on the performance of a membrane-less microbial fuel cell. *Catalysts* **2020**, *10*, 619. [CrossRef]
28. Zhang, X.; Xia, X.; Ivanov, I.; Huang, X.; Logan, B.E. Enhanced activated carbon cathode performance for microbial fuel cell by blending carbon black. *Environ. Sci. Technol.* **2014**, *48*, 2075–2081. [CrossRef]
29. Kamperidis, T.; Tremouli, A.; Pandis, P.K.; Lyberatos, G. In Condensate originating from household food waste as a substrate using microbial fuel cell technology. In Proceedings of the 17th International Conference on Environmental Science and Technology CEST2021, Athens, Greece, 1–4 September 2021.
30. APHA/AWWA/WEF. *Standard Methods for the Examination of Water and Wastewater*; American Public Health Association: Washington, DC, USA, 2012. [CrossRef]

31. Tremouli, A.; Karydogiannis, I.; Pandis, P.K.; Papadopoulou, K.; Argirusis, C.; Stathopoulos, V.N.; Lyberatos, G. Bioelectricity production from fermentable household waste extract using a single chamber microbial fuel cell. *Energy Procedia* **2019**, *161*, 2–9. [CrossRef]
32. Tremouli, A.; Pandis, P.K.; Kamperidis, T.; Stathopoulos, V.N.; Argirusis, C.; Lyberatos, G. Performance assessment of a four-air cathode membraneless microbial fuel cell stack for wastewater treatment and energy extraction. *E3S Web Conf.* **2019**, *116*, 00093. [CrossRef]
33. Tremouli, A.; Pandis, P.K.; Karydogiannis, I.; Stathopoulos, V.N.; Argirusis, C.; Lyberatos, G. Operation and electro(chemical) characterization of a microbial fuel cell stack fed with fermentable household waste extract. *Glob. NEST J.* **2019**, *21*, 253–257.
34. Chalkia, V.; Pandis, P.; Stathopoulos, V. Shape forming of ceramic tubes for electrochemical reactors by gel-casting method. *ECS Trans.* **2015**, *68*, 2339–2348. [CrossRef]
35. Pandis, P.; Kharlamova, T.; Sadykov, V.; Stathopoulos, V. Development of layered anode structures supported over apatite-type solid electrolytes. *MATEC Web Conf.* **2016**, *41*, 04001. [CrossRef]
36. Kung, C.C.; Liu, C.C.; Sun, Y.; Yu, X. Innovative microbial fuel cell for energy harvesting. In Proceedings of the 2012 IEEE Energytech, Cleveland, OH, USA, 29–31 May 2012; pp. 1–4. [CrossRef]
37. Hidalgo, D.; Sacco, A.; Hernandez, S.; Tommasi, T. Electrochemical and impedance characterization of microbial fuel cells based on 2d and 3d anodic electrodes working with seawater microorganisms under continuous operation. *Bioresour. Technol.* **2015**, *195*, 139–146. [CrossRef] [PubMed]
38. Manohar, A.K.; Bretschger, O.; Nealson, K.H.; Mansfeld, F. The use of electrochemical impedance spectroscopy (eis) in the evaluation of the electrochemical properties of a microbial fuel cell. *Bioelectrochemistry* **2008**, *72*, 149–154. [CrossRef]
39. Manohar, A.K.; Mansfeld, F. The internal resistance of a microbial fuel cell and its dependence on cell design and operating conditions. *Electrochim. Acta* **2009**, *54*, 1664–1670. [CrossRef]
40. Lee, M.; Kondaveeti, S.; Jeon, T.; Kim, I.; Min, B. Influence of humidity on performance of single chamber air-cathode microbial fuel cells with different separators. *Processes* **2020**, *8*, 861. [CrossRef]
41. Dominguez-Benetton, X.; Sevda, S.; Vanbroekhoven, K.; Pant, D. The accurate use of impedance analysis for the study of microbial electrochemical systems. *Chem. Soc. Rev.* **2012**, *41*, 7228–7246. [CrossRef]
42. Martin, E.; Savadogo, O.; Guiot, S.R.; Tartakovsky, B. Electrochemical characterization of anodic biofilm development in a microbial fuel cell. *J. Appl. Electrochem.* **2013**, *43*, 533–540. [CrossRef]
43. Ramasamy, R.P.; Sekar, N. Electrochemical impedance spectroscopy for microbial fuel cell characterization. *J. Microb. Biochem. Technol.* **2013**, S6-004. [CrossRef]

*applied sciences*

MDPI

*Article*

# Impact of Biochar Application on Germination Behavior and Early Growth of Maize Seedlings: Insights from a Growth Room Experiment

**Liaqat Ali [1], Wang Xiukang [2,*], Muhammad Naveed [3,*], Sobia Ashraf [3], Sajid Mahmood Nadeem [1], Fasih Ullah Haider [4] and Adnan Mustafa [5,6,7]**

1 Department of Soil Science, University of Agriculture Faisalabad, Sub-Campus Burewala Vehari, Vehari 61100, Pakistan; liaqatali2188@yahoo.com (L.A.); smnadeem@uaf.edu.pk (S.M.N.)
2 College of Life Sciences, Yan'an University, Yan'an 716000, China
3 Institute of Soil and Environmental Sciences, University of Agriculture Faisalabad, Faisalabad 38040, Pakistan; sobiaashraf13@googlemail.com
4 College of Resources and Environmental Sciences, Gansu Agricultural University, Lanzhou 730070, China; fasihullahhaider281@gmail.com
5 Biology Centre CAS, SoWa RI, Na Sádkách 7, 37005 České Budějovice, Czech Republic; adnanmustafa780@gmail.com
6 Department of Agrochemistry, Soil Science, Microbiology and Plant Nutrition, Faculty of AgriSciences, Mendel University in Brno, Zemedelska 1, 613 00 Brno, Czech Republic
7 Institute of Chemistry and Technology of Environmental Protection, Faculty of Chemistry, Brno University of Technology, Purkynova 118, 612 00 Brno, Czech Republic
* Correspondence: wangxiukang@yau.edu.cn (W.X.); muhammad.naveed@uaf.edu.pk (M.N.)

**Citation:** Ali, L.; Xiukang, W.; Naveed, M.; Ashraf, S.; Nadeem, S.M.; Haider, F.U.; Mustafa, A. Impact of Biochar Application on Germination Behavior and Early Growth of Maize Seedlings: Insights from a Growth Room Experiment. *Appl. Sci.* **2021**, *11*, 11666. https://doi.org/10.3390/app112411666

Academic Editor: Carlos Rico de la Hera

Received: 13 September 2021
Accepted: 30 November 2021
Published: 9 December 2021



**Abstract:** Reduced germination and early crop maturity due to soil compaction, nutrients stress, and low moisture are major constraints to achieve optimum crop yield, ultimately resulting in significant economic damages and food shortages. Biochar, having the potential to improve physical and chemical properties of soil, can also improve nutrients and moisture access to plants. In the present study, a growth room experiment was conducted to assess biochar influence on maize seed germination, early growth of seedlings, and its physiological attributes. Corn cob biochar (CCB) was mixed with soil at different rates (0.5%, 1%, 1.5%, 2%, 2.5%, and 3% *w*/*w*) before seed sowing. Results obtained showed that increasing CCB application rate have neutral to positive effects on seed germination and seedling growth of maize. Biochar addition at the rate of 1.5% (*w*/*w*) significantly increased shoot dry biomass (40%), root dry biomass (32%), total chlorophyll content (a and b) (55%), germination percentage (13%), seedling vigor (85%), and relative water content (RWC) (68%), in comparison to un-amended control treatment. In addition to this, it also improved germination rate (GR) by 3% as compared to control treatment, while causing a reduction in mean emergence time (MET). Moreover, application of biochar (3%) also resulted in enhancement of antioxidant enzyme activity, particularly superoxide dismutase (SOD) and catalase (CAT) by 13% and 17%, respectively. Conclusively, biochar application is an attractive approach to improve the initial phase of plant growth and provide better crop stand and essential sustainable high yields.

**Keywords:** biochar; germination; seedling vigor; chlorophyll; germination rate; maize growth

## 1. Introduction

Better crop yield depends on early growth and developmental phases, namely germination, seedling vigor, and plant development. These phases of plant growth are critically related with soil properties, such as porosity, moisture, and nutrient supply. In this way, germination and seedling quality have vital role on crop growth and yield throughout the cropping season [1]. Poor germination of seeds could be enhanced through scarification (mechanical method to reduce seed dormancy) and temperature treatment to seeds [2,3], and the decontaminating or screening of seeds using treatment based on magnetic fields and irradiation with light of various wavelength [4,5]. Germination rate could also be improved by soaking seeds in algal or *Cyanobacteria* aqueous culture prior to sowing [6]. Soil properties perform a key role in seed germination and growth because early crop stand is directly influenced by soil characteristics such as organic matter (OM), soil bulk density (BD), soil porosity, and soil moisture content [7]. In addition to this, improvement in soil properties also increases nutrients and moisture availability to plants, resulting in high germination percentage and improved seedling vigor of crop. Along with this, improving soil organic contents positively manipulates soil physical parameters that ultimately leads to better crop growth and enhanced productivity [8].

Biochar is a highly carbon-containing organic material produced through the process of pyrolysis of several kinds of biomasses [9–11]. Due to its complex aromatic structure, biochar is recalcitrant in nature that makes it more difficult to degrade it in soil [12]. This property of biochar is also an excellent approach to fix organic carbon in soil for longer durations by using crop residues which not only improves soil organic carbon, moisture holding capacity, and physical properties, but also reduces environmental carbon footprints [13,14]. Biochar mixing into the soil directly or indirectly affects different soil indicators such as soil structure, soil porosity [15], cation retention capacity [12], moisture retention [16], soil fertility improvement [17], retention of contaminants and xenobiotics in soil [18], and bacterial diversity [19,20]. It also improves soil physico-chemical characteristics by enhancing plant growth and productivity [21]. Although biochar is widely used in agriculture, its effect varies depending on plant variety and soil type [22], as soil type may also affect the function of biochar [23]. Several other researchers also proposed that biochar application into the soil is an attractive approach for better crop growth [24–26].

Previous research experiments performed by various scientists were more inclined to assess biochar role in soil health and crop yield when applied to soil [27,28]. Whereas the effect of biochar on early growth stages of plants, viz. seedling emergence and growth is seldom studied. Therefore, the present study was conducted with the special aim to understand the effect of biochar application as organic amendment on the germination of maize seeds and early growth of its seedlings.

## 2. Materials and Methods

### 2.1. Production and Characterization of Biochar

Corn cobs were collected from field, crushed, and air dried to $\leq$10% ($v/w$) moisture level for biochar production. Crushed cobs were then subjected to pyrolysis at 350 °C temperature according to Sanchez et al. [29]. Biochar pH and electrical conductivity (ECe) was determined in distilled water in 1:20 [30], while ash and moisture contents of biochar were measured by placing open crucible in muffle furnace at 107 °C and 750 °C, respectively, till constant weight. Wet digestion method was practiced for nitrogen (N), phosphorus (P), and potassium (K) determination in biochar by using the Kjeldahl method, a UV-VIS spectrophotometer, and a flame photometer, respectively [31]. Various physico-chemical properties of biochar are presented in Table 1.

**Table 1.** Physicochemical properties of biochar produced at 350 °C.

| Parameters | Units | Pyrolysis Temperature (350 °C) |
|---|---|---|
| Yield | % | 43–46 |
| $pH_{1:20}$ | - | 7.10 |
| $EC_{1:20}$ | dS $m^{-1}$ | 0.73 |
| Ash content | % | 13.2 |
| Moisture content | % | 2.49 |
| Cation exchange capacity (CEC) | $cmol_c\ kg^{-1}$ | 43–45 |
| Carbon | % | 58.23 |
| Nitrogen | % | 1.33 |
| Phosphorus | % | 0.43 |
| Potassium | % | 1.02 |
| Sulphur | % | 0.93 |

### 2.2. Germination Assay

A germination assay was practiced to evaluate biochar effect on germination of maize seeds and seedling growth. For this purpose, sandy loam soil was filled in plastic trays. Before sowing, developed biochar was manually mixed in sand at the rates of 0%, 0.5%, 1%, 1.5%, 2%, 2.5%, and 3% (*w*/*w*) in triplicates. Twelve seeds of maize (*Zea mays* L.) variety Syngenta NK-6621 were seeded in each tray to calculate germination percentage and rate. Thus, a total of 21 experimental units were assorted using completely randomized design (CRD) and the method described by International Seed Testing Association (ISTA) was used seven days after sowing (DAS). Seed germination percentage (GP), germination rate (GR), mean emergence time (MET), and seed vigor (SV) were calculated using following formulas:

$$\text{Germination Percentage (GP)} = \left({}^{\text{SNG}}/_{\text{SN0}}\right) \times 100$$

where, "SNG" represent germinated seeds in total and "SN0" represents total number of viable seeds [32].

$$\text{Germination Rate} = \frac{\sum \text{N}}{\sum (\text{n} \times \text{g})}$$

where, "N" is the number of germinated seeds, "n" is the seeds germinated on growth day, "g" is the total number of germinated seeds.

$$\text{Mean Emergence Time (MET)} = \frac{\sum \text{Dn}}{\sum \text{n}}$$

where, "n" is the germinated seeds and "D" represents total number of days [33].

$$\text{Seed Vigour (SV)} = \text{GP} \times \text{Seedling length}$$

where, "GP" is germination percentage.

An equation by AOSA [34] was used to calculate emergence index (EI):

$$\text{Emergence index} = \frac{\text{Germinated seeds}}{\text{No of days of first count}} + \frac{\text{Germinated seeds}}{\text{No of days of final count}}$$

Time required for 50% emergence of seeds (E50) was determined using equation given by Farooq et al. [35]:

$$E_{50} = \text{ti} + \left[ \left( \frac{\text{N}}{2} - \text{ni} \right) (\text{tj} - \text{ti}) \setminus \text{nj} - \text{ni} \right]$$

Here, in the equation, "N" is the final number of germinating seeds, while nj and ni are the cumulative number of seeds germinated by adjacent counts at times tj and ti, respectively, when ni < N/2 < nj.

*2.3. Determination of Agronomic Parameters of Maize Seedlings*

After 20 days of seedling growth, the experiment was terminated and data regarding agronomic parameters, such as seedling length, fresh and dry weight, etc., were recorded considering equal number of germinated seeds (six seeds) for all the treatments with and without biochar application.

*2.4. Physiological and Biochemical Attributes*

Freshly harvested maize seedlings were subjected to study various physiological attributes, viz. chlorophyll a and b contents, relative water contents (RWC), superoxide dismutase, and catalase activity.

The Arnon [36] method was used for chlorophyll a and b determination in leaf samples. For this purpose, 0.5 g fresh leaf sample was taken from every treatment and homogenized with 80% acetone, and left for 24 h at 0.4 °C, filtered, and stored for analysis. Chlorophyll a and b contents in leaf extract was determined using a UV-visible spectrometer (Shimadzu UV-1201) at a wavelength of 645 and 663 nm, respectively, and were calculated in mg $g^{-1}$ by using the following formula [37].

$$\text{Chlorophyll}_a = (0.999\ \text{OD}_{663} - 0.099\ \text{OD}_{645})$$

$$\text{Chlorophyll}_b = (-0.328\ \text{OD}_{663} + 1.770\ \text{OD}_{645})$$

Relative water contents (RWC) were also measured by using the formula given below:

$$\text{Relative water content}\ \% = \frac{\text{fresh wt} - \text{dry wt}}{\text{turgid wt} - \text{dry wt}} \times 100$$

The Elavarthi and Martin [38] method was used to determine antioxidant (catalase and superoxide dismutase) activity in leaf samples.

*2.5. Statistical Analysis of Data*

Results of this experiment were statistically analysed using statistical software Statistix 8.1® (Analytical Software, Tallahassee, FL, USA). Tukey's multiple comparison test ($p \leq 0.05$) was used for comparing treatment means at 95% confidence interval. Principal component analysis was performed by using R software packages.

## 3. Results

*3.1. Impact of Biochar on the Germination of Maize Seeds*

Significant ($p \leq 0.05$) improvement in mean emergence time, germination energy, and emergence index were noticed with biochar application, at the rate of 1.5 and 2.5% (Figure 1). Germination of maize seeds started on third day, and the highest number of seeds (eight seeds) were germinated under treatment where biochar was applied at 1 and 1.5% rate, compared to the control, where six seeds germinated without biochar. Seed germination percentage was maximum (97%) with 1%, 1.5%, 2% and 2.5% biochar application (Figure 2). Seedling vigor significantly ($p \leq 0.05$) increased (85%) with 1.5% biochar compared to the control (Figure 3).

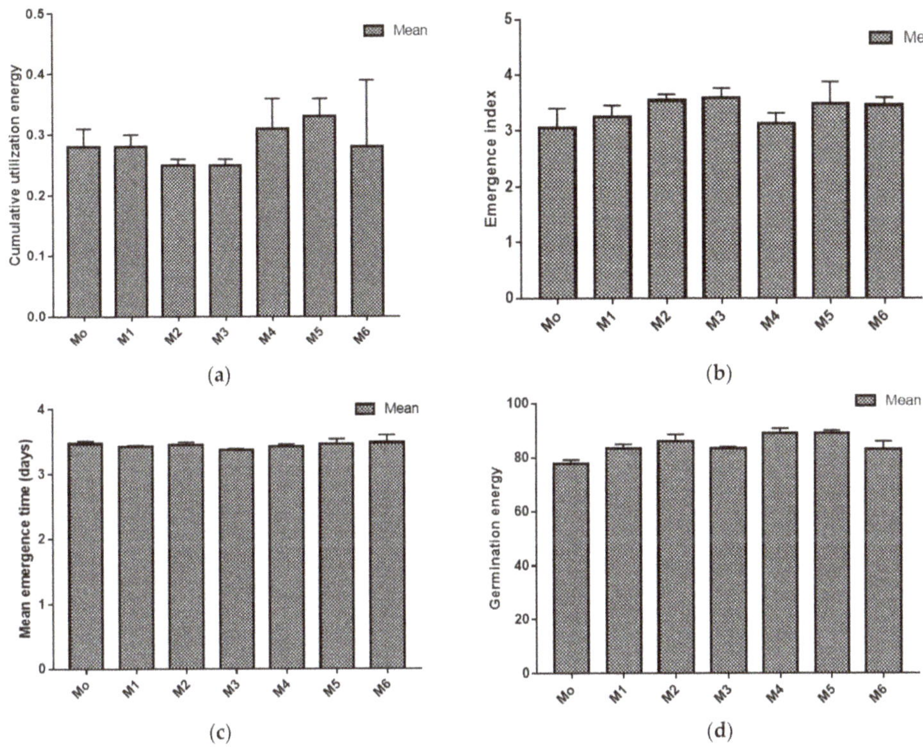


**Figure 1.** Effect of biochar rates on cumulative utilization energy (CUE) (**a**), emergence index (**b**), mean emergence time (**c**), and germination energy (**d**) of maize seedlings. Data presented is average of three repeats. Means sharing the similar letters do not differ significantly ($p \geq 0.05$).

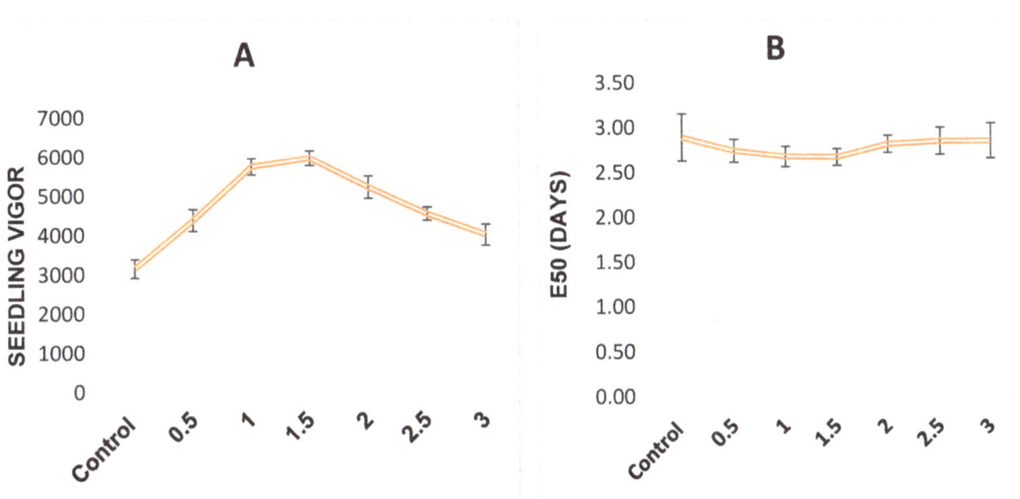


**Figure 2.** Effect of different biochar rates on (**A**) seedling vigor and (**B**) $E_{50}$. Data is the average of three repeats. Error bars showing significance of treatment ($p \leq 0.05$).

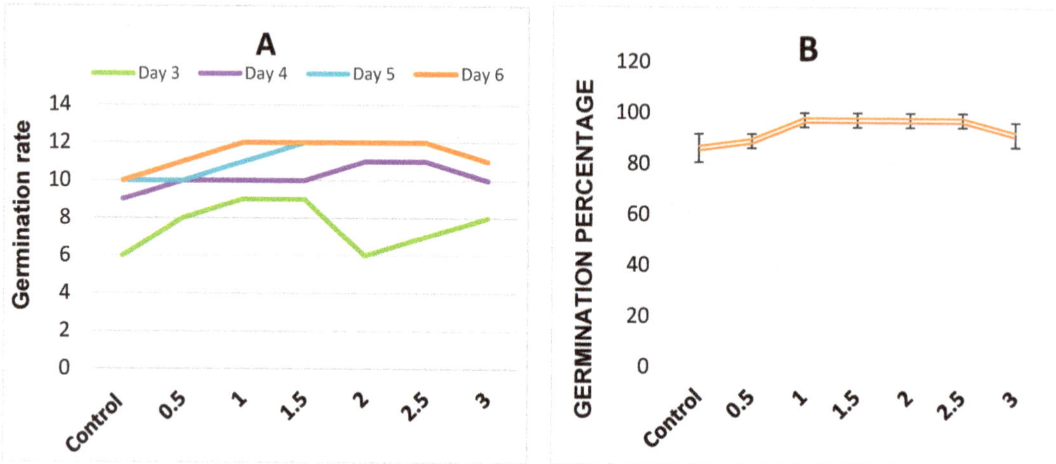


**Figure 3.** (**A**) Effect of different biochar rates on number of germination seeds per day. (**B**) Effect of different biochar rates on germination percentage of maize seeds. Here, the data presented is the average of three replicates of each treatment.

*3.2. Impact of Biochar on Morpho-Physiological Attributes and Enzyme Activity of Maize Seedlings*

3.2.1. Morphological Attributes

Results regarding maize seedlings growth (Table 2) revealed that biochar application to soil influenced maize seedlings positively. Seedling length was increased significantly ($p \leq 0.05$) at a lower rate of biochar, and maximum shoot length was 34 cm at 1.5% biochar, which was 46% more than control; conversely, higher rates of biochar (>1.5%) showed non-significant effects on shoot length, as compared to the control (Table 2). Fresh (1.09 g) and dry weight (0.089 g) of seedlings shoots was the highest with 1.5 and 2% biochar application, while it was 0.67 g and 0.063 g, in control. However, biochar application at higher rates (>2%) inhibited seedlings root length, while lower rates (<2%) increased root length. The maximum root length (27.33 cm) was observed with biochar application at the rate of 1.5%. Moreover, this increased root length was statistically at par where 1% and 2% biochar was applied (Table 2). Significant ($p \leq 0.05$) increase in root fresh and dry mass of maize seedlings was observed with the application of biochar when applied at the rate of 1.5 and 2%, respectively, although higher rate of biochar addition showed no-significant difference between fresh and dry mass of seedling roots, as compared to the control treatment (without biochar).

**Table 2.** Biochar effects on maize seedling growth parameters.

| Biochar Level (%) | Seedlings Shoot | | | Seedlings Root | | |
|---|---|---|---|---|---|---|
| | Length (cm) | Fresh Weight (g) | Dry Weight (g) | Length (cm) | Fresh Weight (g) | Dry Weight (g) |
| Control | 23.33 ± 1.45d | 0.67 ± 0.02d | 0.063 ± 0.002e | 13.33 ± 1.45d | 0.37 ± 0.02d | 0.090 ± 0.004d |
| 0.5 | 28.33 ± 1.67bc | 0.85 ± 0.05c | 0.074 ± 0.004cd | 21.00 ± 1.53bc | 0.47 ± 0.02bc | 0.098 ± 0.005cd |
| 1.0 | 32.33 ± 0.88ab | 1.01 ± 0.02ab | 0.085 ± 0.002ab | 26.67 ± 1.67a | 0.52 ± 0.02ab | 0.114 ± 0.003ab |
| 1.5 | 34.00 ± 0.58a | 1.09 ± 0.05a | 0.088 ± 0.004ab | 27.33 ± 1.20a | 0.57 ± 0.02a | 0.119 ± 0.002ab |
| 2.0 | 28.67 ± 1.76bc | 1.08 ± 0.07ab | 0.089 ± 0.003a | 25.00 ± 1.15ab | 0.52 ± 0.05ab | 0.122 ± 0.003a |
| 2.5 | 26.33 ± 1.20cd | 1.03 ± 0.06ab | 0.078 ± 0.003bc | 20.33 ± 1.33c | 0.42 ± 0.03cd | 0.107 ± 0.004bc |
| 3.0 | 25.67 ± 2.03cd | 0.94 ± 0.03bc | 0.063 ± 0.005de | 18.00 ± 1.53c | 0.40 ± 0.03cd | 0.100 ± 0.005cd |

Values are the mean of three replicates. Means sharing similar letters do not differ significantly (at $p \leq 0.05$).

#### 3.2.2. Physiological and Biochemical Attributes

Data regarding RWC showed that biochar application significantly ($p \leq 0.05$) increased RWC, and the highest (83%) was observed with 3% biochar treatment, as compared to the control treatment (Table 3). Chlorophyll a and b contents were 74% and 43% higher than control with biochar application at the rate 1.5% and 2%, respectively (Table 3). Moreover, a significant increase in total chlorophyll (chlorophyll a + b) was also observed with biochar application over the treatment set as the control.

**Table 3.** Biochar effect on relative water content (RWC), chlorophyll (a and b), and enzyme (catalase (CAT), super oxide dismutase (SOD)) activity of maize seedlings.

| Biochar Level (%) | RWC (%) | Chlorophyll a mg $g^{-1}$ Fresh Weight | Chlorophyll b mg $g^{-1}$ Fresh Weight | Chlorophyll a + b | SOD U $g^{-1}$ fw $min^{-1}$ | CAT U $g^{-1}$ fw $h^{-1}$ |
|---|---|---|---|---|---|---|
| Control | 61 ± 2.08d | 1.40 ± 0.11c | 0.82 ± 0.02e | 2.20 ± 0.13c | 11.93 ± 0.73a | 15.20 ± 1.17ab |
| 0.5 | 63 ± 1.86cd | 1.55 ± 0.07c | 0.87 ± 0.03de | 2.40 ± 0.09c | 11.80 ± 0.98a | 14.68 ± 0.86ab |
| 1.0 | 67 ± 2.31cd | 1.98 ± 0.08b | 0.97 ± 0.04cd | 2.95 ± 0.13b | 11.68 ± 0.88a | 13.68 ± 0.88b |
| 1.5 | 68 ± 3.53bd | 2.43 ± 0.11a | 0.99 ± 0.05bc | 3.41 ± 0.16a | 11.31 ± 0.65a | 14.33 ± 0.34ab |
| 2.0 | 71 ± 2.96bc | 2.31 ± 0.06a | 1.17 ± 0.04a | 3.47 ± 0.10a | 11.66 ± 0.32a | 15.68 ± 0.88ab |
| 2.5 | 76 ± 1.53ab | 2.30 ± 0.13a | 1.10 ± 0.05ab | 3.39 ± 0.18a | 12.34 ± 0.33a | 16.01 ± 0.58ab |
| 3.0 | 84 ± 2.91a | 2.20 ± 0.08ab | 0.92 ± 0.04ce | 3.11 ± 0.07ab | 12.68 ± 0.88a | 16.68 ± 1.33a |

Data is the average of three repeats. Columns sharing similar letters do not differ significantly (at $p \leq 0.05$).

Data regarding measurement of antioxidant enzyme activity showed that biochar application also increased superoxide dismutase (SOD) and catalase (CAT) activity non-significantly ($p \geq 0.05$), although all the treatment results were statistically significant ($p \leq 0.05$) at par with each other (Table 3).

### *3.3. Principal Component and Correlation Analyses*

The principal component analysis (PCA) showed the distribution of different treatments in maize crop as presented in the PCA-biplot (Figure 4). Significant results were obtained from the biplot of PCA performed for two factors (cumulative variance 83.9%), the first factor represents 56.3% variation, while 27.6% of the difference is explained by the second factor. Hence, great variation was found among all the treatments in maize plants. Here, visual impact (Figure 4) highlights the relationship and variation observed among all studied parameters of maize plant. Moreover, significant positive and negative correlations were observed between various studied parameters (Table 4).

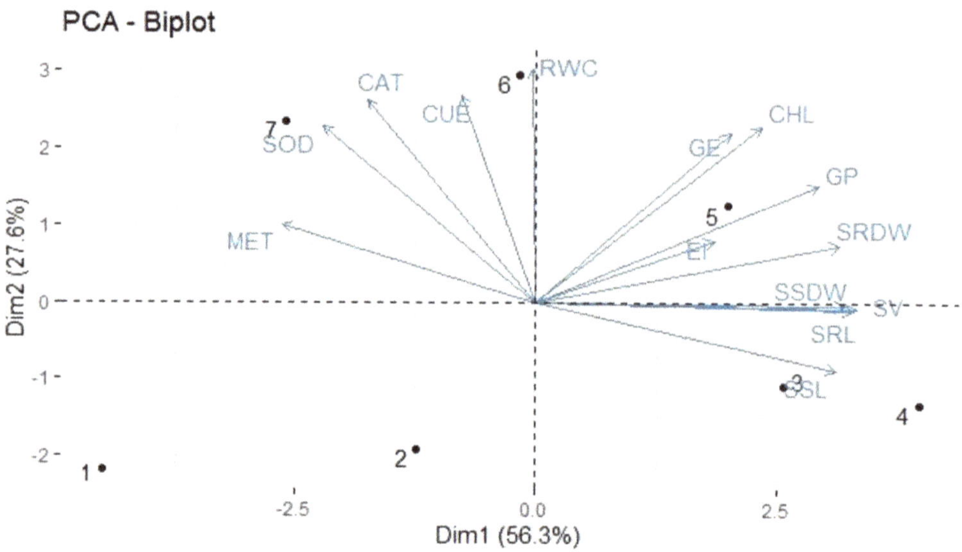


**Figure 4.** Principal component analysis showing the correlation between the impact of biochar application on the germination of maize seeds and early growth of its seedlings, with the, treatments: T1: Control; T2: 0.5% Biochar; T3: 1.0% Biochar; T4: 1.5% Biochar; T5: 2.0% Biochar; T6: 2.5% Biochar; T7: 3.0% Biochar. The abbreviations are: GP, germination percentage; GE, germination energy; MET, mean emergence time; EI, emergence index; SV, seed vigor; RWC, relative water content; CHL, chlorophyll a and b; CAT, catalase; SOD, superoxide dismutase; SSDW, seedling shoot dry weight; SRDW, seedling root dry weight; SRL, seedling root length; SSL, seedling shoot length; CUE, cumulative utilization energy.

**Table 4.** Correlation matrix among different parameters of maize.

| Parameter | GP | GE | MET | EI | SV | RWC | Chl (a + b) | CAT | SOD | SDW |
|---|---|---|---|---|---|---|---|---|---|---|
| GE | 0.632 *** | | | | | | | | | |
| MET | −0.137 ns | −0.121ns | | | | | | | | |
| EI | 0.598 *** | 0.249 ns | −0.173 ns | | | | | | | |
| SV | 0.756 *** | 0.563 ** | −0.432 ** | 0.509** | | | | | | |
| RWC | 0.261 ns | 0.302 ns | 0.287 ns | 0.329 ns | 0.022 ns | | | | | |
| Chl (a + b) | 0.600 *** | 0.636 *** | −0.214 ns | 0.372 ns | 0.583 ** | 0.598 *** | | | | |
| CAT | −0.105 ns | 0.099 ns | −0.170 ns | −0.132 ns | −0.238 ns | 0.257 ns | 0.178 ns | | | |
| SOD | −0.099 ns | 0.079 ns | −0.166 ns | −0.236 ns | −0.148 ns | 0.068 ns | −0.050 ns | 0.567 ** | | |
| SDW | 0.480 ** | 0.648 *** | −0.407 ns | 0.216 ns | 0.800 *** | −0.179 ns | 0.502 ** | −0.144 ns | −0.215 ns | |
| RDW | 0.604 *** | 0.599 *** | −0.101 ns | 0.257 ns | 0.711 *** | 0.160 ns | 0.647 *** | −0.253 ns | −0.410 ns | 0.787 *** |

*** shows highly significant at $p \leq 0.01$, ** shows significant at $p \leq 0.05$, ns shows non-significant at $p \geq 0.05$. The abbreviations are: GP, germination percentage; GE, germination energy; MET, mean emergence time; EI, emergence index; SV, seed vigor; RWC, relative water content; Chl (a + b), chlorophyll a and b; CAT, catalase; SOD, superoxide dismutase; SDW, shoot dry weight; and RDW, root dry weight.

## 4. Discussion

Biochar feedstock type, rate, and application method could play an important role in better plant response [17,39–41], due to its specific characteristics which are favorable to the soil health and could improve soil properties along with crop productivity [42,43]. However, low or high nutrient content present in biochar that could influence germination of seeds depends upon properties of biochar feedstock [44]. As biochar has high nutrient retention and water holding capacity [45–47], we supposed that soil amendment with biochar may play a potential role in germination and early seedling development. The present study showed that biochar at a lower rate (Table 2) enhanced the early growth of maize seedlings that is evident from earlier studies [15,41,46], and improved plant macro- (N, P, K) and micronutrients (Zn, Mn, Fe, B, Cu, etc.), as well as moisture retention in the soil [48]. Our findings are consistent with [49] who reported an improvement in the initial growth of shoot and root length in soil amended with biochar at a lower rate. Increase in seedling growth with biochar application is also documented by Yang et al. [50]. Similarly, rice seedling lengths of roots and shoots increased with biochar nanoparticles [51]. Marzouk [52] documented that olive pomace wastes biochar improved the root growth of date palm seedlings. Shoot and root length increment in *Coreopsis, Eschscholzia,* and *Leucanthemum* was also reported by Benjamin [53]. This is because of a positive plant growth response to applied biochar [41,54]; although, the increase in plant growth could be due to the synergistic effect of biochar and fertilizer [55], more nutrient availability [56], balanced microbes [57], and substrate with better properties [58].

In the present study, the biomass of root and shoot seedlings of maize increased under the influence of biochar addition (Table 2), which is in accordance with the previous findings of Lehmann et al. [59]. They reviewed that biochar application increased shoot and root biomass in biochar amended soil. Similarly, significant enhancement in root and shoot biomass has also been reported previously by Bu et al. [60] in *Robinia pseudoacacia* L. seedlings. However, studies conducted by [59,60] also reported reduced root to shoot ratio due to biochar addition into the soil. Conversely, here in our study, an increase in root to shoot ratio was observed, as reported by Samuel et al. [61].

Significant increase in total chlorophyll, and a and b chlorophyll, was observed in this experiment with biochar amendment as compared to the control treatment (Table 3). Reduced chlorophyll content in control treatment might be due to the reduction in specific enzyme suppression necessary for chlorophyll synthesis and reduced mineral uptake, e.g., Mg required for pigment synthesis [62]. Chlorophyll content of *Malus hupehensis Rehd.* seedlings was improved with biochar application into the soil [63]. Afsharipoor and Roosta [64] also reported improved chlorophyll content (chlorophyll a, b and total chlorophyll) in vegetables grown in organics-amended soil [65], and in *Amaranthus* as well [66].

Biochar application also played a very important role in increasing the enzyme activity; a non-significant increase in superoxide dismutase (SOD) and catalase (CAT) expression in leaves is shown in Table 3. Enhanced antioxidant activity (SOD and CAT) in seedling leaves after biochar addition to soil is also reported previously by Wang et al. [63], which could be due to the improved uptake of essential minerals or biochar trigger activities of some specific enzymes [67].

Mean emergence time (MET) also decreased in the current experiment as evident by (Figure 1). A similar finding, i.e., reduced MET, was also observed by Qayyum et al. [68], due to the utilization of vegetable waste biochar as soil amendment. Low moisture level in soil reduces the seed germination as well [69], and biochar application in our study helped to overcome this shortcoming because biochar enhances the water holding capacity of the soil and maintains the soil moisture level which promotes germination rate of seeds.

In the present study, biochar application increased seed germination and germination rate or mean emergence time (MET) in days (Figures 3 and 4). The possible reason for better germination is more water content held by biochar, as reported by Kammann et al. [46] and Basso et al. [70]. Use of organic amendments put into the soil has greater influence on moisture in growing medium that ultimately results in improved seed germination [71]; although, soluble plant growth regulators and native bacteria present in the amendment also lead to healthier germination of seeds. Here, better seed germination of maize plants could also be due to the mineral nutrients present in biochar which are released slowly, thus maintaining the fertility level of the soil to promote growth attributes [50]. Another factor that may also be the reason of better germination is the adsorption capacity of biochar. This study resembled with [72,73], who reported that biochar addition into the soil increased germination parameters. Seed germination rate was enhanced by biochar addition as organic amendment, specifically when applied at a lower rate [74]. Improved germination of *Robinia pseudoacacia* L. with biochar addition into the calcareous soil has also been reported [60]. Germination rate was increased significantly with biochar and other treatments compared to the control [52] at a lower rate of applied biochar because a higher rate of biochar may have some unwanted substances in olive pomace-derived biochar, leading to a reduced germination rate [75]. Overall, biochar addition as organic amendment in soil showed a positive effect on the growth of maize seedlings by making a continuous supply of required nutrients [76–78]. Along with this, biochar also reduces the level of contaminants present in soil that otherwise hamper plant growth by causing negative impacts on both soil and plant health [78].

## 5. Conclusions

The present investigation concluded that application of biochar has a positive to neutral effect on maize seed germination and growth of its seedling. Increased seedling growth and biomass through biochar application reflected that biochar has the capacity to hold more water, and also has necessary minerals in it which are released slowly to promote plant growth or crop yield. Moreover, chlorophyll (a and b) and relative water content were also increased significantly in maize seedlings, but antioxidant activity (SOD and CAT) increased non-significantly with biochar application. Hence, biochar addition to soil as an organic amendment could be a better, easier, and more affordable option to enhance germination and early growth of maize seedlings to improve crop development and production in an environmentally friendly "green" way.

**Author Contributions:** Conceptualization, L.A. and A.M.; methodology, W.X.; software, F.U.H.; validation, M.N., S.A. and S.M.N.; formal analysis, A.M.; resources, W.X.; writing—original draft preparation, L.A. and M.N.; writing—review and editing, A.M., M.N. and W.X.; funding acquisition, W.X., and M.N. All authors have read and agreed to the published version of the manuscript.

**Funding:** This work received no external funding.

**Institutional Review Board Statement:** Not applicable.

**Informed Consent Statement:** Not applicable.

**Data Availability Statement:** All the data can be accessible from the corresponding author on request.

**Acknowledgments:** The authors extend their appreciation to the Institute of Soil and Environmental Sciences, University of Agriculture, Faisalabad, Pakistan.

**Conflicts of Interest:** The authors declare no conflict of interest.

## References

1. Khajeh-Hosseini, M.; Lomholt, A.; Matthews, S. Mean germination time in the laboratory estimates the relative vigour and field performance of commercial seed lots of maize (*Zea mays* L.). *Seed Sci. Technol.* **2009**, *37*, 446–456. [CrossRef]
2. Duczmal, K.; Tucholska, H. *Nasiennictwo [Seed Production]*; PWRiL: Poznań, Poland, 2000; Volume 1, pp. 205–234.
3. Cavallaro, V.; Barbera, A.C.; Maucieri, C.; Gimma, G.; Scalisi, C.; Patanè, C. Evaluation of variability to drought and saline stress through the germination of different ecotypes of carob (*Ceratonia siliqua* L.) using a hydrotime model. *Ecol. Eng.* **2016**, *95*, 557–566. [CrossRef]
4. Ciupak, A.; Szczurowska, I.; Gładyszewska, B.; Pietruszewski, S. Impact of laser light and magnetic field stimulation on the process of buckwheat seed germination. *Tech. Sci.* **2007**, *10*, 1–10. [CrossRef]
5. Grzesik, M.; Janas, R.; Górnik, K.; Romanowska-Duda, Z. Biological and physical methods of seed production and processing. *J. Res. Appl. Agric. Eng.* **2012**, *57*, 147–152.
6. Karthikeyanb, N.; Prasannaa, R.; Nainb, L.; Kaushik, B.D. Evaluating the potential of plant growth promoting *Cyanobacteria* as inoculants for wheat. *Eur. J. Soil Biol.* **2007**, *43*, 23–30. [CrossRef]
7. Oliet, J.A.; Planelles, R.; Segura, M.L.; Artero, F.; Jacobs, D.F. Mineral nutrition and growth of containerized *Pinus halepensis* seedlings under controlled-release fertilizer. *Sci. Hortic.* **2004**, *103*, 113–129. [CrossRef]
8. Hassanzadeh, G. A Study on the Effects of Organic, Inorganic and Integrated Fertilizers on the Quantitative and Qualitative Traits of Different Sunflower (*Helianthus annuus* L.) Cultivars in West Azerbaijan, Iran. Ph.D. Thesis, Tarbiat Modares University, Tehran, Iran, 2000; 195p. (In Persian).
9. Major, J.; Rondon, M.; Molina, D.; Riha, S.J.; Lehmann, J. Nutrient leaching in a Colombian savanna Oxisol amended with biochar. *J. Environ. Qual.* **2012**, *41*, 1076–1086. [CrossRef]
10. Ahmad, M.; Rajapaksha, A.U.; Lim, J.E.; Zhang, M.; Bolan, N.; Mohan, D.; Vithanage, M.; Lee, S.S.; Ok, Y.S. Biochar as a sorbent for contaminant management in soil and water: A review. *Chemosphere* **2019**, *99*, 19–33. [CrossRef]
11. Lu, K.; Yang, X.; Shen, J.; Robinson, B.; Huang, H.; Liu, D.; Bolan, N.; Pei, J.; Wang, H. Effect of bamboo and rice straw biochars on the bioavailability of Cd, Cu, Pb and Zn to *Sedum plumbizincicola*. *Agric. Ecosyst. Environ.* **2014**, *191*, 124–132. [CrossRef]

12. Jindo, K.; Monedero, M.A.S.; Hernández, T.; García, C.; Furukawa, T.; Matsumoto, K. Biochar influences the microbial community structure during manure composting with agricultural wastes. *Sci. Total Environ.* **2012**, *416*, 476–481. [CrossRef]
13. Glaser, B.; Lehmann, J.; Zech, W. Ameliorating physical and chemical properties of highly weathered soils in the tropics with charcoal—A review. *Biol. Fertile Soils* **2002**, *35*, 219–230. [CrossRef]
14. Parsons, A.J.; Rowarth, J.S.; Newman, P. Managing pastures for animals and soil carbon. *Proc. N. Z. Grassland Assoc.* **2009**, *71*, 77–84. [CrossRef]
15. Chan, K.Y.; Van Zwieten, L.; Meszaros, I.; Downie, A.; Joseph, S. Using poultry litter biochars as soil amendments. *Aust. J. Soil Res.* **2008**, *46*, 437–444. [CrossRef]
16. Jeong, C.Y.; Dodla, S.K.; Wang, J.J. Fundamental and molecular composition characteristics of biochars produced from sugarcane and rice crop residues and by-products. *Chemosphere* **2016**, *142*, 4–13. [CrossRef] [PubMed]
17. Lehmann, J. Bioenergy in the black. *Front. Ecol. Environ.* **2007**, *5*, 381–387. [CrossRef]
18. Teixido, M.; Hurtado, C.; Pignatello, J.J.; Beltran, J.L.; Granados, M.; Peccia, J. Predicting contaminant adsorption in black carbon (biochar)-amended soil for the veterinary antimicrobial sulfamethazine. *Environ. Sci. Technol.* **2013**, *47*, 6197–6205. [CrossRef]
19. Kolton, M.; Harel, Y.M.; Pasternak, Z.; Graber, E.R.; Elad, Y.; Cytryn, E. Impact of biochar application to soil on the root-associated bacterial community structure of fully developed greenhouse pepper plants. *Appl. Environ. Microbiol.* **2011**, *77*, 4924–4930. [CrossRef]
20. Makoto, K.; Hirobe, M.; DeLuca, T.H.; Bryanin, S.V.; Procopchuk, V.F.; Koike, T. Effects of fire-derived charcoal on soil properties and seedling regeneration in a recently burned *Larix gmelinii/Pinus sylvestris* forest. *J. Soils Sediments* **2011**, *11*, 1317–1322. [CrossRef]
21. Asai, H.; Samson, B.K.; Stephan, H.M.; Songyikhangsuthor, K.; Homma, K.; Kiyono, Y.; Horie, T. Biochar amendment techniques for upland rice production in Northern Laos 1. Soil physical properties, leaf SPAD and grain yield. *Field Crops Res.* **2009**, *111*, 81–84. [CrossRef]
22. Noguera, D.; Barot, S.; Laossi, K.R.; Cardoso, J.; Lavelle, P.; Cruz de Carvalho, M. Biochar but not earthworms enhances rice growth through increased protein turnover. *Soil Biol. Biochem.* **2012**, *52*, 13–20. [CrossRef]
23. Macdonald, L.M.; Farrell, M.; Van Zwieten, L.; Krull, E.S. Plant growth responses to biochar addition: An Australian soils perspective. *Biol. Fertil. Soils* **2014**, *50*, 1035–1045. [CrossRef]
24. Wu, H.; Lai, C.; Zeng, G.; Liang, J.; Chen, J.; Xu, J.; Dai, J.; Li, X.; Liu, J.; Chen, M.; et al. The interactions of composting and biochar and their implications for soil amendment and pollution remediation: A review. *Crit. Rev. Biotechnol.* **2017**, *37*, 754–764. [CrossRef] [PubMed]
25. Ye, S.; Zeng, G.; Wu, H.; Zhang, C.; Dai, J.; Liang, J.; Yu, J.; Ren, X.; Yi, H.; Cheng, M.; et al. Biological technologies for the remediation of co-contaminated soil. *Crit. Rev. Biotechnol.* **2017**, *37*, 1062–1076. [CrossRef] [PubMed]
26. Zama, E.F.; Reid, B.J.; Arp, H.P.H.; Sun, G.-X.; Yuan, H.Y.; Zhu, Y.-G. Advances in research on the use of biochar in soil for remediation: A review. *J. Soils Sediments* **2018**, *18*, 2433–2450. [CrossRef]
27. Oh, T.K.; Shinogi, Y.; Chikushi, J.; Lee, Y.H.; Choi, B. Effect of aqueous extract of biochar on germination and seedling growth of lettuce (*Lactuca sativa* L.). *J. Fac. Agric. Kyushu Univ.* **2012**, *57*, 55–60. [CrossRef]
28. Alburquerque, J.A.; Salazar, P.; Barroon, V.; Torrent, J.; Campillo, M.D.C.; Gallardo, A.; Villar, R. Enhanced wheat yield by biochar addition under different mineral fertilization levels. *Agron. Sustain. Develop.* **2013**, *33*, 475–484. [CrossRef]
29. Sanchez, M.E.; Lindao, E.; Margaleff, D.; Martınez, O.; Moran, A. Pyrolysis of agricultural residues from rape and sunflowers: Production and characterization of bio-fuels and biochar soil management. *J. Anal. Appl. Pyrolysis* **2009**, *85*, 142–144. [CrossRef]
30. Nelson, D.W.; Sommers, L.E. Total carbon, organic carbon and organic matter. In *Methods of Soil Analysis, Part 2: Chemical and Microbiological Properties*; Klute, A., Ed.; American Society of Agronomy: Madison, WI, USA, 1982; pp. 570–571.
31. Wolf, B. The comprehensive system of leaf analysis and its use for diagnosing crop nutrient status. *Commun. Soil Sci. Plant Anal.* **1982**, *13*, 1035–1059. [CrossRef]
32. Scott, S.J.; Jones, R.A.; Williams, W.A. Review of data analysis methods for seed germination. *Crop Sci.* **1984**, *24*, 1192–1199. [CrossRef]
33. Ellis, R.A.; Roberts, E.H. The quantification of ageing and survival in orthodox seeds. *Seed Sci. Technol.* **1981**, *9*, 373–409.
34. Association of Official Seed Analysis (AOSA). Rules for testing seeds. *J. Seed Technol.* **1990**, *12*, 1–112.
35. Farooq, M.; Basra, S.M.A.; Hafeez, K.; Ahmad, N. Thermal hardening: A new seed vigor enhancing tool in rice. *J. Integr. Plant Biol.* **2005**, *47*, 187–193. [CrossRef]
36. Arnon, D.I. Copper enzymes in isolated chloroplasts, polyphenoxidase in *Beta vulgaris*. *Plant Physiol.* **1949**, *24*, 1–15. [CrossRef]
37. Nagata, M.; Yamashta, I. Simple method for simultaneous determination of chlorophyll and carotenoides in tomato fruit. *J. Jpn. Soc. Food Sci. Technol.* **1992**, *39*, 925–928. [CrossRef]
38. Elavarthi, S.; Martin, B. Spectrophotometric assays for antioxidant enzymes in plants. In *Plant Stress Tolerance*; Humana Press: New York, NY, USA, 2010; pp. 273–280.
39. Van Zwieten, L.; Kimber, S.; Morris, S.; Chan, K.Y.; Downie, A.; Rust, J.; Joseph, S.; Cowie, A. Effects of biochar from slow pyrolysis of papermill waste on agronomic performance and soil fertility. *Plant Soil* **2010**, *327*, 235–246. [CrossRef]
40. Barrow, C.J. Biochar: Potential for countering land degradation and for improving agriculture. *Appl. Geogr.* **2012**, *34*, 21–28. [CrossRef]
41. Solaiman, Z.M.; Murphy, D.V.; Abbott, L.K. Biochars influence seed germination and early growth of seedlings. *Plant Soil* **2012**, *353*, 273–287. [CrossRef]

42. Garnett, E.; Jonsson, L.M.; Dighton, J.; Murnen, K. Control of pitch pine seed germination and initial growth exerted by leaf litters and polyphenolic compounds. *Biol. Fertile Soils* **2004**, *40*, 421–426. [CrossRef]
43. Hille, M.; den Quden, J. Charcoal and activated carbon as adsorbate to phytotoxic compounds—A comparative study. *Oikos* **2005**, *108*, 202–207. [CrossRef]
44. Gaskin, J.W.; Steiner, C.; Harris, K.; Das, K.C.; Bibens, B. Effect of low-temperature pyrolysis conditions on biochar for agricultural use. *Am. Soc. Agric. Biol. Eng.* **2008**, *51*, 2061–2069.
45. Laird, D.A.; Fleming, P.; Davis, D.D.; Horton, R.; Wang, B.; Karlen, D.I. Impact of biochar amendments on the quality of a typical Midwestern agricultural soil. *Geoderma* **2010**, *158*, 443–449. [CrossRef]
46. Kammann, C.I.; Linsel, S.; Geobling, J.W.; Koyro, H.W. Influence of biochar on drought tolerance of *Chenopodium quinoa* Wild and on soil-plant relations. *Plant Soil* **2011**, *345*, 195–210. [CrossRef]
47. Page-Dumroese, D.; Robichaud, P.R.; Brown, R.E.; Tirocke, J.M. Water repellency of two forest soils after biochar addition. *Am. Soc. Agric. Biol. Eng.* **2015**, *58*, 335–342.
48. Rogovska, N.; Laird, D.; Cruse, R.; Trabue, S.; Heaton, E. Germination tests for assessing biochar quality. *J. Environ. Qual.* **2012**, *41*, 1014–1022. [CrossRef]
49. Shaohua, L.; Pan, B.; Li, H.; Zhang, D.; Xing, B. Detecting free radicals in biochars and determining their ability to inhibit the germination and growth of corn, wheat and rice seedlings. *Environ. Sci. Technol.* **2014**, *48*, 8581–8587.
50. Yang, L.; Liao, F.; Huang, M.; Yang, L.; Li, Y. Biochar improves sugarcane seedling root and soil properties under a pot experiment. *Sugar Tech.* **2015**, *17*, 36–40. [CrossRef]
51. Zhang, L.; Wang, Y.; Mao, J.; Chen, B. Effects of biochar nanoparticles on seed germination and seedling growth. *Environ. Pollut.* **2020**, *256*, 113409. [CrossRef] [PubMed]
52. Marzouk, E. Soil-less seed germination and root growth of date palm affected by biochar and metal nanoparticles. *J. Soil Sci. Agric. Eng.* **2017**, *8*, 77–84. [CrossRef]
53. Benjamin, K.H. Herbaceous Perennial Seed Germination and Seedling Growth in Biochar-amended Propagation Substrates. *HortScience* **2018**, *53*, 236–241.
54. Vaughn, S.F.; Kenar, J.A.; Thompson, A.R.; Peterson, S.C. Comparison of biochars derived from wood pellets and pelletized wheat straw as replacements for peat in potting substrates. *Ind. Crops Prod.* **2013**, *511*, 437–443. [CrossRef]
55. Deenik, J.L.; McClellan, T.; Uehara, G.; Antal, M.J., Jr.; Campbell, S. Charcoal volatile matter content influences plant growth and soil nitrogen transformations. *Soil Sci. Soc. Am. J.* **2010**, *744*, 1259–1270. [CrossRef]
56. Headlee, W.; Brewer, C.E.; Hall, R.B. Biochar as a substitute for vermiculite in potting mix for hybrid poplar. *BioEnergy Res.* **2014**, *71*, 120–131. [CrossRef]
57. Graber, E.R.; Harel, Y.M.; Kolton, M.; Cytryn, E.; Silber, A.; David, D.R.; Tsechansky, L.; Borenshtein, M.; Elad, Y. Biochar impact on development and productivity of pepper and tomato grown in fertigated soilless media. *Plant Soil* **2010**, *337*, 481–496. [CrossRef]
58. Alburquerque, J.A.; Calero, J.M.; Barron, V.; Torrent, J.; del Campillo, M.C.; Gallardo, A.; Villar, R. Effects of biochars produced from different feedstocks on soil properties and sunflower growth. *J. Plant Nutr. Soil Sci.* **2014**, *177*, 16–25. [CrossRef]
59. Lehmann, J.; Rillig, M.C.; Thies, J.; Masiello, C.A.; Hockaday, W.C.; Crowley, D. Biochar effects on soil biota-a review. *Soil Biol. Biochem.* **2011**, *43*, 1812–1836. [CrossRef]
60. Bu, X.L.; Xue, J.H.; Wu, Y.B.; Ma, W.B. Effect of Biochar on Seed Germination and Seedling Growth of *Robinia pseudoacacia* L. In Karst Calcareous Soils. *Commun. Soil Sci. Plant Anal.* **2020**, *51*, 352–363. [CrossRef]
61. Samuel, A.; Hund, A.; Martinsen, V.; Cornelissen, G. Biochar amendment increases maize root surface areas and branching: A shovelomics study in Zambia. *Plant Soil* **2015**, *395*, 45–55.
62. Kazemi, N.; Nejad, R.A.K.; ahimi, H.F.; Saadatmand, S.; Sattari, T.N. Effects of exogenous salicylic acid and nitric oxide on lipid peroxidation and antioxidant enzyme activities in leaves of *Brassica napus* L. under nickel stress. *Sci. Hortic.* **2010**, *126*, 402–407. [CrossRef]
63. Wang, Y.; Pan, F.; Wang, G.; Zhang, G.; Wang, Y.; Chen, X.; Mao, Z. Effects of biochar on photosynthesis and antioxidative system of *Malus hupehensis* Rehd. seedlings under replant conditions. *Sci. Hortic.* **2014**, *175*, 9–15. [CrossRef]
64. Afsharipoor, S.; Roosta, H. Effect of different planting beds on growth and development of strawberry in hydroponic and aquaponic cultivation systems. *Plant Ecophysiol.* **2011**, *2*, 61–66.
65. Narkhede, S.D.; Attarde, S.B.; Ingle, S.T. Study on effect of chemical fertilizer and vermicompost on growth of chilli pepper plant (*Capsicum annum*). *J. Environ. Health* **2011**, *6*, 327–332.
66. Malathi, M.; Uma, B. Vermicompost as a soil supplement to improve growth and yield of *Amaranthus* species. *Res. J. Agric. Biol. Sci.* **2009**, *5*, 1054–1060.
67. Rondon, M.A.; Lehmann, J.; Ramirez, J.; Hurtado, M. Biological nitrogen fixation by common beans (*Phaseolus vulgaris* L.) increases with bio-char additions. *Biol. Fertil. Soils* **2007**, *43*, 699–708. [CrossRef]
68. Qayyum, M.F.; Abid, M.; Danish, S.; Saeed, M.K.; Ali, M.A. Effect of various biochars on seed germination and carbon mineralization in an alkaline soil. *Pak. J. Agric. Sci.* **2015**, *51*, 977–982.
69. Wang, J.; Pan, X.; Liu, Y.; Zhang, X.; Xiong, Z. Effects of biochar amendment in two soils on greenhouse gas emissions and crop production. *Plant Soil* **2012**, *360*, 287–298. [CrossRef]
70. Leubner-Metzger, G. Hormonal interactions during seed dormancy release and germination. In *Handbook of Seed Science and Technology*; Basra, A., Ed.; The Haworth Press: Binghamton, NY, USA, 2006; pp. 303–342.

71. Basso, A.S.; Miguez, F.E.; Laird, D.A.; Horton, R. Westgate, M. Assessing potential of biochar for increasing water-holding capacity of sandy soils. *GCB Bioenergy* **2013**, *5*, 132–143. [CrossRef]
72. Lazcano, C.; Sampedro, L.; Zas, R.; Dominguez, J. Vermicompost enhances germination of the maritime pine (*Pinus pinaster* Ait.). *N. For.* **2010**, *39*, 387–400. [CrossRef]
73. Agboola, K.; Moses, S.A. Effect of Biochar and Cowdung on Nodulation, Growth and Yield of Soybean (*Glycine max* L. Merrill). *Int. J. Agric. Biosci.* **2015**, *4*, 154–160.
74. Alie, k.; Abibatu, K.; Mary, M.; Mansaray, P.; Sawyerr, A. Effects of biochar derived from maize stover and rice straw on the germination of their seeds. *Am. J. Agric. For.* **2014**, *2*, 246–249.
75. Li, Y.; Shen, F.; Guo, H.; Wang, Z.; Yang, G.; Wang, L.; Zhang, Y.; Zeng, Y.; Deng, S. Phytotoxicity assessment on corn stover biochar, derived from fast pyrolysis, based on seed germination, early growth, and potential plant cell damage. *Environ. Sci. Pollut. Res.* **2015**, *22*, 9534–9543. [CrossRef] [PubMed]
76. Thies, J.E.; Rillig, M.C. Characteristics of biochar: Biological properties. In *Biochar Environmental Management: Science and Technology*; Routledge: London, UK, 2009; pp. 85–105.
77. Liu, Z.; Chen, X.; Jing, Y.; Li, Q.; Zhang, J.; Huang, Q. Effects of biochar amendment on rapeseed and sweet potato yields and water stable aggregate in upland red soil. *Catena* **2014**, *123*, 45–50. [CrossRef]
78. Ding, Y.; Liu, Y.; Liu, S.; Li, Z.; Tan, X.; Huang, X.; Zeng, G.; Zhou, L.; Zheng, B. Biochar to improve soil fertility. A review. *Agron. Sustain. Dev.* **2016**, *36*, 1–18. [CrossRef]

applied sciences

MDPI

Article

# Energy Balance of Turbocharged Engines Operating in a WWTP with Thermal Hydrolysis. Co-Digestion Provides the Full Plant Energy Demand

José García-Cascallana [1], Daniela Carrillo-Peña [2], Antonio Morán [2], Richard Smith [3] and Xiomar Gómez [2,*]

1 Area of Chemical Engineering, Department of Applied Chemistry and Physics, University of León, Campus de Vegazana, 24071 León, Spain; jgc0504@yahoo.es

2 Chemical and Environmental Bioprocess Engineering Group, Natural Resources Institute (IRENA), University of León, Avenida de Portugal 41, 24009 León, Spain; dcarp@unileon.es (D.C.-P.); amorp@unileon.es (A.M.)

3 Department of Chemical and Environmental Engineering, University of Nottingham, Coates Building B12, Nottingham NG7 2RD, UK; r.smith@nottingham.ac.uk

* Correspondence: xagomb@unileon.es

**Abstract:** The energy balance of lean-burn turbocharged engines using biogas as fuel is reported. Digestion data were obtained from the wastewater treatment plant (WWTP) of the city of Burgos (Spain), operating with a thermal hydrolysis unit for sludge pre-treatment. Operational performance of the plant was studied by considering the treatment of sludge as a comparative base for analyzing global plant performance if co-digestion is implemented for increasing biogas production. The calculation methodology was based on equations derived from the engine efficiency parameters provided by the manufacturer. Results from real data engine performance when evaluated in isolation as a unique control volume, reported an electrical efficiency of 38.2% and a thermal efficiency of 49.8% leading to a global efficiency of 88% at the operating point. The gross electrical power generated amounted to 1039 kW, which translates into 9102 MWh/year, with an economic value of 837,384 €/year which was completely consumed at the plant. It also represents 55.1% of self-consumption regarding the total electricity demand of the plant. The analysis of the system considering the use of the total installed capacity by adding a co-substrate, such as cheese whey or microalgae, reveals that total electrical self-consumption is attained when the co-substrate is directly fed into the digester (cheese whey case), obtaining 16,517 MWh/year equivalent to 1,519,160 €/year. The application of thermal hydrolysis as pre-treatment to the co-substrate (microalgae case study) leads to lower electricity production, but still attains a better performance than a mono-digestion baseline scenario.

**Keywords:** biogas valorization; heat recovery; engine cooling circuit; electricity self-consumption; global efficiency

**Citation:** García-Cascallana, J.; Carrillo-Peña, D.; Morán, A.; Smith, R.; Gómez, X. Energy Balance of Turbocharged Engines Operating in a WWTP with Thermal Hydrolysis. Co-Digestion Provides the Full Plant Energy Demand. *Appl. Sci.* **2021**, *11*, 11103. https://doi.org/10.3390/app112311103

Academic Editor: Xiaolin Wang

Received: 31 October 2021
Accepted: 19 November 2021
Published: 23 November 2021



## 1. Introduction

The use of biogas as a fuel comprises three essential aspects. Biogas is a renewable fuel that is produced from the anaerobic digestion of wastes such as animal manure, food, agricultural, domestic wastes, and sewage sludge. This gas contains methane ($CH_4$) and carbon dioxide ($CO_2$) as major components, with the latter reaching concentrations up to 40% or even more depending on reactor conditions. Methane content in biogas has average values between 51–55% [1]—data reported from the evaluation of Italian biogas plants treating different organics—the high $CO_2$ levels adversely affect the heating value of biogas, combustion characteristics and composition of emitted gases [2]. The valorization of this gas to produce energy aids in climate change mitigation and reduces the use of fossil fuels. The digestion process also generates a digestate, that can be treated as a residue or as a valuable by-product used as an organic amendment [3,4], allowing nutrient recycling.

A great emphasis has been placed on increasing the efficiency in biogas production as a second-generation biofuel [5,6]. The continuous increase in the global energy demand and the fact that the primary energy resources available, such as oil, natural gas, coal and nuclear fuel are not renewables, forces the scientific community to search for alternative sources and increase the efficiency of conversion processes for producing bioenergy. Anaerobic digestion can also contribute to reducing the environmental impact and emission of $CH_4$ as it allows targeting conversion processes, eliminating uncontrolled degradation and dispersed emission of pollutants. In this way, the valorization of biogas can be achieved in waste treatment plants for producing energy or upgraded to achieve a quality similar to that of natural gas [7].

Several studies have focused on improving digestion performance of sewage sludge by the application of different techniques, through the addition of supplements [8–10], the application of pre-treatments [11–13], and co-digestion [14–16] as a way of enhancing biogas production and benefiting from the common utilization of plant equipment. The introduction of a pre-treatment unit in an existing digestion plant causes changes in the electrical and thermal demand because of the additional equipment installed.

There is an increasing concern about the energy demand and emissions derived from the use of fossil fuels, creating social pressure for making alternative fuels more accessible and economically feasible in accordance with sustainability criteria. Renewable sources offer a wide range of attractive options, as they are unlimited, but the degree of dispersion that this type of energy production has may condition their feasibility [17]. Digestion plants for treating sewage sludge, livestock, food and agricultural wastes to produce biogas have been installed worldwide. However, several studies reported that the plant scale closely affects techno-economic viability, which is associated with the availability of financial incentives such as feed-in tariffs and electricity selling price [18,19]. Levelized costs related to electricity production are lower for large-scale plants because large-scale devices have more efficient conversion coefficients and lower capital costs per unit of electricity produced [20]. Additionally, energy performance studies should consider the whole plant configuration and the energy demand linked with each specific process unit. Otherwise, any attempt to increase biogas yields may lead to a disappointing global performance because of the extra energy demand [21].

The use of biogas in engines and boilers has become the most common way of valorizing biogas. The use of combined heat and power (CHP) engines for heat and electricity production shows better performance in terms of greenhouse gas (GHG) emission reduction due to the on-site use of electricity [22]. However, this option is not always feasible based on the geographical location of the plant and the performance of the digestion process [23]. Biogas composition affects engine efficiency in different ways. Feroskhan and Ismail [24] studied an engine in dual fuel mode with diesel as pilot fuel and biogas as the main fuel for which different methane compositions were evaluated. These authors suggested that exhaust gas temperature and air–fuel ratio were the key engine parameters affected, with the latter exhibiting a marginal increase with methane enrichment. However, when evaluated as a single fuel in compression engines, the presence of $CO_2$ showed detrimental engine performance with increasing content.

Gupta and Mittal [25] indicated that increasing levels of $CO_2$ in biogas leads to a decrease in the brake thermal efficiency, whereas the flame initiation and combustion durations were increased. Nitric oxide (NOx) emissions were reduced, but as a detrimental effect, these authors reported an increase in hydrocarbon and carbon monoxide (CO) emissions. On the other hand, the high $CO_2$ content in biogas gives it a greater resistance to detonation, the latter being a phenomenon that, if it occurs during combustion, adversely affects the efficiency, emissions and durability of engines [26]. Hydrogen sulfide ($H_2S$) is another common component of biogas. Its presence may have negative consequences to human health and impacts equipment durability [27]. Therefore, it is usually removed by different methods such as reaction with iron oxides or by biological techniques [28,29].

Turbocharging is an efficient technology available for increasing the effectiveness of engines and output power [30]. Turbochargers are widely used in internal combustion reciprocating engines, as they increase power thanks to the compression effect, allowing a greater mass of air to flow into the cylinder whilst recovering a proportion of the exhaust gas energy [31,32]. Since cylinder capacity is increased, keeping size constant, this action is known as downsizing [33]. The basis of turbocharging is to increase engine power by introducing a greater amount of air and fuel into the cylinders, as the air–fuel ratio is restricted to a very narrow range to achieve complete combustion. Standard petrol engines operate with compression ratios between 6:1 and 10:1 [34], to avoid self-ignition of the air–fuel mixture. Increasing the compression ratio to 16–18 improves the efficiency and power output. This can be attained when biogas is used as fuel due to its good detonation resistance [35]. The use of biogas using a dual-fuel turbocharged engine was evaluated by Ahmed et al. [36]. These authors concluded that the presence of biogas did not adversely affect brake thermal efficiency.

The present manuscript aimed to analyze the energy balance of turbocharged biogas engines operating at the Burgos wastewater treatment plant (WWTP), where thermal hydrolysis sludge pre-treatment units are installed. Biogas is used for producing electricity and thermal energy. The energy efficiency of the plant was evaluated considering the addition of a co-substrate to enhance biogas production and analyze its effect on the energy balance. The engine thermal circuits were studied to establish the efficiency of the global energy system and its improvement. A sensitivity analysis of the main operating parameters of the engines and cooling circuits was carried out.

The analysis performed considered the operating point of the WWTP and assumed the production of biogas from thermally pre-treated sludge as a baseline scenario. A comparison regarding electricity production was made if a co-substrate was available to enhance digestion performance and take advantage of the high treatment capacity of the plant. It was essential to consider the hypothesis that the characteristics of the co-substrate regarding digestibility could prevent the system from reaching energy autarky.

## 2. Materials and Methods

### 2.1. General Operating Conditions

Figure 1 shows the operating diagram of biogas engines in the WWTP of Burgos (Spain). The WWTP treats domestic wastewater from the surrounding area. The treatment capacity is 156,504 $m^3/d$ and 1,056,692 e.i. (equivalent inhabitant). The plant has the following sections: (1) pretreatment, where greater size objects, sand and grease are removed, (2) primary decanters, (3) secondary treatment consisting of the conventional activated sludge system, and (4) tertiary treatment consisting of lamellar decanters, ozone and ultraviolet rays.

The primary sludge is sent to a gravity thickener to increase solid concentration. Waste activated sludge (WAS) is concentrated by air flotation thickeners. The sludge line is composed of the thickening, anaerobic digestion, conditioning and digestate dewatering section. Biogas is produced from mesophilic digesters. The sludge is a mixture of primary sludge and WAS. Sludge is preheated by thermal hydrolysis (165 °C) and then cool to 50 °C. This temperature takes into account heat losses by thermal transmission in pipework and digesters. The plant uses Cambi thermal pre-treatment technology, which is a batch-type process using flash steam preheating [21]. The hydraulic retention time (HRT) of the digestion reactor is 17 days and the temperature is 41 °C. Four rotary vane biogas compressors are used for mixing the digester content, with a flow capacity of 448 $m^3/h$ each (measured at standard temperature and pressure conditions). Four gasometers with a capacity of 703 $m^3$/each are available for biogas storage and four biogas engines for electricity production.

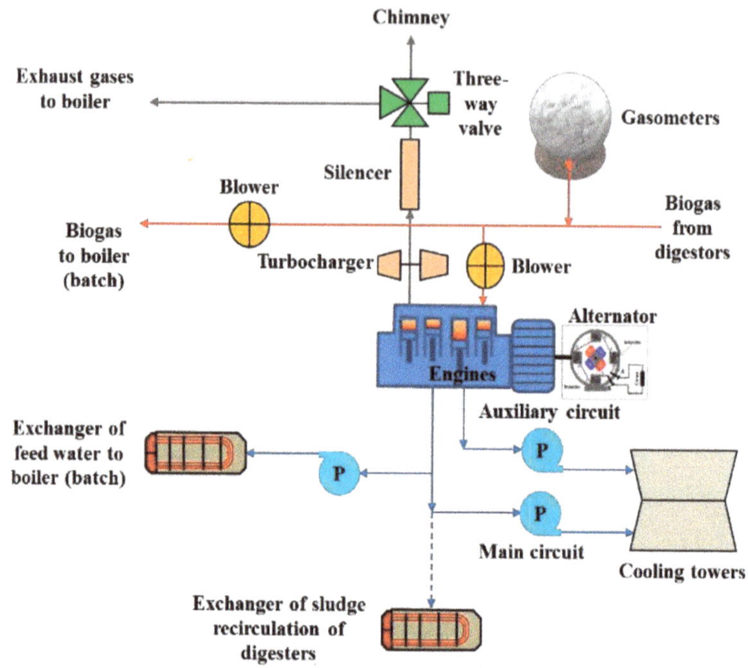


**Figure 1.** Scheme describing the general operating circuit of the biogas engine.

The mixed recovery boiler also consumes a small quantity of biogas to generate steam for the thermal hydrolysis unit. The exhaust gas from the engines at the turbocharger's outlet is sent to the silencers, then to the recovery boiler and finally, to the stack. The thermal energy from the auxiliary cooling circuit is dissipated in cooling towers. A fraction of the energy from the main circuit is sent to the boiler heat exchanger to preheat the feeding water. Energy surplus is dissipated in cooling towers. The thermal energy of the main circuit is only used for heating the digesters when thermal hydrolysis is not operating. Gasometers are used to store biogas, maintain the system pressure and supply biogas to the boiler—when steam injection stage takes place—and to the gas engines.

*2.2. Wastewater Treatment Plant (WWTP) Parameters*

Table 1 presents data regarding biogas production and consumption of Burgos WWTP. Values correspond to the operating point of the plant reported in García-Cascallana et al. [21], where a detailed description is given. This operating point is based on biogas consumption of the boiler and engines. This point also considers the use of engine exhaust gases in the mixed recovery boiler during an average operating cycle, where injection and no injection stages of steam are evaluated.

The theoretical analysis of the biogas line was also evaluated considering the addition of a co-substrate for increasing electricity production. The recovery of thermal energy in the plant is also analyzed under this configuration. Mass and energy balances were performed for the co-digestion scenario assuming the availability of readily degradable co-substrate and also the application of thermal hydrolysis to a complex co-substrate.

**Table 1.** Biogas parameters, operating point and electricity demand of the Burgos WWTP. Biogas reported at standard temperature and pressure (STP) conditions.

| Parameter | Value |
|---|---|
| Biogas characteristics | - |
| LHV [1] (kJ/m$^3$) | 21,240 |
| Density (kg/m$^3$) | 1.15 |
| Specific heat capacity ($c_p$, (kJ/kg °C)) | 1.56 |
| $CH_4$ (%) | 62.4 |
| $CO_2$ (%) | 34.1 |
| $H_2$ (%) | 2.5 |
| $H_2O$ (%) | 1.0 |
| WWTP operating conditions | - |
| Biogas production (kW) | 3590 |
| Biogas consumption (kW) | 2720 |
| Boiler biogas average consumption (kW) [2] | 870 |
| WWTP electrical power demand (kW) | 1885 |

[1] LHV: Lower heating value. [2] Values obtained from WWTP operating with a thermal hydrolysis treatment unit: Cambi THP B6.2 [20].

### 2.3. Engines

The plant has four internal combustion engines, namely spark ignition, Otto cycle, four strokes, cooled by a water circuit. The engines are Guascor type SFGLD360s one step turbocharging and intercooling, having a V two-rows of six-cylinder disposition and operating under lean-burn conditions—(λ) of 1.3–1.7—. The engines are of the square type with a stroke/bore ratio close to unity. Carburetion is electronic, blowing the lean mixture into the turbocharger and inlet manifolds to the cylinders. The biogas is sent from the gasometers or digester head to the engines—at 100.7 kPa and 30 °C—by four side channel blowers equipped with frequency converters. The blower nominal flow rate is 300 m$^3$/h at 2800 rpm, raising the pressure to 115 kPa and the temperature to 55 °C, before being introduced into the biogas ramp and carburetor, where it is mixed with the admission air, compressor and inlet manifolds before entering the combustion chamber. The operating principle of the carburetor is a Venturi effect. Air flows through the carburetor creating a vacuum proportional to the air velocity, causing biogas to enter and homogenously mix with air. The pressure in the carburetor cannot be higher than atmospheric pressure 101 kPa, a restriction achieved by an electronic controller.

According to the indications of the load/speed controller, a throttle valve at the outlet of the intercooler regulates the air–biogas ratio and the engine power. The intercooler lowers the temperature of the cylinder inlet air, thus preventing uncontrolled detonation and increasing its density. The compression ratio of turbocharging is usually 2–4. The exhaust gases enter the turbine radially and leave it axially, whereas, it is the other way around for the compressor—the air inlet is axial and the outlet radial—. The centripetal turbine is driven by the exhaust gases and is mechanically coupled at the same centrifugal compressor shaft. The engines are cooled by water in closed circuits using plate-type heat exchangers.

The engine is a four-stroke and every two crankshaft turns (angle of rotation 2 × 360°) causes one drive stroke in each of the cylinders. Therefore, there is a driving stroke in the cylinders each 120° (720°/6 = 120°), with a firing order of sequence 1-8-5-10-3-7-6-11-2-9-4-12, achieving good smoothness when running. The alternator converts mechanical energy into electrical energy by a synchronous generator of 753 kVA, 400 V. A pair of poles keep synchronism rotating at 1500 rpm, frequency 50 Hz and allows a power variation factor of 0.8–1.

The residual thermal energy of the engine exhaust gases is used in the mixed recovery boiler. A fraction of the thermal energy from the main cooling circuits is used for preheating the boiler feed water. At the Burgos WWTP, the maximum degree of loading at which engines operate is 90% of the rated power as a preventive maintenance requirement to extend engine service life and reduce the frequency and number of maintenance stops for spark plug replacements. Figure 2 shows a diagram of one of the six cylinders' lines of the engines. Each engine has two lines of this type accounting for 12 cylinders. The electrical power generated depends on the conditions of the inlet air, such as temperature, pressure, humidity and the backpressure of the exhaust gas circuit and biogas composition.

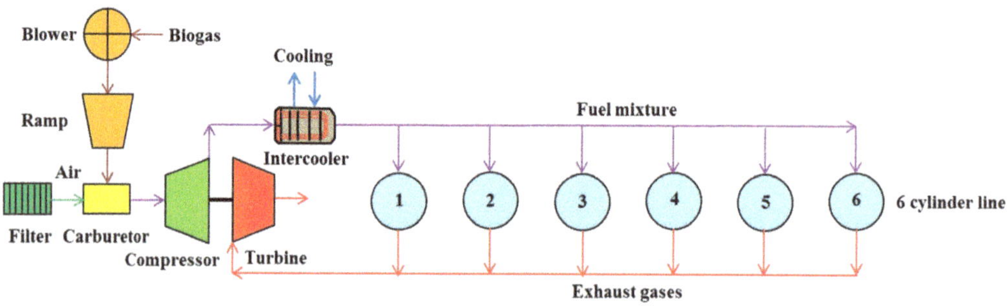


**Figure 2.** Scheme of one of the six cylinders' lines of the engine. The engine has two equivalent lines.

The reference conditions for the engine parameters ISO 3046/1 are: 100 kPa ambient pressure, 25 °C ambient temperature and 30% relative humidity. The tolerances at 100% of degree of loading are: biogas consumption ±5%, main, auxiliary and emergency cooling circuits and exhaust gases ±15%, mechanical losses and radiation ±25%, exhaust gas temperature ±20 °C, and air and gas mass flow ±10%. Table 2 shows the key parameters of the biogas engines analyzed.

**Table 2.** Parameters of biogas engines in Burgos WWTP.

| Parameter | Value |
|---|---|
| $V_t$, total volume displacement ($cm^3$) | 35,900 |
| Unitary volume displacement ($cm^3$) | 2993 |
| Number of cylinders (units) in V disposition | 12 |
| Number of cylinders lines (units) | 2 |
| Bore (cm) | 15.2 |
| $S_t$, stroke (cm) | 16.5 |
| Stroke/bore ratio | 1.08 |
| Piston area ($cm^2$) | 181 |
| Compression ratio (volumetric) | 11.6 |
| Combustion chamber volume ($cm^3$) | 282 |
| Ignition advance angle (°) | 20 |
| Combustion chamber volume/displacement per cylinder ratio (%) | 9.4 |
| Total weight (kg) | 7100 |

The biogas ramp for controlling the cylinder biogas supply has a pressure regulation between 112.8 and 116.3 kPa. The engines are equipped with an exhaust gas silencer with 30 dB sound attenuation and an exhaust gas distribution valve (three-ways). This valve sends the gas flow to the boiler heat recovery zone or directly to the atmosphere through the chimney. Table 3 shows the different values obtained for the operating parameters of the engines based on the loading degree. The engine rotation speed is constant for any degree of loading to maintain synchronism with the alternator speed.

**Table 3.** Engine operating parameters obtained at different loading degrees.

| Parameters | Degree of Loading | | | |
|---|---|---|---|---|
| | 100 | 80 | 60 | 40 |
| n, rotation speed (rpm) | 1500 | | | |
| Effective power | | | | |
| $MP_{eng}$, mechanical power (kW) | 617 | 494 | 370 | 247 |
| $EP_{eng}$, electrical power (cos(φ) = 1) (kW) | 598 | 478 | 357 | 236 |
| Effective average pressure (kPa) | 1370 | 1100 | 820 | 550 |
| Efficiency and fuel consumption | | | | |
| $\eta_{mec\text{-}eng}$, mechanical engine efficiency (%) | 40.3 | 38.8 | 36.6 | 32.9 |
| $\eta_{el\text{-}eng}$, electrical engine efficiency (cosφ = 1) (%) | 39.1 | 37.6 | 35.3 | 31.4 |
| $TP_{biogas\text{-}eng}$, biogas consumption (kW) | 1530 | 1272 | 1012 | 752 |
| Thermal losses | | | | |
| Radiation power losses (kW) | 28 | 25 | 21 | 17 |
| Mechanical power losses (kW) | 19 | 16 | 13 | 11 |
| Exhaust gas power losses (25 °C) (kW) | 476 | 406 | 330 | 252 |
| Exhaust gas power losses (120 °C) (kW) | 375 | 323 | 264 | 203 |
| Thermal power losses in the main water circuit (kW) | 299 | 244 | 195 | 155 |
| Thermal power losses in the auxiliary water circuit (kW) | 110 | 103 | 96 | 81 |
| $TP_{IC}$, thermal power losses in the intercooler (kW) | 36 | 33 | 30 | 20 |
| Thermal power losses in the oil exchanger (kW) | 74 | 70 | 66 | 61 |
| Combustion and emissions [1] | | | | |
| Exhaust gases dry oxygen (%) | 9.0 | 8.8 | 8.5 | 8.2 |
| NOx (mg/m$^3$) | 500 | | | |
| CO (mg/m$^3$) | <1000 | | | |
| NMHC [2] (mg/m$^3$) | <300 | | | |
| Other parameters | | | | |
| $T_4$, outlet temperature of the turbine exhaust gases (°C) | 474 | 488 | 503 | 518 |
| $M_{air}$, inlet air mass flow (kg/h) | 3030 | 2510 | 1970 | 1460 |
| $M_{eg}$, exhaust gases mass flow (kg/h) | 3320 | 2740 | 2160 | 1600 |
| $p_{bp\_max}$, maximum exhaust back-pressure (kPa) | 4.5 | | | |

[1] Standardized emission at 5% oxygen. [2] NMHC stands for non-methane hydrocarbons. Mechanical and electrical power are referred to 830 m of altitude above sea level (a.s.l.).

The mechanical power is affected by the inlet air density, which varies with the altitude and ambient temperature. Equation (1) calculates the engine mechanical efficiency losses ($Losses_{\eta\text{-}a}$) due to altitude. Equation (2) estimates losses ($Losses_{\eta\text{-}T}$) due to the inlet air temperature [37] and finally, Equation (3) is used to calculate the air atmospheric pressure ($p_{atm}$) as a function of altitude:

$$Losses_{\eta-a} = (z-500)/150 \quad (1)$$

$$Losses_{\eta-T} = (T_{env}-25)/5 \quad (2)$$

$$p_{atm} = p_{atm-0} \times (1 - a \times z)^b \quad (3)$$

where $z$ represents the altitude measured at sea level and expressed in m, $T_{env}$ is the temperature of the location, and $p_{atm\text{-}0}$ stands for the atmospheric pressure at sea level. $a$ and $b$ are constants (a = 2.2557 × $10^{-5}$ m$^{-1}$, b = 5.2559). The altitude of Burgos is 859 m above the sea level, but the engines of the WWTP are at 830 m above sea level, so the real atmospheric pressure considered in the present manuscript for evaluating engine performance was 91.7 kPa with an air density of 1.07 kg/m$^3$ at 25 °C.

### 2.4. Engine Cooling System

Each engine has three different cooling systems. The main system is comprised of the block, crankcase and cylinder cooling jacket and it is refrigerated by two circuits, the primary and the secondary circuit, separated by a plate heat exchanger. The secondary cooling system supplies the energy needed for preheating the boiler feed water (batch operation) and the surplus heat is dissipated in cooling towers. The second cooling system is known as Auxiliary. This circuit is composed of the intercooler and lubricating oil circuit, also having two internal circuits, as in the previous case. This cooling system keeps the air–fuel mixture temperature at 65 °C to limit the risk of detonation before combustion takes place in cylinders. The heat absorbed from the air–fuel mixture is dissipated in the cooling towers. The oil outlet temperature of the heat exchanger is between 75 and 90 °C.

The third cooling system is the Emergency system. This system aids in cooling the block, crankcase and cylinder jackets, and it is deployed when the main circuit is not capable of removing all the heat. The primary circuit of this system is located in series with the main cooling system, and the secondary circuit is located in series with the auxiliary circuit. Since each engine has two lines of six V-cylinders, there are two blocks of cooling systems for the crankcase and cylinder jacket, and two others for the intercooler and oil exchangers. In calculations, they were considered to be arranged in parallel. The three circuits consist of plate heat exchangers, expansion vessels, three-way thermostatic valves (to allow a regulation of 8–10 °C temperature), temperature probes, manometers, and circulation pumps for the primary and secondary circuits. The fluid used in the heat exchanger of the primary circuit is a mixture of water and 20% glycol. Secondary cooling systems and cooling towers use water as fluid. The efficiency of the cooling exchangers was assumed as unity.

The WWTP has a total of seven cooling towers with forced draught, having a nominal thermal power of 274 kW each, making a total of 1918 kW of install nominal cooling capacity. Four towers are used to cool the main system, two for the auxiliary and one in reserve. Table 4 shows the parameters of the three cooling systems.

**Table 4.** Parameters of the three cooling circuits of the engine.

| Cooling Circuit | Temperature (°C) | Volumetric Flow ($m^3$/h) | | Nominal Thermal Power of the Heat Plate Exchangers (kW) |
|---|---|---|---|---|
| | Primary | Primary | Secondary | |
| Main | 75–90 (engine outlet) | 80 | 40 | 299 |
| Emergency | – | – | – | 99 |
| Auxiliary | 55 (engine inlet) | 25 | 30 | 110 |

## 2.5. Engine Parameters and Efficiency

Equation (4) is used to calculate the average linear piston speed ($C_{eng}$), Equation (5) for the effective biogas consumption ($g_{eff}$), and Equation (6) for the volumetric efficiency of the engine ($\eta_v$):

$$C_{eng} = 2 \times S_t \times \frac{n}{60} \tag{4}$$

$$g_{eff} = \frac{3.6 \times \rho_{biogas}}{\eta_{mec-eng} \times LHV} \tag{5}$$

$$\eta_v = \frac{M_{mix}}{V_t \times \left(\frac{2 \times n}{60 \times n_{stroke}}\right) \times \rho_{ref-mix}} \tag{6}$$

where $S_t$ is the stroke length (m), $n$ represents the rotation speed in revolutions per minute (rpm) at which the engine is running. The density of biogas ($\rho_{biogas}$) was assumed as 1.15 kg/$m^3$ at standard temperature and pressure conditions. The mechanical efficiency of the engine ($\eta_{mec\text{-}eng}$) is a parameter considered as a function of the degree of loading. The mass of air–fuel mixture ($M_{mix}$) is used for estimating $\eta_v$, using as density ($\rho_{ref\text{-}mix}$) a variable value calculated based on the impulsion pressure, and outlet temperature of the mixture exiting the turbocharger. $V_t$ corresponds to the cylinder displacement volume and $n_{stroke}$ is the number of strokes per cycle.

The electrical efficiency ($\eta_{\text{el-eng}}$) is calculated using Equation (7). *EP* and *TP* in the formula stand for electrical and thermal power. Equation (8) calculates the thermal efficiency ($\eta_{\text{thermal-eng}}$) based on the ratio between the useful thermal power of the engine ($TP_{\text{useful-eng}}$) and the thermal power contained in biogas and consumed by the engine ($TP_{\text{biogas-eng}}$). Equation (9) estimates the energy efficiency ($\eta_{\text{energy-eng}}$), and Equation (10) estimates the usable thermal energy factor of the engine (*UTEF*) considering the thermal power that has application in the WWTP, that is, the power used ($TP_{\text{usable-eng}}$) against the value of the thermal power susceptible of application. Equation (11) estimates the work/heat ratio of engine $(W/Q)_{\text{eng}}$, as the ratio between the mechanical power of the engine ($MP_{\text{eng}}$) and $TP_{\text{useful-eng}}$:

$$\eta_{\text{el-eng}} = EP_{\text{eng}} \times 100/\ TP_{\text{biogas-eng}} \tag{7}$$

$$\eta_{\text{thermal-eng}} = TP_{\text{useful-eng}} \times 100/\ TP_{\text{biogas-eng}} \tag{8}$$

$$\eta_{\text{energy-eng}} = (EP_{\text{eng}} + TP_{\text{useful-eng}}) \times 100/\ TP_{\text{biogas-eng}} \tag{9}$$

$$UTEF = TP_{\text{usabel-eng}} \times 100/\ TP_{\text{useful-eng}} \tag{10}$$

$$(W/Q)_{\text{eng}} = MP_{\text{eng}} \times 100/\ TP_{\text{useful-eng}} \tag{11}$$

Equation (12) estimates the ratio of the stoichiometric mass of air ($A_{\text{stoich}}$) for biogas combustion considering molar fraction of methane in biogas ($X_{CH4}$) and the molecular weight of biogas and air ($MW_{\text{biogas}}$, $MW_{\text{air}}$). Equation (13) estimates the stoichiometric ratio ($F_{\text{stoich}}$) for biogas–air:

$$A_{\text{stoich}} = \frac{X_{CH4} \times 2 \times MW_{\text{air}}}{MW_{\text{biogas}}} \tag{12}$$

$$F_{\text{stoich}} = 1/Air_{\text{stoich}} \tag{13}$$

The ratio of biogas–air (F) at the engine operating conditions is obtained from Equation (14) in terms of the mass flow of exhaust gases ($M_{\text{eg}}$) and the mass of air ($M_{\text{air}}$). The ratio of Air per unit of fuel ($A_{\text{real}}$) is given by Equation (15) and the mass of excess air is estimated using Equation (16):

$$F = \frac{(M_{\text{eg}} - M_{\text{air}})}{M_{air}} \tag{14}$$

$$A_{\text{real}} = 1/F \tag{15}$$

$$E_{air} = [(A_{\text{stoich}}/A_{\text{real}}) - 1] \times 100 \tag{16}$$

The Guascor engines work with a biogas–air ratio of approximately 9.6%. This translates into a large excess of combustion air over stoichiometric conditions, resulting in an average $O_2$ content in the exhaust gas of about 9%. It is assumed that the exhaust gas enthalpy is about the same as that of air under the same conditions, as these gases have, as main components, $N_2$ followed by $CO_2$, $O_2$ and traces of water vapor. All these compounds have specific heat values very similar to air for the temperatures considered, except for water vapor, but this has a negligible effect because of its small content in exhaust gases.

Equation (17) is a NASA polynomial equation valid for temperatures between 200–1000 K [38] and allows calculating specific heat values for air ($C_{\text{p-air}}$) as a function of the temperature (K) ($T_{\text{air}}$). In this equation, $R_g$ is an ideal gas constant with a value of 0.287 kJ/kg K. Equation (18) and Equation (19) are used to calculate the specific enthalpy of air ($h_{\text{air}}$), equivalent to that of the mixture air–biogas ($h_{\text{mix}}$), and the enthalpy of exhaust gases ($h_{\text{eg}}$). Equation (20) establishes the equivalence for $C_{\text{p-air}}$ and the specific heat values of gas mixture and exhaust gases ($C_{\text{p-mix}}$, $C_{\text{p-eg}}$). Equation (21) is used to calculate the specific entropy:

$$C_{p-air} = R_g \times \left(3.56839 - 6.788729 \times 10^{-4} \times T_{air} + 1.5537 \times 10^{-6} \times T_{air}^2 - 3.29937 \times 10^{-12} \times T_{air}^3 - 466.395 \times 10^{-15} \times T_{air}^4\right) \quad (17)$$

$$h_{air} = h_{mix} = C_{p-air} \times T_{air} = C_{p-mix} \times T_{mix} \quad (18)$$

$$h_{eg} = C_{p-eg} \times T_{eg} \quad (19)$$

$$C_{p-air} = C_{p-mix} = C_{p-eg} \quad (20)$$

$$s = C_{p-air} \times \ln(T) - R_g \times \ln(p) \quad (21)$$

### 2.6. Electricity Production

The degree of loading is calculated from Equation (22) considering the ratio between the mechanical power of the engine ($MP_{eng}$) and its nominal value ($MP_{eng\text{-}nom}$) and expressed as percentage. Equation (23) estimates the installed electrical capacity utilization factor (*IECUF*) as the ratio between the electrical power generated by the engines ($EP_{eng}$) and the electrical power installed ($EP_{inst}$). Equation (24) estimates the degree of electrical power self-consumption (*DSC*) in the WWTP as the ratio between the $EP_{eng}$ and the electrical power consumed in the plant ($EP_{consumed}$):

$$DL = MP_{eng} \times 100/MP_{eng-nom} \quad (22)$$

$$IECUF = EP_{eng} \times 100/EP_{ins} \quad (23)$$

$$DSC = EP_{eng} \times 100/EP_{consumed} \quad (24)$$

Savings obtained from electricity self-consumption were estimated by considering the price of electricity in Spain for medium voltage. A value of 92 €/MWh was used for electricity for the range of 2000–20,000 MWh/year [39].

### 2.7. Turbocharging

Each of the two rows of cylinders has a turbocharger with a centrifugal compressor driven by a centripetal turbine from the exhaust gases and an intercooler. Equations (25) and (26) calculate $MP_{eng}$ developed by the internal combustion engine:

$$MP_{eng} = \frac{1}{3600} \times M_{air} \times F \times \left(\frac{LHV}{\rho_{biogas}}\right) \times \eta_{et} \quad (25)$$

$$MP_{eng} = V_t \times \frac{2n}{60\, n_{stroke}} \times \rho_{air} \times \eta_v \times F \times \left(\frac{LHV}{\rho_{biogas}}\right) \times \eta_{et} \quad (26)$$

where $\eta_{et}$ is the engine effective efficiency. Equation (27) calculates the pressure ratio of the compressor ($r_c$), neglecting the pressure losses in filter, intercooler and inlet manifold to cylinders. This equation considers the outlet pressure of the compressor ($p_1$) and the isentropic outlet pressure as equivalent ($p_{1S}$):

$$r_c = \frac{p_1 + p_{atm-0}}{p_0} = \frac{p_{1S} + p_{atm-0}}{p_0} \quad (27)$$

Equation (28) is used to calculate the temperature of the isentropic mixture at the compressor outlet ($T_{1S}$), and the isentropic efficiency ($\eta_{is\text{-}C}$) is calculated from Equation (29), considering the compressor outlet temperature ($T_1$) and the inlet temperature ($T_0$). $\gamma$ is the adiabatic index:

$$T_{1S} = T_0 \left(\frac{1}{r_c}\right)^{\frac{(1-\gamma)}{\gamma}} \quad (28)$$

$$\eta_{is-C} = \frac{(T_{1S} - T_0)}{(T_1 - T_0)} \times 100 \quad (29)$$

The thermal power of the compressor ($TP_C$) is estimated using the temperature difference between the outlet and inlet temperature as well as the specific heat value for the gas mixture ($C_{p\text{-mix}}$) as shown in Equation (30):

$$TP_C = \frac{M_{mix}(1+F)}{3600} C_{p-mix}(T_1 - T_0) \quad (30)$$

In a similar way, the thermal power of the intercooler ($TP_{IC}$) is also estimated using Equation (31) with the outlet theoretical temperature ($T_{2S}$) included in the temperature difference term:

$$TP_{IC} = \frac{M_{mix}(1+F)}{3600} C_{p-mix}(T_1 - T_{2S}) \quad (31)$$

The efficiency of the intercooler ($\eta_{IC}$) is calculated as:

$$\eta_{IC} = \frac{(T_1 - T_2)}{(T_1 - T_{2S})} \times 100 \quad (32)$$

where $T_2$ is the temperature of the mixture at the intercooler outlet. The mechanical power of the turbine ($MP_T$) is calculated from Equation (33) using the temperature difference between the exhaust gases' inlet ($T_3$) and outlet ($T_4$) stream. The isentropic efficiency of the turbine ($\eta_{IS\text{-}T}$) is given by Equation (34) denoting the outlet isentropic temperature of turbine exhaust gases as ($T_{4S}$):

$$MP_T = \frac{M_{mix}(1+F)}{3600} C_{p-mix}(T_3 - T_4) \quad (33)$$

$$\eta_{is-T} = \frac{(T_3 - T_4)}{(T_3 - T_{4S})} \times 100 \quad (34)$$

Equation (35) expresses the pressure relationship for the turbine under isentropic conditions, with $p_{4S}$ denoting the pressure of the outlet exhaust gases' stream under isentropic conditions and $p_3$ representing the pressure of the gas stream entering the turbine. The expansion of the exhaust gases in the turbine is assumed as an irreversible flow due to the adiabatic assumptions based on the small contact surface and the short residence time, considering also steady flow at the inlet and outlet zone. Compressors have three boundary zones, pumping, blocking and over speed zones whilst turbines have none. To achieve the necessary exhaust back-pressure, the turbine opposes to the flow of gases. For this reason, its effective passage area must be small enough, and this feature is called permeability. The turbine discharge pressure is obtained from summing up the reference pressure of the inlet gas mixture to the compressor ($p_0$) and the outlet back-pressure of exhaust gases ($p_{bp}$), see Equation (36). Equation (37) represents the turbine expansion ratio ($r_{expansion}$):

$$p_3 = p_{4S} \left( \frac{T_3}{T_{4S}} \right)^{\frac{\gamma}{(1-\gamma)}} \quad (35)$$

$$p_4 = p_0 + p_{bp} \quad (36)$$

$$r_{expansion} = \frac{p_3}{p_{4S}} = \frac{p_3}{p_4} \quad (37)$$

The turbine is subjected to pulses from the cylinder gases, and a shaped volute configuration is used to attain better efficiency. The twinning between compressor-turbine must achieve the equality for mechanical powers considering performance losses. Equation (38) represents the relationship between the two mechanical powers of the compressor and turbine given by the mechanical efficiency $\eta_{mec\text{-}T}$. The equality for the rotation speed of the two turbomachines ($n_C$, $n_T$) is given by Equation (39):

$$MP_T = \frac{MP_C}{\eta_{mec-T}} \quad (38)$$

$$n_C = n_T \tag{39}$$

The increase in back pressure in a turbocharged system causes a decrease in pressure ratio across the turbine. Therefore, the engine has to pump the gases out of the cylinder against a higher pressure, reducing the efficiency and increasing fuel consumption [40]. The back pressure must be kept to the minimum value necessary to achieve higher turbine efficiency and thus higher engine efficiency [41]. Figure 3 shows a diagram of a turbocharger. The compression of the inlet air is driven by the turbine through the exhaust gases. The main components are: air filter, compressor, intercooler, common coupling shaft, and turbine. Table 5 shows the values required for calculating the compressor and the turbine parameters for the turbocharger. The thermodynamic parameters of a turbocharger are calculated by means of performance maps, which are usually generated experimentally on test benches and are used as a look-up table for engine models [42].

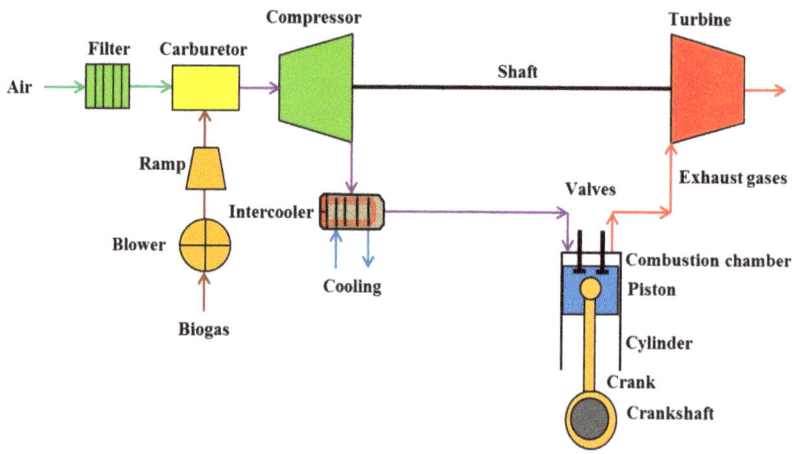


**Figure 3.** Turbocharger schematic diagram.

**Table 5.** Compressor and turbine calculation parameters.

| Parameter | Value |
|---|---|
| $T_0$, inlet air-mixture temperature (°C) | 25 |
| $p_0$, inlet air-mixture pressure (830 m a.s.l) (kPa) | 91.7 |
| ,, adiabatic coefficient of air-mixture and exhaust gases | 1.4 |
| $\eta_{\text{mec-T}}$, turbine mechanical efficiency (%) | 97 |
| $\eta_{\text{IC}}$, intercooler efficiency (%) | 95 |
| $\eta_{\text{is-C}}$, isentropic efficiency of the compressor (%) | 75 |
| $\eta_{\text{is-T}}$, isentropic efficiency of the turbine (%) | 75 |

## 2.8. Sensitivity Analysis

The energy parameters of the engines were evaluated by considering the effect on the output at varying values of the input variable. The degree of loading was selected as the independent input variable varying in a range between 40 and 100%. Sensitivity analysis is an instrument for the assessment of the input parameters and their impact on the model output [43]. Sensitivity analysis and model validation are linked in that these techniques attempt to assess the appropriateness of a particular model specification and to establish the strength of the conclusions drawn from the model [44]. Different variables measuring engine performance were selected as the response variable of the model. Regression methods are usually used to replace high complex models and regression coefficients provide a means of applying sensitivity rankings to input parameters [45]. The $R^2$ value was calculated to estimate the adequacy of the model to data. This parameter is a summary statistic that represents the proportion of total variance of the response. The $R^2$ has a range from 0 to 1, where values close to 0 indicate that the model does not appreciably summarize the behavior of the response, and values close to 1 indicate that a high proportion of the variance is explained by the regression model [44].

## 3. Results and Discussion

The engine energy parameters were evaluated considering the range associated with the minimum degree of loading and maximum operating value. Once the response of the system is established, it is subsequently analyzed by the effect of coupling multiple engines as the availability of biogas increases. The engine energy balance was also studied in detail along with the associated cooling circuit to establish particular points where energy recovery should be improved. Finally, the hypothetical addition of a co-substrate is studied considering energy equations previously formulated to analyze if the electrical and thermal demand of the overall plant can be fully covered by energy valorization of biogas.

### 3.1. Sensitivity Analysis of Biogas Engines

The energy parameters of the engines are shown in the following figures. Values were calculated from the minimum operating point corresponding to a 40% degree of loading at working conditions established at the WWTP, until the maximum value of 100% (full load). Electrical and thermal efficiencies were obtained based on these trend lines. Figure 4a–f shows the key parameters of the engines. All curves were approximated to a linear behavior and $R^2$ (coefficient of determination) was estimated along with the resulting equation, which is indicated in each graph.

The increase in the degree of loading brings with it an associated steep rise in biogas consumption, and therefore, in the input energy. However, the electrical power does not follow the same slope, experiencing a minor increase due to the engine efficiency, which increases with the degree of loading, attaining 40.3% as the maximum value. The maximum power and torque regime in this type of engine are developed at full load. The difference in rotational speed between these two parameters is zero since the engine always rotates at a constant speed to maintain synchronism, so these engines have no elasticity. Explosion ignition engines allow minimal variations of the air–fuel ratio between minimum and full load. Quantitative regulation of this ratio is necessary and this is achieved by throttling the inlet at the carburetor butterfly valve. For this reason, the pumping losses are increased at a low degree of loading as the mixture inlet pressure is reduced and therefore, the efficiency decreases. The power at a full degree of loading is generally proportional to the mass of air admitted per unit of time. The efficiency of this type of engine is a function of the filling or renewal volume into the cylinders. The engine volumetric efficiency thus decreases with the decrease in the degree of loading, as well as the power output.

The turbocharger increases the mass of the mixture (air–biogas) entering into the cylinders compared to atmospheric aspirated engines, attaining a volumetric efficiency of 102.2%. However, the exhaust gas back-pressure increases the turbine inlet pressure from 140.2 to 144.7 kPa (3.2%), lowering the average effective pressure and causing a loss of power. Losses are those derived from mechanical and radiation losses. The first one corresponds mainly to those caused by friction and derived from the various auxiliary elements (oil and water pumps, fans, etc.). They present slight variations between minimum and full load because the engine rotates at constant speed. Radiation losses follow an upward trend with the degree of loading. However, they do not vary significantly because the temperature of the engine block is controlled by the cooling system.

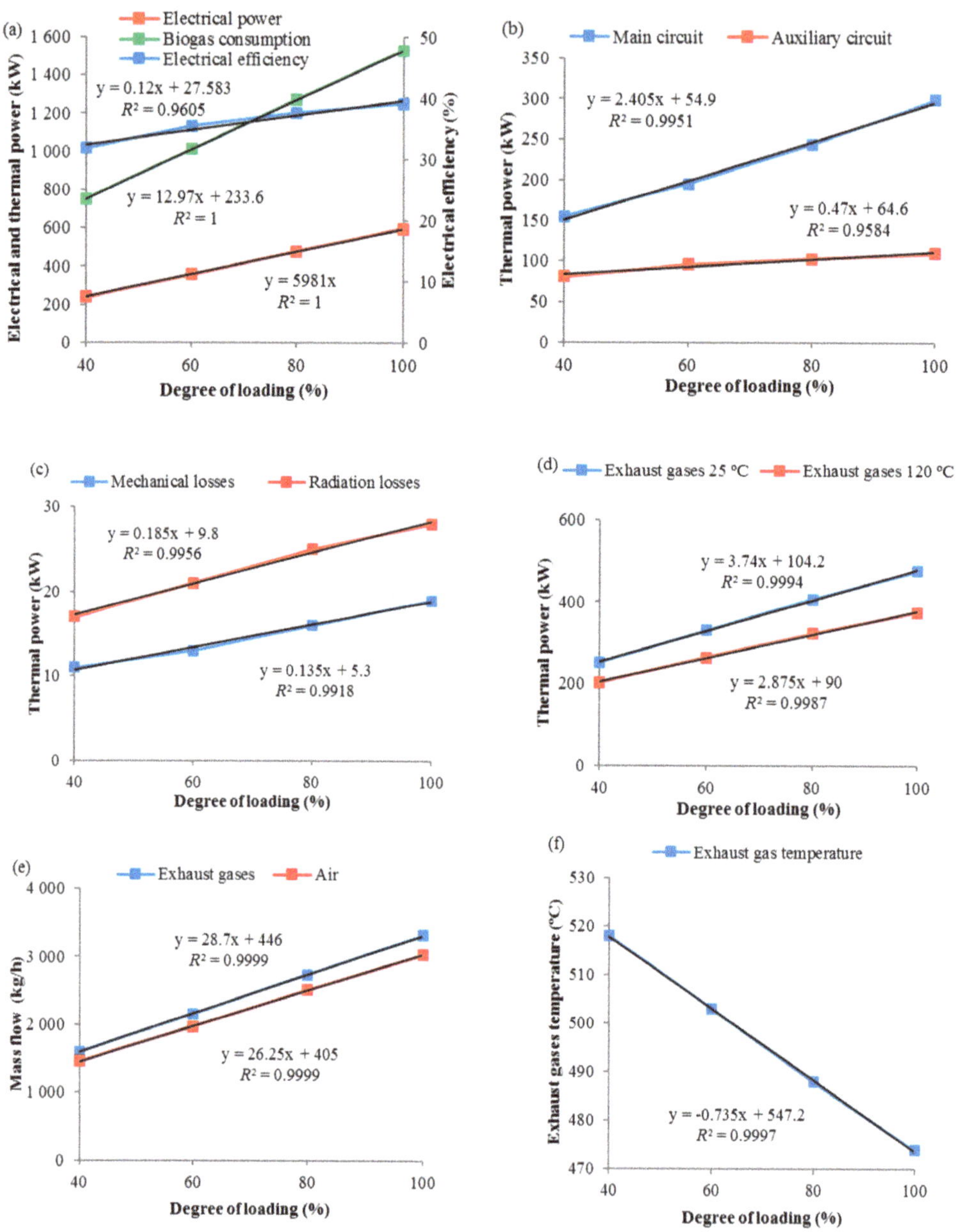


**Figure 4.** Sensitivity analysis of the engines' parameters as a function of the degree of loading. (**a**) Electrical power, biogas consumption and electrical efficiency; (**b**) Thermal power in cooling circuits; (**c**) Mechanical and radiation losses; (**d**) Thermal power in exhaust gases at 25 and 120 °C; (**e**) Mass flow of exhaust gases and air; (**f**) Exhaust gas temperature.

The thermal power associated with cooling circuits has different behavior. The auxiliary circuit has cooling components related to the air compression system and the lubricating oil circuits. Thus, it is not affected as much as the main circuit, which is directly linked to the performance of the engine. Increasing biogas consumption brings, as a consequence, a greater release of heat, which the main circuit should absorb to keep the crankcase and cylinders at the proper temperature.

The intercooler cools down the mixture, increasing its density in an isobaric way (neglecting the small pressure losses). The compressor-turbine assembly is independent of the drive shaft, which gives a greater degree of freedom in terms of efficiency. The rise in exhaust gas back-pressure causes a decrease in the pressure ratio across the turbocharger, resulting in lower mass flow through the components and less air available for the engine [46], adversely affecting the performance.

The density of the mixture inside a combustion chamber of a naturally aspirated engine is 1.07 kg/m$^3$ at 25 °C, whereas, for a turbocharged engine, the density of the mixture leaving the intercooler is 1.78 kg/m$^3$ (65 °C). This higher density translates into a greater mass, increasing by 66.3%. This is a crucial aspect of large power engines for centralized treatment of wastes. The conversion of biogas into energy is more favorable for large-scale systems given the greater power and efficiency CHP units can provide, due to the compression of the inlet gas mixture and the recovery of heat from different engine stages. On the contrary, decentralized digestion plants have to deal with greater thermal losses and higher electrical demand [47,48]. If biogas is valorized in situ, small engines should be used with lower electrical efficiency, generally using atmospheric air intake systems and having limited capacity for heat recovery.

Mechanical requirements impose engine cooling to limit the temperature of the different engine elements, and cooling should be compatible with correct lubrication, allowing a moderate expansion. As the degree of loading increases, so does the heat transmission per unit area, the temperature difference between gases and cooling fluid, so that cooling requirements are increased, and also mechanical losses but to a minor degree.

The cylinder head is the part of the engine that has the most exigent needs for cooling. Cooling by lubricating oil is necessary to keep engine temperature within the correct values. This task is performed along with the lubrication of piston-sleeve, piston-connecting rod, connecting rod-crank, camshaft-pusher, and exhaust valve assemblies without the oil losing its properties. Cooling requirements by lubricating oil vary little from minimum to full load in absolute value. The cooling of the intake mixture to the cylinders at the turbo outlet is small at either minimum or maximum load due to the low energy content of the exhaust gases because gases have low density and low $Cp$ value.

All parameters, previously analyzed, increase in value with the degree of loading. However, the exhaust gas temperature follows the opposite trend. Efficiencies at low loading levels are lower for the reasons indicated above, which causes the exhaust gas outlet temperature to be higher than that expected at full load. This effect increases the exhaust gas losses due to the higher temperature gradient. Table 6 presents other calculated parameters for the engines having relevance in the global analysis. Reducing the degree of loading significantly affects engine performance. Therefore, installing several engines operating close to full loading is better than having a single one with low flexibility when biogas production is significantly reduced.

### 3.2. Coupling of Biogas Engines

Maximum efficiency is achieved when all engines are working at 100% loading. However, this is not always the case since the coupling of engines is based on biogas availability and boiler and engine demand. In addition, coupling of engines takes place after one of them reaches 100% loading. When the degree of loading slightly exceeds the nominal value, an additional engine comes into operation. Thus, the fuel is distributed between the operating set at a much lower loading, decreasing the efficiency significantly. In this way, a higher biogas availability causes an additional engine to start up and reduces the efficiency of the whole set.

**Table 6.** Key parameters of biogas engines calculated at different degree of loading.

| Parameter | Degree of Loading | | | |
|---|---|---|---|---|
| | 100 | 80 | 60 | 40 |
| Average linear piston speed (m/s) | 8.3 | | | |
| Effective torque (N m) | 3807 | 3043 | 2273 | 1502 |
| Specific electrical power (kW/t) | 84.2 | 67.3 | 50.6 | 33.7 |
| Effective specific mechanical power (kW/L) | 17.2 | 13.8 | 10.3 | 6.9 |
| Thermal load or power per unit of piston area (kW/cm$^2$) | 3.3 | 2.6 | 2 | 1.3 |
| Efficiency and consumption | | | | |
| Specific biogas consumption (g/kWh) | 470 | 466 | 514 | 567 |
| $\eta_{thermal\text{-}eng}$, engine thermal efficiency (%) | 48.6 | 50.4 | 52.6 | 55.7 |
| $\eta_{energy\text{-}eng}$, engine energy efficiency (%) | 87.7 | 88.0 | 87.9 | 87.1 |
| *UTEF* (%) | 85.5 | 84.2 | 82.2 | 81.1 |
| $(W/Q)_{eng}$, work/heat ratio | 81.1 | 75.9 | 68.5 | 57.7 |
| Combustion and emissions | | | | |
| Stoichiometric air–biogas ratio (%) | 17.1 | | | |
| *A*, air–biogas ratio (%) | 10.8 | 10.9 | 10.4 | 10.4 |
| $E_{air}$, excess air (%) | 58.3 | 56.7 | 64.9 | 64.0 |

Figure 5a,b demonstrates the abovementioned behavior. In these figures, the electrical power produced and biogas consumption are represented. The steep increase in biogas consumption is associated with the incoming operation of the additional engine. This behavior is indicated in Figure 5b, where a saw-teeth diagram represents the electrical efficiency. The steep decrease in efficiency when reducing the loading is explained by how biogas is distributed between the engines. The first point of operation in these diagrams corresponds with the engine working at a 40% degree of loading, which is the minimum possible for these engines.

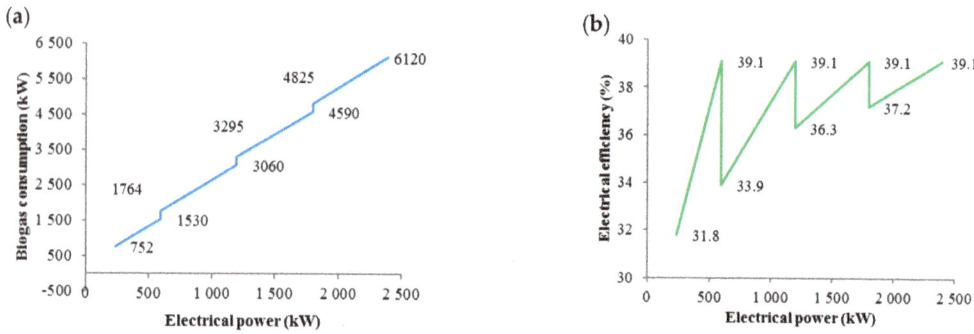


**Figure 5.** Coupling of the biogas engines, variation of parameters depending on the electrical power generated. (**a**) Biogas consumption; (**b**) Electrical efficiency.

The efficiency decreases as the number of operating engines increases until coupling the last one, leading to an efficiency decrease of 5.2, 2.8 and 1.9%. This is explained by the values of the engine loading moving from 50, 66.7 and 75% in each engine. This feature is a critical operational characteristic of the daily WWTP operation. For specific electrical power values, starting up a new engine may decrease the global efficiency sharply. Better efficiency is in some cases obtained by burning the excess biogas in a flare. If engines work at 90% loading, as recommended by the manufacturer, the problem may be aggravated at these so-called singular operating points.

### 3.3. Engine Energy Balance

Figure 6a shows the energy balance of the engines at the operating point (see Table 1). Supplementary Material contains the thermodynamics diagram of the turbocharger as a function of specific entropy (Figure S1). Figure S2 represents mass and energy balance of one cylinder in biogas engines. Heat capacity and specific enthalpy of the fluid flowing through the compressor, intercooler and turbine are shown in Table S1. Table S2 contains energy parameters of these same elements. A loading of 86.9% was assumed for each engine (two units coupled). A fraction of the thermal power from the main cooling circuit, 19.9%, is used to raise the temperature of the boiler feed water from 15 to 80 °C. This boiler produces the steam needed in the thermal pre-treatment unit. The remaining thermal power of this circuit is dissipated in the cooling towers, 80.1%, and there is no possibility of using this heat in the sludge recirculation exchanger. However, when the digesters operated without hydrolysis pre-treatment, this heat was used to keep the mesophilic temperature of digesters. Introducing a thermal pre-treatment unit into the digestion section creates different heat demands. Therefore, sludge is heated up during thermal pre-treatment and subsequently cooled down to the desired temperature to cover for any thermal losses associated with digester conditions and sludge piping line, being no longer necessary to adjust the temperature by the heat sludge recirculation exchanger. This waste heat has a negative impact on the energy efficiency of the process. The demand of the thermal pre-treatment unit is covered by the boiler and engine exhaust gases, giving rise to a different heating circuit where no other use is given to that of the engine main cooling circuit.

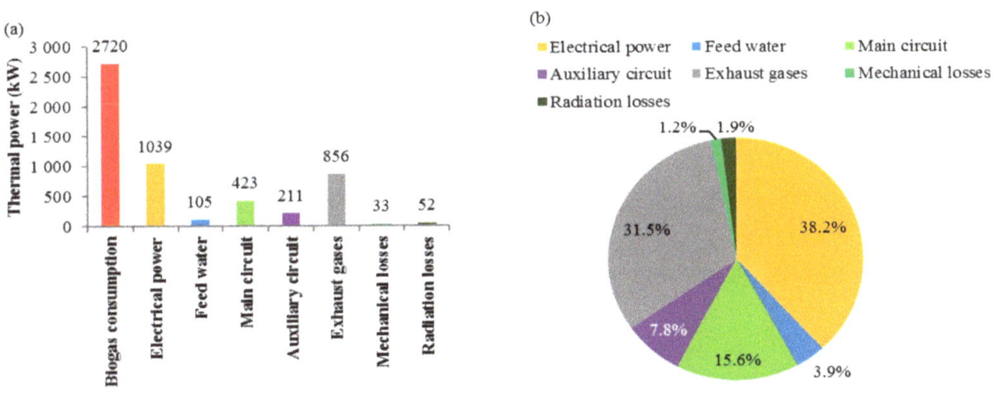


**Figure 6.** Energy balance of the biogas engines. (**a**) Parameters in absolute value; (**b**) Parameters expressed as percentage.

Figure 6b shows the energy balance represented as a percentage. Biogas consumption was used as base (expressed as thermal power). The available thermal energy of the main cooling circuit represents a significant share of the thermal distribution. Therefore, the gain associated with the higher efficiency of the turbocharged engine should translate into full use of the available thermal energy to take full advantage of the energy contained in biogas. In this graph, about 31% of the energy is in exhaust gases and about 23% in the cooling circuits. Thus, finding full use of this energy in the WWTP is imperative for increasing the efficiency of the global process.

Table 7 shows the values obtained for the useful thermal power associated with engine operation, broken down into different efficiencies. Absolute and relative values are reported. The values obtained from nominal engine operation were used as base. Excess air at 59.4% is observed, which gives an indication of the operating performance of these engines under lean-burn conditions. This condition allows increasing the mass flow of gases and, as a consequence, their energy. The useful thermal power represents the power capable of finding use in different equipment. Therefore, mechanical losses and radiation losses are discarded, and it is considered for calculation heat transfer from exhaust gases until a minimum temperature of 142 °C is reached. A value for *UTEF* of 84.4% indicates that this type of engine achieves a high level of energy recovery.

**Table 7.** Parameters of biogas engines at the operating point.

| Parameter | Value |
|---|---|
| $MP_{eng}$, mechanical power (kW) | 1072 |
| *DL*, degree of loading (%) (2 engines) | 86.9 |
| Effective average pressure (kPa) | 1193 |
| *Specific* $TP_{biogas\text{-}eng}$, engine specific biogas consumption (g/kWh) | 468 |
| $\eta_{mec\text{-}eng}$, engine mechanical efficiency (%) | 39.3 |
| $\eta_{el\text{-}eng}$, engine electrical efficiency (%) | 38.2 |
| $\eta_{thermal\text{-}eng}$, engine thermal efficiency (%) | 49.8 |
| $\eta_{energy\text{-}eng}$, engine energy efficiency (%) | 88 |
| *UTEF*, usable thermal energy factor of the engines (%) | 84.4 |
| $(W/Q)_{eng}$, work/heat ratio | 0.77 |
| Thermal losses | |
| Exhaust gases thermal power at 25 °C (kW) | 858 |
| Exhaust gases thermal power at 120 °C (kW) | 679 |
| Thermal power losses in the auxiliary water circuit (kW) | 211 |
| Oil exchanger (kW) | 143 |
| $TP_{IC}$, thermal power of the intercooler (kW) | 68 |
| Combustion and emissions | |
| Exhaust gases dry oxygen (%) | 8.9 |
| Stoichiometric air–biogas ratio (%) | 17.1 |
| *A*, air–biogas ratio (%) | 10.7 |
| $E_{air}$, excess air (%) | 59.4 |
| $T_4$, exhaust gas temperature (°C) | 483 |
| $M_{air}$, inlet air mass flow (kg/h) | 5371 |
| $M_{eg}$, exhaust gas mass flow (kg/h) | 5878 |

### 3.4. Cooling Circuit Energy Balance

Figure 7 shows the diagram of the main, emergency and auxiliary cooling circuits of the engines. Each circuit consists of a primary and a secondary section as previously described, both separated by a plate heat exchanger. When the digestion section worked as a conventional mesophilic system, the energy from this circuit was used to increase the sludge temperature and keep the digester at 37 °C. However, after installing the thermal pre-treatment unit, this heat recovery was no longer needed. Part of the energy available in the main circuit—secondary section—is now used to preheat the feeding water of the recovery mixed boiler. The exceeding heat is dissipated in cooling towers. Due to the physical proximity between the exchangers of the main cooling circuit and the boiler feed water preheating circuit, the minimum approach point considered for calculations was 5 °C. A significant amount of low-grade energy does not find any application in the WWTP when the thermal hydrolysis unit is operating. This assessment highlights the relevance when evaluating biogas production enhancement. All sets of additional equipment must be considered, along with changes in heat recovery and the quality of the energy needed after implemented modifications.

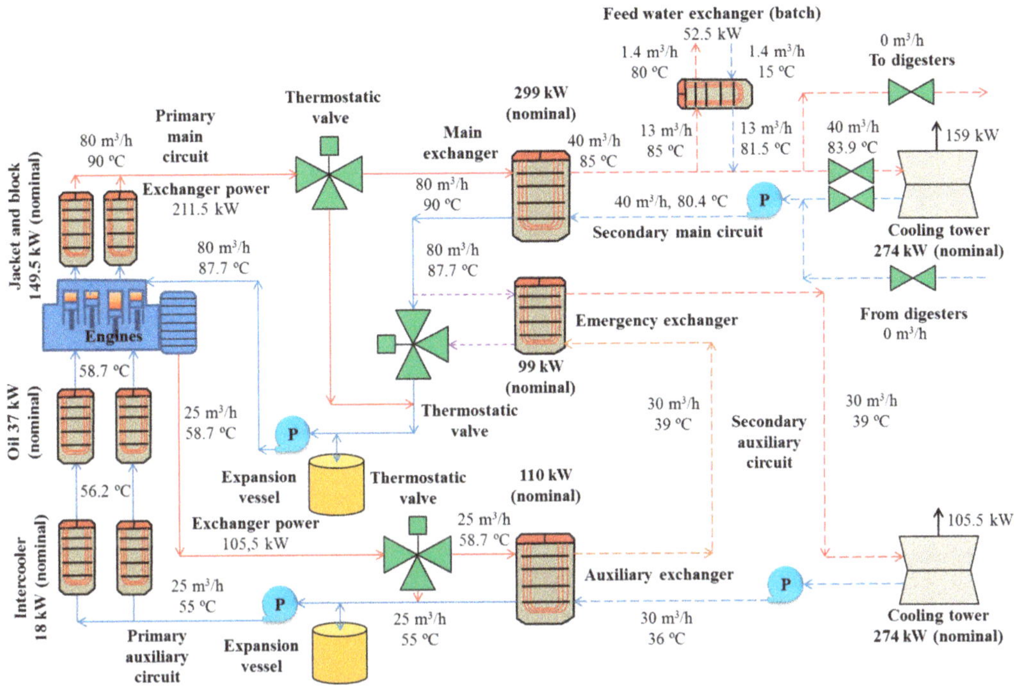


**Figure 7.** Mass and energy balance of the engine cooling circuits at the operating point.

The emergency exchanger only comes into service in the event of insufficient power, breakdown, or failure of the primary exchanger. In contrast, its secondary circuit is cooled by water coming out of the secondary exchanger from the auxiliary circuit, both being positioned in series and fully cooled in cooling towers. The secondary section of the auxiliary circuit also dissipates its energy in cooling towers.

The three-way thermostatic valves in both the main and auxiliary circuits allow the water temperature to be kept at 90 °C at the outlet and 55 °C at the inlet of the engine, by recirculating water through the existing bypasses without going through any of the heat plate exchangers located in the two cooling circuits. The heat exchangers of the block, crankcase and cylinder jackets are shown in Figure 7 as external independent units for the sake of simplicity to facilitate process description. This representation eases heat transfer calculation and aids in understanding the different circuits considered. The cooling water is fed at the bottom part of the engine, where the crankcase is located, and it circulates up through the cylinder jackets, exiting through the cylinder heads. All these components are located inside the engine. The thermal efficiency of plate heat exchangers was assumed to be unity.

Table 8 shows results from the energy balance, considering electrical energy production. The gross electrical power generated was 1039 kW, which translates into 9102 MWh/year, equivalent to 837,384 €/year. This energy is completely consumed at the plant. The value of the resulting utilization factor is 43.4%, not reaching 50% of the installed capacity, which is too low to guarantee the installation reaches profitability limits. The plant installed capacity for producing electricity is much greater than that produced by the engines. The inclusion of thermal hydrolysis for enhancing sludge stabilization and reducing sludge volume modifies the global energy balance because steam is now needed along with the additional equipment operating at higher temperatures. This modification, in an already operating WWTP, results in enhanced performance of digestion, but it does not necessarily lead to a better energy balance given all the additional equipment needed for the extra units and the lower capacity for finding use of low-temperature streams. However, the extra digester capacity is advantageous due to the shorter time required for sludge stabilization. This improvement in reactor capacity allows the treatment of co-substrates with high methane potential, thus producing more biogas to be valorized in the existent engines.

**Table 8.** Results from the energy balance of the WWTP.

| Parameters | Values |
|---|---|
| Number of operating engines | 2 |
| Gross electrical power generated (kW) | 1039 |
| $EP_{inst}$, installed electrical power of the engines (kW) | 2392 |
| *IECUF*, installed electrical capacity utilization factor (%) | - |
| *DSC*, degree of self-consumption (%) | 55.1 |
| Gross electrical energy generation (MWh/d) | 24.9 |
| Gross electrical energy generation (MWh/year) | 9102 |
| Value of gross electrical energy generation (€/year) | 837,384 |

High-strength organics considered suitable co-substrates include waste cooking oil, residual glycerin, or cheese whey from the food industry [49–54]. The organic fraction of municipal solid wastes has been extensively studied as co-substrate [55–57], but it has the risk of increasing the amount of inert material [58], deteriorating sludge agronomic quality.

### 3.5. Energy Balance: Co-Digestion Case

Cheese whey and residual glycerol both have high methane yields (446 mL $CH_4$/g VS, reported by Papirio et al. [59] for cheese whey and 561.3 mL $CH_4$/g VS reported by González et al. [60] for glycerol). In the present case, 2302 kW of extra biogas can be valorized based on the plant installed capacity. Therefore, if considering characteristics for these substrates similar to those reported by de Albuquerque et al. [61], with whey having a concentration of 70 g COD/L and glycerol a concentration of 1520 g COD/L (COD: chemical oxygen demand), the WWTP could treat a stream of 366 $m^3$/d in the case of cheese whey, but for glycerol, this amount is reduced to 16 $m^3$/d. A similar analysis can be performed for the use of waste cooking oil, also characterized by high methane yield (922 mL $CH_4$/g VS, methane yield reported by Marchetti et al. [62]). However, in this latter case, complexities regarding the appearance of possible inhibitory conditions due to long-chain fatty acid accumulation need to be carefully evaluated. In addition, storage and handling of waste cooking oil may be more problematic because of additional heating systems needed to avoid pipe clogging.

Figure 8 shows the co-digestion configuration. The degree of electrical self-consumption is now 55.1%, still a low value. However, it should be taken into account that the WWTP has a tertiary treatment system based on lamellar decantation and the use of $O_3$ and ultraviolet rays for final polishing of the treated water, significantly rising the energy demand of the whole plant. Co-digestion with cheese whey could add an extra 2302 kW of biogas, summing to 5892 kW. A fraction of this power, 870 kW, would be used by the recovery boiler for producing steam to cover the demand of the thermal hydrolysis pre-treatment (Cambi THP B6.2, two reactors of 6 $m^3$). The remaining amount (5022 kW) could be used as fuel in engines.

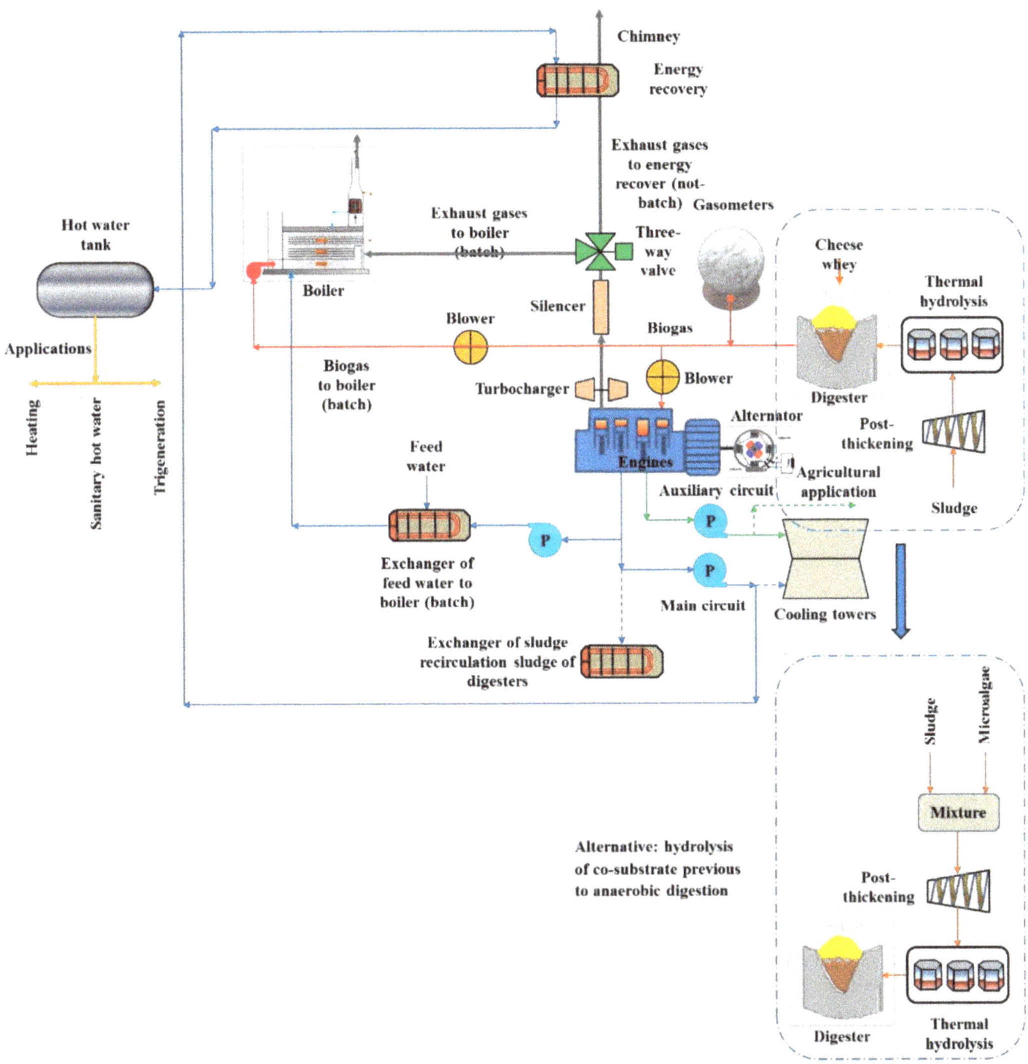


**Figure 8.** Proposed scheme for cheese whey co-digestion and hydrolyzed sludge. As alternative to this process is represented in the case where a thermally pre-treated co-substrate is introduced into the digestion line. The scheme also represents biogas valorization, the recovery energy system, and thermal hydrolysis.

Co-digestion is a suitable option for increasing biogas production and reactor productivity. The energy balance is positively affected in the hypothetical case of adding cheese whey as co-substrate since this material is readily degraded and does not require any pre-treatments. The greater availability of biogas affects engines' performance, increasing the mass of exhaust gases and improving the recovery boiler energy balance.

Table 9 shows the results of the energy balance under co-digestion configuration. This table also shows results from the hypothetical case of adding a hydrolyzed co-substrate. This latter configuration is also schematized in Figure 8 as an alternative treatment option. For this second valorization scheme, a volumetric flow of microalgae biomass was hydrolyzed with sludge. The methane yield for hydrolyzed microalgae biomass reported by Ometto et al. [63] was 223 mL $CH_4$/g VS, whereas a value of 339.1 mL $CH_4$/g VS was reported by Méndez et al. [64]. Thus an average value of 281 mL $CH_4$/g VS was used for evaluating the energy balance. A microalgae volumetric flow of 410 $m^3$/d (TS content of 7%, VS percentage of 91%) was considered to keep previous assumptions consistent. Biogas produced would be, in this case, equivalent to 6253 kW. The thermal hydrolysis unit would demand 1231 kW, leading to the same amount of biogas available as fuel in engines (5022 kW). Electricity production would be increased until reaching energy self-sufficiency (1885 kW), obtaining savings valued at 1,519,160 €/year. However, the application of the microalgae co-digestion strategy requires the installation of an additional hydrolysis unit to cover the higher flow.

**Table 9.** Energy balance of biogas engines considering co-digestion with cheese whey and with hydrolyzed microalgae. For simplicity, the same biogas yield after thermal pre-treatment was assumed.

| Parameter | Sludge + Cheese Whey | Sludge + Microalgae Biomass |
|---|---|---|
| Biogas production (kW) | 5892 | 6523 |
| Biogas boiler (kW) | 870 | 1231 |
| Electrical power (kW) | 1885 | 1885 |
| Feed water (kW) | 105 | 184 |
| Main circuit (kW) | 873 | 778 |
| Auxiliary circuit (kW) | 406 | 417 |
| Exhaust gases (kW) | 1593 | 1596 |
| Mechanical losses (kW) | 63 | 63 |
| Radiation losses (kW) | 97 | 99 |

For both cases studied, the hydrolyzed substrate contains enough energy to provide the thermal demand of digestion. Therefore, the thermal energy available from the engine cooling circuit is not needed. The main cooling circuit has a thermal power of 873 kW in the cheese whey co-digestion case and 778 kW in the micro-algae case with a temperature of the hot fluid stream of 84.2 °C. This stream could supply domestic hot water and heating to the office building of the WWTP, or be used in a trigeneration system by installing an absorption chiller to cover cooling needs in summer. The auxiliary cooling circuit has similar power for both co-digestion cases, with a hot fluid maximum temperature of only 49 °C (the approach temperature between fluids was 10 °C). This value is too low to produce any work. The low temperature of this stream and, therefore, the low amount of energy stored prevent its use as residual thermal energy in any process having direct application in the WWTP. However, given the large quantity of these low sources in urban areas, future research efforts should be performed to find efficient ways for this type of heat recovery [65].

The residual energy of exhaust gases is 981 kW, considering an outlet temperature of 142 °C, from which just 323 kW are used in the recovery boiler, with the remaining being wasted by emission directly into the atmosphere. If this energy is used to raise the water temperature in the main cooling circuit, the usable power for heating, domestic hot water, or trigeneration, could be considerably increased to 1531 kW. Table 10 presents the results from the energy balance. A similar analysis for the hydrolysis case leads to a slightly lower amount of residual energy contained in exhaust gases. In this latter case, the value is 978 kW, but a greater amount is needed in the recovery boiler because of the increment in steam demand. Thus, 567 kW are used in this unit. The remaining amount (411 kW) is wasted. Just as previously analyzed, if this power were used, the amount available for domestic hot water applications or trigeneration would be 1189 kW.

**Table 10.** Results from the cooling circuit energy balance. Quantities refer to thermal power having application in other uses. The low temperature of the auxiliary circuit stream makes its application unfeasible at the current WWTP configuration (losses in plate and exhaust gas heat exchangers were not taken into account for the sake of simplicity).

| Process | Useful Energy Recovery from Exhaust Gases (kW) | Main Circuit (kW) (Useful) | Auxiliary Circuit (kW) (Not Useful) | Total Energy (kW) (Useful) |
|---|---|---|---|---|
| Sludge digestion | 351 | 423 | 211 | 774 |
| Cheese whey co-digestion | 658 | 873 | 406 | 1531 |
| Microalgae co-digestion | 411 | 778 | 417 | 1189 |

## 4. Conclusions

This research studied the production of biogas in a WWTPs treating hydrolyzed sewage sludge. The pre-treatment increases the energy demand of the plant, but it also increases the digester's treatment capacity, providing an opportunity for valorizing different waste streams. A detailed energy balance of biogas engines was carried out to establish specific points where energy recovery needs improvement. The hypothetical addition of a co-substrate to the digestion line was also analyzed, considering a readily degradable substrate and another one needing thermal hydrolysis. The engine energy balance demonstrated a high electrical and thermal efficiency. The addition of a co-substrate attains the full utilization of the electrical installed capacity of the WWTP (1885 kW). Therefore, co-digestion provides a significant enhancement in plant overall performance because of the greater amount of electricity produced reaching energy autarky when a co-substrate such as cheese whey is co-digested with thermal pre-treated sludge. Further research is necessary to find suitable ways of recovering energy from low temperature streams to avoid dissipating this energy into the atmosphere.

**Supplementary Materials:** The following are available online at https://www.mdpi.com/article/10.3390/app112311103/s1, Figure S1: Thermodynamics diagram of the turbocharger as a function of specific entropy. Figure S2: Mass and energy balance of one cylinder in biogas engines Table S1: Heat capacity and specific enthalpy of the fluid flowing through the compressor, intercooler and turbine. Table S2: Energy parameters of compressor, intercooler and turbine.

**Author Contributions:** Conceptualization, J.G.-C. and X.G.; methodology, J.G.-C.; software, J.G.-C.; validation, J.G.-C. and X.G.; formal analysis, D.C.-P.; investigation, J.G.-C.; resources, A.M.; data curation, D.C.-P.; writing—original draft preparation, J.G.-C. and X.G.; writing—review and editing, X.G. and R.S.; visualization, R.S.; supervision, X.G. All authors have read and agreed to the published version of the manuscript.

**Funding:** This research received no external funding.

**Institutional Review Board Statement:** Not applicable.

**Informed Consent Statement:** Not applicable.

**Conflicts of Interest:** The authors declare no conflict of interest.

## References


1. Benato, A.; Macor, A. Italian biogas plants: Trend, subsidies, cost, biogas composition and engine emissions. *Energies* **2019**, *12*, 979. [CrossRef]
2. Karagöz, M.; Sarıdemir, S.; Deniz, E.; Çiftçi, B. The effect of the $CO_2$ ratio in biogas on the vibration and performance of a spark ignited engine. *Fuel* **2018**, *214*, 634–639. [CrossRef]
3. Ngumah, C.C.; Ogbulie, J.N.; Orji, J.C.; Amadi, E.S. Biogas potential of organic waste in Nigeria. *J. Urban Environ. Eng.* **2013**, *7*, 110–116. [CrossRef]
4. Morales-Polo, C.; Cledera-Castro, M.D.M.; Moratilla Soria, B.Y. Reviewing the anaerobic digestion of food waste: From waste generation and anaerobic process to its perspectives. *Appl. Sci.* **2018**, *8*, 1804. [CrossRef]

5. Singh, A.; Olsen, S.I.; Nigam, P.S. A viable technology to generate third-generation biofuel. *J. Chem. Technol. Biotechnol.* **2011**, *86*, 1349–1353. [CrossRef]
6. Sarangi, P.K.; Nayak, M.M. Agro-Waste for Second-Generation Biofuels. *Liq. Biofuels Fundam. Charact. Appl.* **2021**, 697–709. [CrossRef]
7. Ellacuriaga, M.; Cascallana, J.G.; González, R.; Gómez, X. High-Solid Anaerobic Digestion: Reviewing Strategies for Increasing Reactor Performance. *Environments* **2021**, *8*, 80. [CrossRef]
8. Thiruselvi, D.; Yuvarani, M.; Salma, A.; Arafath, Y.; Jagadiswary, D.; Kumar, M.A.; Anuradha, D.; Shanmugam, P.; Sivanesan, S. Enhanced biogas from sewage sludge digestion using iron nanocatalyst from Vitex negundo leaf extract: Response surface modeling. *Int. J. Environ. Sci. Technol.* **2021**, *18*, 2161–2172. [CrossRef]
9. Calabrò, P.S.; Fazzino, F.; Limonti, C.; Siciliano, A. Enhancement of Anaerobic Digestion of Waste-Activated Sludge by Conductive Materials under High Volatile Fatty Acids-to-Alkalinity Ratios. *Water* **2021**, *13*, 391. [CrossRef]
10. Sakaveli, F.; Petala, M.; Tsiridis, V.; Darakas, E. Enhanced Mesophilic Anaerobic Digestion of Primary Sewage Sludge. *Water* **2021**, *13*, 348. [CrossRef]
11. Tian, X.; Trzcinski, A. Effects of physico-chemical post-treatments on the semi-continuous anaerobic digestion of sewage sludge. *Environments* **2017**, *4*, 49. [CrossRef]
12. Wang, T.; Xu, B.; Zhang, X.; Yang, Q.; Xu, B.; Yang, P. Enhanced biogas production and dewaterability from sewage sludge with alkaline pretreatment at mesophilic and thermophilic temperatures. *Water Air Soil Pollut.* **2018**, *229*, 1–10. [CrossRef]
13. Le, T.M.; Vo, P.T.; Do, T.A.; Tran, L.T.; Truong, H.T.; Xuan Le, T.T.; Chen, Y.-H.; Chang, C.C.; Chang, C.C.; Tran, Q.C.; et al. Effect of assisted ultrasonication and ozone pretreatments on sludge characteristics and yield of biogas production. *Processes* **2019**, *7*, 743. [CrossRef]
14. Martínez, E.J.; Gil, M.V.; Fernández, C.; Rosas, J.G.; Gómez, X. Anaerobic codigestion of sludge: Addition of butcher's fat waste as a cosubstrate for increasing biogas production. *PLoS ONE* **2016**, *11*, e0153139. [CrossRef]
15. Bardi, M.J.; Aminirad, H. Synergistic effects of co-trace elements on anaerobic co-digestion of food waste and sewage sludge at high organic load. *Environ. Sci. Pollut. Res.* **2020**, *27*, 1–16. [CrossRef]
16. Coura, R.; Alonso, J.M.; Rodrigues, A.C.; Ferraz, A.I.; Mouta, N.; Silva, R.; Brito, A.G.D. Spatially Explicit Model for Anaerobic Co-Digestion Facilities Location and Pre-Dimensioning Considering Spatial Distribution of Resource Supply and Biogas Yield in Northwest Portugal. *Appl. Sci.* **2021**, *11*, 1841. [CrossRef]
17. Deheri, C.; Acharya, S.; Thatoi, D.; Mohanty, A. A review on efficiency of biogas and hydrogen on diesel engine in dual fuel mode. *Fuel* **2020**, *260*, 116337. [CrossRef]
18. Imeni, S.M.; Puy, N.; Ovejero, J.; Busquets, A.M.; Bartroli, J.; Pelaz, L.; Ponsá, S.; Colón, J. Techno-economic assessment of anaerobic co-digestion of cattle manure and wheat straw (raw and pre-treated) at small to medium dairy cattle farms. *Waste Biomass Valoriz.* **2020**, *11*, 4035–4051. [CrossRef]
19. Zhou, L.; Hülsemann, B.; Cui, Z.; Merkle, W.; Sponagel, C.; Zhou, Y.; Guo, J.; Dong, R.; Müller, J.; Oechsner, H. Operating performance of full-scale agricultural biogas plants in Germany and China: Results of a year-round monitoring program. *Appl. Sci.* **2021**, *11*, 1271. [CrossRef]
20. Oreggioni, G.D.; Gowreesunker, B.L.; Tassou, S.A.; Bianchi, G.; Reilly, M.; Kirby, M.E.; Toop, T.A.; Theodorou, M.K. Potential for energy production from farm wastes using anaerobic digestion in the UK: An economic comparison of different size plants. *Energies* **2017**, *10*, 1396. [CrossRef]
21. García-Cascallana, J.; Borge-Díez, D.; Gómez, X. Enhancing the efficiency of thermal hydrolysis process in wastewater treatment plants by the use of steam accumulation. *Int. J. Environ. Sci. Technol.* **2019**, *16*, 3403–3418. [CrossRef]
22. Kim, D.; Kim, K.T.; Park, Y.K. A Comparative Study on the Reduction Effect in Greenhouse Gas Emissions between the Combined Heat and Power Plant and Boiler. *Sustainability* **2020**, *12*, 5144. [CrossRef]
23. Pérez, V.; Lebrero, R.; Muñoz, R. Comparative evaluation of biogas valorization into electricity/heat and poly (hydroxyalkanoates) in waste treatment plants: Assessing the influence of local commodity prices and current biotechnological limitations. *ACS Sustain. Chem. Eng.* **2020**, *8*, 7701–7709. [CrossRef]
24. Feroskhan, M.; Ismail, S. A review on the purification and use of biogas in compression ignition engines. *Int. J. Automot. Mech. Eng.* **2017**, *14*, 4383–4400. [CrossRef]
25. Gupta, S.K.; Mittal, M. Effect of biogas composition variations on engine characteristics including operational limits of a spark-ignition engine. *J. Eng. Gas Turbines Power* **2019**, *141*, 10100. [CrossRef]
26. Gómez-Montoya, J.P.; Amell, A.A.; Olsen, D.B. Prediction and measurement of the critical compression ratio and methane number for blends of biogas with methane, propane and hydrogen. *Fuel* **2016**, *186*, 168–175. [CrossRef]
27. Okoro, O.V.; Sun, Z. Desulphurisation of biogas: A systematic qualitative and economic-based quantitative review of alternative strategies. *ChemEngineering* **2019**, *3*, 76. [CrossRef]
28. Sadegh-Vaziri, R.; Babler, M.U. Removal of Hydrogen Sulfide with Metal Oxides in Packed Bed Reactors—A Review from a Modeling Perspective with Practical Implications. *Appl. Sci.* **2019**, *9*, 5316. [CrossRef]
29. Ellacuriaga, M.; García-Cascallana, J.; Gómez, X. Biogas Production from Organic Wastes: Integrating Concepts of Circular Economy. *Fuels* **2021**, *2*, 9. [CrossRef]
30. Ketata, A.; Driss, Z.; Abid, M.S. Experimental analysis for performance evaluation and unsteadiness quantification for one turbocharger vane-less radial turbine operating on a gasoline engine. *Arab. J. Sci. Eng.* **2018**, *43*, 4763–4781. [CrossRef]

31. Thompson, I.G.M.; Spence, S.; McCartan, C.; Talbot-Weiss, J.; Thornhill, D. One dimensional modeling of a turbogenerating spark ignition engine operating on biogas. *SAE Int. J. Engines* **2011**, *4*, 1354–1364. [CrossRef]
32. Lee, W.; Schubert, E.; Li, Y.; Li, S.; Bobba, D.; Sarlioglu, B. Electrification of turbocharger and supercharger for downsized internal combustion engines and hybrid electric vehicles-benefits and challenges. In Proceedings of the 2016 IEEE Transportation Electrification Conference and Expo (ITEC), Dearborn, MI, USA, 27–29 June 2016. [CrossRef]
33. Kusztelan, A.; Yao, Y.; Marchant, D.; Wang, Y. A Review of Novel Turbocharger Concepts for Enhancements in Energy Efficiency. *IJTEE* **2010**, *2*, 75–82. [CrossRef]
34. Malaquias, A.C.T.; Netto, N.A.D.; da Costa, R.B.R.; Teixeira, A.F.; Costa, S.A.P.; Baêta, J.G.C. An evaluation of combustion aspects with different compression ratios, fuel types and injection systems in a single-cylinder research engine. *J. Braz. Soc. Mech. Sci.* **2020**, *42*, 1–16. [CrossRef]
35. Bora, B.J.; Saha, U.K. Theoretical performance limits of a biogas–diesel powered dual fuel diesel engine for different combinations of compression ratio and injection timing. *J. Energy Eng.* **2016**, *142*, E4015001. [CrossRef]
36. Ahmed, S.A.; Zhou, S.; Zhu, Y.; Tsegay, A.S.; Feng, Y.; Ahmad, N.; Malik, A. Effects of Pig Manure and Corn Straw Generated Biogas and Methane Enriched Biogas on Performance and Emission Characteristics of Dual Fuel Diesel Engines. *Energies* **2020**, *13*, 889. [CrossRef]
37. Guascor. Gas Engines Power Rating Correction. 2021. Available online: http://www.guascor.ruinfoIC-G-B-00-001e_B.pdf (accessed on 21 October 2021).
38. Abu-Nada, E.; Al-Hinti, I.; Al-Sarkhi, A.; Akash, B. Thermodynamic modeling of spark-ignition engine: Effect of temperature dependent specific heats. *Int. Commun. Heat Mass Transf.* **2016**, *33*, 1264–1272. [CrossRef]
39. EUROSTAT. Electricity Prices for Non-Household Consumers—Bi-Annual Data (from 2007 Onwards). Available online: https://appsso.eurostat.ec.europa.eu/nui/submitViewTableAction.do (accessed on 1 April 2021).
40. Di Battista, D.; Mauriello, M.; Cipollone, R. Waste heat recovery of an ORC-based power unit in a turbocharged diesel engine propelling a light duty vehicle. *Appl. Energy* **2015**, *152*, 109–120. [CrossRef]
41. Kesgin, U. Effect of turbocharging system on the performance of a natural gas engine. *Energy Convers. Manag.* **2005**, *46*, 11–32. [CrossRef]
42. Romagnoli, A.; Manivannan, A.; Rajoo, A.S.; Chiong, M.S.; Feneley, A.; Pesiridis, A.; Martinez-Botas, R.F. A review of heat transfer in turbochargers. *Renew. Sust. Energ. Rev.* **2017**, *79*, 1442–1460. [CrossRef]
43. Lenhart, T.; Eckhardt, K.; Fohrer, N.; Frede, H.G. Comparison of two different approaches of sensitivity analysis. *Phys. Chem. Earth* **2002**, *27*, 645–654. [CrossRef]
44. Salciccioli, J.D.; Crutain, Y.; Komorowski, M.; Marshall, D.C. Sensitivity Analysis and Model Validation. In *Secondary Analysis of Electronic Health Records*; Springer: Cham, Switzerland, 2016. [CrossRef]
45. Hamby, D.M. A review of techniques for parameter sensitivity analysis of environmental models. *Environ. Monit. Assess.* **1994**, *32*, 135–154. [CrossRef] [PubMed]
46. Kocsis, L.B.; Moldovanu, D.; Băldean, D.L. The Influence of Exhaust Backpressure Upon the Turbocharger's Boost Pressure. In *Proceedings of the European Automotive Congress EAEC-ESFA 2015*; Andreescu, C., Clenci, A., Eds.; Springer: Cham, Switzerland, 2016. [CrossRef]
47. González, R.; Blanco, D.; González-Arias, J.; García-Cascallana, J.; Gómez, X. Description of a decentralized small scale digester for treating organic wastes. *Environments* **2020**, *7*, 78. [CrossRef]
48. González, R.; Hernández, J.E.; Gómez, X.; Smith, R.; Arias, J.G.; Martínez, E.J.; Blanco, D. Performance evaluation of a small-scale digester for achieving decentralised management of waste. *Waste Manag.* **2020**, *118*, 99–109. [CrossRef] [PubMed]
49. Fierro, J.; Martínez, E.J.; Morán, A.; Gómez, X. Valorisation of used cooking oil sludge by codigestion with swine manure. *Waste Manag.* **2014**, *34*, 1537–1545. [CrossRef]
50. Chou, Y.C.; Su, J.J. Biogas production by anaerobic co-digestion of dairy wastewater with the crude glycerol from slaughterhouse sludge cake transesterification. *Animals* **2019**, *9*, 618. [CrossRef]
51. Hallaji, S.M.; Kuroshkarim, M.; Moussavi, S.P. Enhancing methane production using anaerobic co-digestion of waste activated sludge with combined fruit waste and cheese whey. *BMC Biotechnol.* **2019**, *19*, 19. [CrossRef]
52. Sprafke, J.; Shettigondahalli Ekanthalu, V.; Nelles, M. Continuous Anaerobic Co-Digestion of Biowaste with Crude Glycerol under Mesophilic Conditions. *Sustainability* **2020**, *12*, 9512. [CrossRef]
53. Kassongo, J.; Shahsavari, E.; Ball, A.S. Co-Digestion of Grape Marc and Cheese Whey at High Total Solids Holds Potential for Sustained Bioenergy Generation. *Molecules* **2020**, *25*, 5754. [CrossRef]
54. Yan, W.; Vadivelu, V.; Maspolim, Y.; Zhou, Y. In-situ alkaline enhanced two-stage anaerobic digestion system for waste cooking oil and sewage sludge co-digestion. *Waste Manag.* **2021**, *120*, 221–229. [CrossRef]
55. Bolzonella, D.; Battistoni, P.; Susini, C.; Cecchi, F. Anaerobic codigestion of waste activated sludge and OFMSW: The experiences of Viareggio and Treviso plants (Italy). *Water Sci. Technol.* **2006**, *53*, 203–211. [CrossRef]
56. Alrawashdeh, K.A.B.; Pugliese, A.; Slopiecka, K.; Pistolesi, V.; Massoli, S.; Bartocci, P.; Bidini, G.; Fantozzi, F. Codigestion of Untreated and Treated Sewage Sludge with the Organic Fraction of Municipal Solid Wastes. *Fermentation* **2017**, *3*, 35. [CrossRef]
57. Liu, Y.; Huang, T.; Li, X.; Huang, J.; Peng, D.; Maurer, C.; Kranert, M. Experiments and modeling for flexible biogas production by co-digestion of food waste and sewage sludge. *Energies* **2020**, *13*, 818. [CrossRef]

58. Park, N.D.; Thring, R.W.; Garton, R.P.; Rutherford, M.P.; Helle, S.S. Increased biogas production in a wastewater treatment plant by anaerobic co-digestion of fruit and vegetable waste and sewer sludge—A full scale study. *Water Sci. Technol.* **2011**, *64*, 1851–1856. [CrossRef]
59. Papirio, S.; Matassa, S.; Pirozzi, F.; Esposito, G. Anaerobic co-digestion of cheese whey and industrial hemp residues opens new perspectives for the valorization of agri-food waste. *Energies* **2020**, *13*, 2820. [CrossRef]
60. González, R.; Smith, R.; Blanco, D.; Fierro, J.; Gómez, X. Application of thermal analysis for evaluating the effect of glycer-ine addition on the digestion of swine manure. *J. Therm. Anal. Calorim.* **2018**, *135*, 2277–2286. [CrossRef]
61. De Albuquerque, J.N.; Paulinetti, A.P.; Lovato, G.; Albanez, R.; Ratusznei, S.M.; Rodrigues, J.A.D. Anaerobic Sequencing Batch Reactors Co-digesting Whey and Glycerin as a Possible Solution for Small and Mid-size Dairy Industries: Environmental Compliance and Methane Production. *Appl. Biochem. Biotechnol.* **2020**, *192*, 979–998. [CrossRef]
62. Marchetti, R.; Vasmara, C.; Fiume, F. Pig slurry improves the anaerobic digestion of waste cooking oil. *Appl. Microbiol. Biotechnol.* **2019**, *103*, 8267–8279. [CrossRef]
63. Ometto, F.; Quiroga, G.; Pšenička, P.; Whitton, R.; Jefferson, B.; Villa, R. Impacts of microalgae pre-treatments for improved anaerobic digestion: Thermal treatment, thermal hydrolysis, ultrasound and enzymatic hydrolysis. *Water Res.* **2014**, *65*, 350–361. [CrossRef]
64. Méndez, L.; Mahdy, A.; Demuez, M.; Ballesteros, M.; González-Fernández, C. Effect of high pressure thermal pretreatment on Chlorella vulgaris biomass: Organic matter solubilisation and biochemical methane potential. *Fuel* **2014**, *117*, 674–679. [CrossRef]
65. Wheatcroft, E.; Wynn, H.; Lygnerud, K.; Bonvicini, G.; Leonte, D. The role of low temperature waste heat recovery in achieving 2050 goals: A policy positioning paper. *Energies* **2020**, *13*, 2107. [CrossRef]

*applied sciences*

MDPI

*Article*

# Energetic Valorization of Solid Wastes from the Alcoholic Beverage Production Industry: Distilled Gin Spent Botanicals and Brewers' Spent Grains

**Jesús A. Montes * and Carlos Rico**

Department of Water and Environmental Science and Technologies, University of Cantabria, Avda. Los Castros, s/n, 39005 Santander, Spain; carlos.rico@unican.es
* Correspondence: jesus-andres.montes@alumnos.unican.es; Tel.: +34-942-201-848; Fax: +34-942-201-703

**Featured Application: Valorization of wastes from agri-food industries by their use as fuels and sources of power.**

**Abstract:** In this paper, the authors assess the possibilities of energetic valorization for two solid wastes from alcoholic beverage production. Distilled gin spent botanicals (DGSB) and brewers' spent grains (BSG) are tested, both by themselves and as co-substrates, for their possibilities as substrates for anaerobic digestion in a system of box-type digesters, suited for the process. While BSGs show a good performance for anaerobic digestion, DGSBs, despite showing an acceptable biomethanogenic potential result as not suitable for the process. Experiments using DGSBs as substrate in the reactors result in failure. And, as a co-substrate, the biomethanogenic digestion process appears to be hampered and lagged. Possible explanations for this behavior are explored, as well as other possibilities for the use of the material as a power source given its high heating value.

**Keywords:** biochemical methane potential (BMP); biogas; by-products; higher heating value; alcoholic beverage production

**Citation:** Montes, J.A.; Rico, C. Energetic Valorization of Solid Wastes from the Alcoholic Beverage Production Industry: Distilled Gin Spent Botanicals and Brewers' Spent Grains. *Appl. Sci.* **2021**, *11*, 10158. https://doi.org/10.3390/app112110158

Academic Editor: Francisco Jesus Fernandez-Morales

Received: 22 September 2021
Accepted: 26 October 2021
Published: 29 October 2021



## 1. Introduction

Alcoholic beverage production is, globally, one of the most important industries in the agri-food sector. Both fermented and distillate beverage productions have several things in common, among which we can cite a high usage of resources, both in terms of raw materials and energy. On the other hand, the alcohol-making process produces a flow of waste and by-products that have to be dealt with due to their high pollution potential [1]. A number of countries have established regulations and guidelines for this specific sector. However, the problem is that the differences in the raw materials and processes used for the specific production of the different beverages demand specific analyses and studies for each particular case [2,3], making it difficult to establish unified ways to tackle the particular problems for the differenttypes of waste.

The current socio-economical situation worldwide, with an increased global concern about the environment and the effects of climate change, together with awareness about the scarcity of resources and energy, has resulted in a demand for new solutions to the problems caused by human activities in generaland processing industries in particular.Following the principles of circular economy and using the tools of life cycle analyses and the carbon footprint of the products is a way to ensure the welfare of the planet and its inhabitants, increasing the resilience of society and global economy.

The case of the alcoholic beverage production among other industrial fields is significant. Breweries have been among the first industries which have raised concerns about their pollution creation [4], which has led to the adoption of different domestic regulations (we can cite as examples the Australian Effluent management guidelines for Australian wineries and distilleries [5,6] or the U.K. Guidance for Pollution Prevention GPP29 for Wales, Scotland, and Northern Ireland [7]) and, in other cases, rules of good practice. On the other hand, it is a demanding industrial field, not only in terms of raw materials and water consumption [8] but also in energy power, mainly used for heating in several operations of the alcoholic beverage production process (alcoholic fermentation, boiling of the worts, distillation, etc.). It is noteworthy to say that the sector combines a mixture of tradition (for example, the use of century-old copper pot stills or yeast strains fully developed in the particular industrial premises) with an open mind for development and improvement of those processes; aiming to optimize results both in terms of economy and product quality as well as image improvement towards society [9,10]. Thus, the industrial sector is eager to adopt measures to reduce both their inputs of raw matters, water and power, as well as to improve their production processes and reduce their waste [11]. This leads in a reasonable way to the interest of industries in the reduction and/or valorization of their waste. And an obvious way of dealing with such waste is as a powersource [12], achieving the double benefit of their reduction or elimination, as well as the reduction of power obtained through external sources.

Wastes in alcoholic beverage production can vary in form and nature, depending on the particular beverage produced. However, they usually havea series of things in common, as their origin is from natural matters (almost always from plant material) which usually makes them biodegradable and thus susceptible to biologic treatments. Their aforementioned diversity creates the necessity for the research of characterization and treatment options suited for each kind. And among the different possibilities available to deal with these wastes, anaerobic digestion with biomethane production has been a subject of study for a long time. Currently, there is literature available about anaerobic treatment and valorization in general for waste of almost any kind of alcoholic beverage, from beer or wine production [13–15], whisky and other grain distillates [16,17], to even tequila and other agave-based beverages [18,19]. However, there's a void in studies dealing with the wasteof gin production.

The making process of distilled gin as defined in the European legislation involves the mixing of botanicals and aromatics with ethyl alcohol of agricultural origin and water [20] and the distillation of that mixture. The botanicals used may be a mixture ofdifferent components in different proportions according to the producer's recipe (which is almost always a company secret) but the use of juniper berries (*Juniperus communis* L.) is mandatory, as its taste must be predominant by definition [21]. Other common flavoring agents that may be used or not are angelica (*Angelica archangelica* L.) roots, cinnamon (*Cinnamomum verum P.*) bark, caraway (*Carum carvi* L.) and coriander (*Coriandrum sativum* L.) seeds, and orris root (the rhizomes of *Iris germanica L.* and/or *Iris pallida Lam.*). The distillation product is mixed with water to achieve the desired alcoholic strength by volume, which has to be a 37.5% minimum.

A minimum of two flows of wastes can be identified in the making process. One of solid matter, consisting of the spent botanicals and aromatics, which will be wet and impregnated with a certain amount of alcohols, and another of liquids. At first sight, it appears as a mass of mixed items, which can be distinguished under close inspection, showing their nature. The liquid waste can consist of the heads and tails of the distillation process as well as the filtered remains in the pot still and spent wash. Process and cleaning waters can be added to this flow, increasing its volume.

Through the course of their research career, the authors have had previous experience with distilled gin waste, having dealt with the liquid waste of distilled gin production [22]. However, and knowing the working system of a gin distillery, there's no knowledge available on the treatment of the solid waste flow of distilled gin spent botanicals (DGSBs). Thus, it can be considered necessary to add research for the possible treatment and energetic valorization of these wastes to achieve a total overview of the ways to deal with them.

Experience with distilled gin liquid waste showed the possible need for a co-substrate for anaerobic digestion treatment, as the biological process can be hampered by the nature and characteristics of the substrate. The use of brewer's spent grains (BSGs) resulted in an obvious choice for several reasons. The gin distillery was planning to expand their business, creating their own grain alcohols from malts (and thus, reducing their dependence from external sources of ethyl alcohol of agricultural origin), and the characteristics and composition of distiller's spent grains (DSGs) are very similar to those of BSGs, if not identical.

As the authors had previously proven the feasibility of anaerobic digestion treatment for distilled gin liquid waste, it was decided to try anaerobic digestion for solid wastein box-type reactors for both substrates, alone and in co-digestion. Some other research lines showed the influence of the use of a conductive material in the biochemical methane potential for some substrates from alcoholic beverage production [23], so it was decided to use granulated activated charcoal in some experiments.

## 2. Materials and Methods

### 2.1. Substrates

Solid waste from dry gin production was obtained from the distillery of Destilería Siderit S.L. in Puente Arce (Cantabria, Spain). Some of the different kinds of botanicals could be distinguished under the naked eye, as is the case for juniper berries, cinnamon bark or citric peel. Others were not easily distinguishable in the waste mass. The company is reluctant to provide the full recipe of the used flavours and their origin though, if possible, they preferentially acquire them from sources taking in account geographic proximity criteria.

Spent brewer's grains were collected at the facilities of Cervezas Artesanales de Cantabria S.L. (which operates under the brand "Dougall's") located in the municipality of Liérganes (Cantabria, Spain). Those were a Maris Otter barley malt (the usual malt type to brew "british style pale ale" beers; provider not detailed, but the characteristics of these malts are well established, being a product with known and predictable attributes for its use and for the produced beer qualities) with whole or coarsely ground grains.The characteristics of these substrates are shown in Table 1.

**Table 1.** Characteristics of wastes and by-products analyzed in the study (TS and VS are expressed as a percentage of the total mass).

| Wastes and By-Products | TS (%) | VS (%) | TKN (g/kg TS) | P (g/kg TS) |
|---|---|---|---|---|
| Gin spent botanicals | 35.6 | 34.3 | 1.56 | 0.17 |
| Brewers' spent grains | 25.2 | 24.2 | 3.59 | 0.56 |

### 2.2. Inoculum and Conductive Material

The inoculum used was, originally, the anaerobic effluent from a lab-scale digester treating liquid dairy manure and food waste (manure inoculum-MI). Afterward, it was used in the experimental setup for the box-type assays using food waste as a substrate described in Rico et al., 2020 [24]. Thus, it was assumed that the inoculum was well adapted to the experimental conditions in the laboratory and to substrates with a high degree of heterogeneity, as the ones for this study. This inoculum was used for both BMP tests and box-type reactor assays. Characteristics of the inoculum are shown in Table 2.

**Table 2.** Characteristics of the inoculum used in the study (TS and VS are expressed as a percentage of the total mass).

| | Inoculum |
|---|---|
| TS (%) | 2.23 |
| VS (%) | 1.16 |
| pH | 7.9 |
| Alkalinity (g $CaCO_3$ $L^{-1}$) | 12.1 |
| TAN (g $NH_4^+$-N $L^{-1}$) | 2.7 |

As commented in the introduction, the use of a conductive material in addition to the inoculums has been shown as a significant factor maximizing the biomethanogenic performance of the processes for some substrates as shown in Valero et al. (2019) [23]. Thus, Granular Activated Charcoal (GAC, 20 × 40) was used as a conductive material in some experiments.

### 2.3. Experimental Set-Up

The research part focused in the feasibility of waste treatment and their energetic valorization through anaerobic digestion follows the methodology of obtaining the maximum possible methane production for the tested substrates in optimum conditions and for as much time as the necessary for the depletion of the nutrients in the substrate by means of BMP tests. Biogas and biomethane production in similar conditions to real scale functioning digesters are attained through experimentation with lab-scale box digesters. The closest the attained values are to the ones reached in the BMP tests for the substrate, the more suited the substrate is for anaerobic digestion. A biogas production above 75% of the attained BMP values can be considered as very good.

#### 2.3.1. BMP Experiments

All batch experiments were conducted in triplicate in anaerobic 250 mL serum bottles capped with rubber septum sleeve stoppers. Bottles were filled with the amount of substrate containing 0.5 g VS and the amount of inoculum to provide an inoculum to substrate ratio of 2 (based on volatile solids). Blanks were also tested. After filling the bottles, nitrogen was flushed to remove the oxygen in the headspace of the bottles and thereafter placed in an incubator at 38 °C. All the reactors were manually stirred once a day. The test was stopped for each substrate when methane production was negligible in all the samples. Results are expressed as means subtracting methane production from the blanks. Once the experiment was stopped, the reactors were opened to measure the pH, redox potential and VFA in the effluents.

#### 2.3.2. Box Digesters

Two sets of box digesters, laboratory scale, were used. They were made of 304 stainless steel and consisted of two separate airtight-sealed compartments, one of them containing the feedstock (box tank) and the other for the liquid inoculums (percolate tank). Both tanks were provided with thermostat-controlled electric heating blankets that covered the externalwalls and floor of the box tank and the externalwalls of the percolate tank in order to keep stable operation temperatures in the optimum range for the selected biomethanogenic processes (in this case and as the experiments were performed at the mesophilic range, 36–38 °C).

The box tank compartment works as a solid substrate digester. It is 25 cm wide, 50 cm long, and 25 cm high, with a volume of 31.2 L (21.0 L of useful volume). The bottom has a 1.5% slope to facilitate percolate drainage and the compartment is provided with a physical barrier to contain the solid feedstock. The physical barrier has several openings at the bottom to allow the flow of the percolate, which is gathered at an opening in the bottom that is the opening of the percolation line.

The percolate tank is 25 cm wide, 25 cm long, and 23.5 cm high, provided with a conical bottom, having a volume of 16 L (12 L of useful volume). It works as an inoculum storage tank as well as a liquid digester for the nutrients and chemical compounds washed from the feedstock in recirculation operations.

The box and the percolate tanks were constructed with the indicated size in such a way that in the case of needing a higher or lower liquid inoculum to feedstock ratio, both the tanks could be filled with variable amounts of both materials. Both the compartments are equipped with temperature sensors (bimetallic thermometer, 15 cm stem length, 0–80 °C).

Both tanks in each set were connected by two different lines: a percolation line, through which the liquid in excess in the box tank could percolate by gravity to the percolate tank and which could be closed allowing the box tank and feedstock to be flooded; and a recirculation and distribution line, through which the liquid inoculum stored in the percolate tank could be pumped (by a time controlled peristaltic pumpwhich was set to pump an instant flow of 2 L $min^{-1}$)and distributed at will over the feedstock in the box tank through a sprinkler system(which consists in three perforated pipes in parallel). Both lines were set in a way that allowed only liquids to go from one compartment to the other while making impossible a transmission of the biogas produced in each tank. The compartments were provided with independent openings through which created biogas could be collected in gas bags in order to assess thegas composition and production amount in each chamber.

A detailed explanation and sketch of the system, as well as its mode of operation, can be found in Rico et al., 2020 [24].

### 2.4. Analytical Techniques

The biogas and methane production in the BMP tests was measured by the manometric method as described in Valero et al. (2016) [25]. Headspace pressure was measured in the headspace of the reactors through the septum with a syringe connected to a digital pressure transducer with silicon measuring cell (ifm, Germany-type PN78, up to 2000 mbar). Biogas production in each box digester set was measured connecting the different gas bags to a liquid displacement system device. Biogas samples from the BMP testsand box digesters were analyzed on a 2m Poropak T column in an HP 6890 gas chromatograph (GC) system (Agilent Technologies, Inc. 2850 Centerville Road Wilmington, DE 19808-1610 USA) with helium as the carrier gas and a TCD detector. The methane volumes are expressed at 0 °C and 1 atm in dry conditions. VFA's were determined using an HP6890 GC (fitted with a 2 m 1/8-in glass column, liquid phase 10% AT 1000, packed with solid-support Chromosorb W-AW 80/100 mesh. Nitrogen was used as the carrier gas at a flow rate of 14 mL/min, and a FID detector was installed. The Higher Calorific Value (HCV) of both substrates was evaluated using a Parr calorimeter model 1341EE (Parr Instrument Company211 Fifty-Third StreetMoline, IL 61265-1770 USA) provided with a 1108 Oxygen bomb and a 2901 Ignition unit, both from the same manufacturer. Total Solids (TS), Volatile Solids (VS), chemical oxygen demand (COD), total ammonia nitrogen (TAN) and bicarbonate alkalinity were analyzed according to Standard Methods (APHA, 1998).

### 2.5. Data Analysis

Statistical significance was tested by ANOVA analysis, complemented with mean value comparison using Tukey's HSD tests. Statistical significant difference was analyzed for data related to the relationship between methane production and substrate composition (proportion in weight of each co-substrate in the sample) was determined at a threshold $p$-value of less than 5%.

## 3. Results and Discussion

### 3.1. BMP Tests

A series of biomethanogenic potential tests (BMP) was performed, following the procedures and parameters proposed in Holliger et al. (2016) [26]. Keeping an inoculum to substrate ratio of 2 based on volatile solids, both substrates were tested, as well as a number of mixtures of the substrates.

Due to the heterogeneous nature of DGSBs, a significant sample of the matter (1000 g, making sure by visual inspection that the different components were present, in the same apparent proportion as shown in the bulk sample) was ground and blended until its particles were small and uniform enough to the naked eye. BSGs were also ground coarsely. Milling of the samples was performed taking into account the need of not overheating the samples, with the consequent loss of volatile compounds and water and alcohol contents in the original substrates.

The different mixes of substrates used in this series of tests were performed keeping in mind that both substrates (but mainly DGSBs) have a complex chemical composition, and so interactions and reactions could be expected throughout the biological process, not only because of the original composition but also because of the different nature of by-products created through the process in the decomposition of the original ones. Thus, the series was planned from samples 100% DGSBs (which we named M1) to 100% BSGs (which we named M6), with intermediate steps changing the mixture proportions in 20% (by original weight) each. This way, the tests performed with a mixture of 80% DGSBs and 20% BSGs were the M2 samples, the ones with 60% DGSBs and 40% BSGs were the M2, and so on. The results of the different series of BMP tests can be seen in Figure 1.

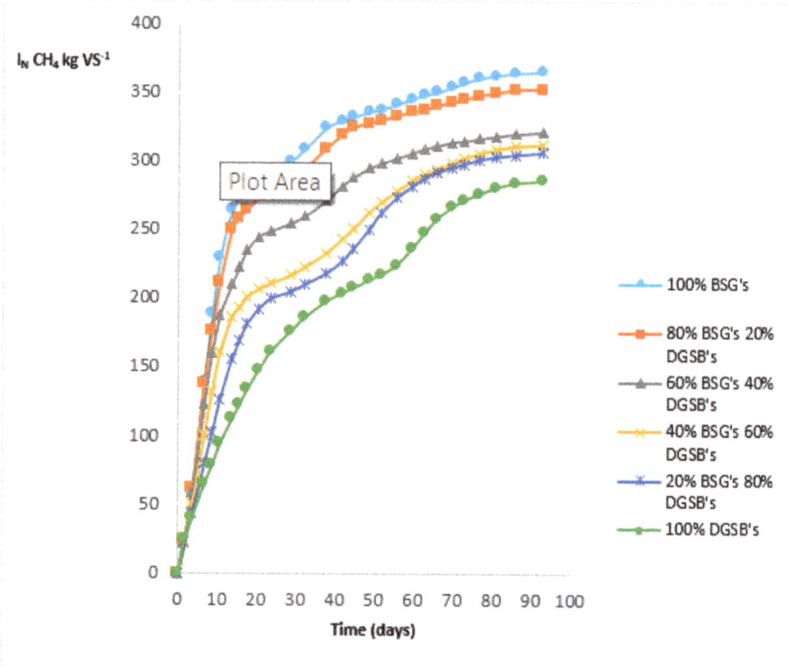


**Figure 1.** Total specific methane production in $l_N$ $CH_4kgVS^{-1}$ from the BMP tests (SD for each sample: M1 (100% DGSBs), 12; M2 (80% DGSBs, 20% BSGs), 15; M3 (60% DGSBs, 40% BSGs), 19; M4 (40% DGSBs, 60% BSGs), 23; M5 (20% DGSBs, 80% BSGs), 15; M6 (100% BSGs), 22.

Two things become apparent from the evolution of the accumulated biogas yield. Firstly, the time needed for the total exhaustion of the biological process is very high. This might be the result of the usage of substrates with a complex nature, a high content in cellulose and lignin as well as other substances, and with a complex degradation through which intermediate compounds (which can pose difficulties for the process) are created. There is a correlation ($p < 0.05$) between the speed of the process and the final biogas yield with the content of DGSBs. In effect, the higher the proportion of DGSBs in the sample, the lower the slope of the yield curve on the first stages of the process (on the first two weeks) and the lesser the final biogas production (it should be reminded that the total content in VS in each sample was the same, regardless of the proportions of the co-substrates used). It was expected a high degree of variability in samples with a high DGSBs content due to the heterogeneous nature of the substrate and the small size of the sample, which implies differences in the presence and proportion of some components in each sample.

Secondly, in relation to the shape of the curves throughout their evolution with time, there is a first phase in which biogas is created normally, with a daily yield that reduces with time as the available substances for anaerobic digestion are reduced in the batch process. However, in all cases and after a lag in the process appears, there is a second relative increase in biogas production, appearing as an inflection point. This second relative increase appears in all cases regardless of the composition of the sample and proportions in the mixture of DGSBs and BSGs. However, it is to be noted that, the higher the content of DGSBs, the later this inflection point appears throughout the experiment. It is also noteworthy that this phenomenon is more remarkablewith the higher content in DGSBs of the sample, at least apparently. Following the inflection point, the biogas yield follows a similar path to the previous state, with its slope decreasing with time and the depletion of available feed substances for the methanogenic biomass.

Several hypotheses can be used to explain this behavior. One of them could be the aforementioned one of the production of intermediate by-products throughout the process, which could lag it with their accumulation. Eventually, specialized biomass that could deal with those by-products could develop and start decomposing them. For BSGs it is known that inhibition can appear due to the creation of p-cresol and other phenolic by-products throughout the process [27]. Due to the complex nature of DGSBs, a similar behavior could be also expected. Another hypothesis is that, due to the nature of the substrates, the hydrolysis phase in the anaerobic digestion process could take longer for substrates high in lignin and other complex matter as DGSBs but, once the mechanism is finally accomplished and activated, it translates in a sudden increase in available feed for biogas production. Finally, it can't be discarded that the inoculum, nevertheless it is supposed to be well adapted and able to cope with almost any feedstock, while starting the process with good efficiency dealing with the easiest to digest substances, at a certain point needs some time and biomass adjustment to cope with other more complex substrates (especially with DGSBs). Once the adaptation is adequate and the biomass and microbial communities are adjusted, there's an increase in biogas production due to the increase in the ability to use the feedstock by the rearranged biomass community.

In the end, it is far beyond the possibilities of the researchers and the scope of this paper to determine which is the correct hypothesis (or combination of them) to explain the attained results. From a practical point of view, the BMP results give a hint that it can be expected that the biological process, applied to laboratory conditions and standard digestion systems, could be difficult and take a longer time than what could be desirable.

The specific methane production values attained of 3659 $l_N$ $CH_4kgVS^{-1}$ for BSGs and 2863 $l_NCH_4$ $kgVS^{-1}$ for DGSBs are in line with those attained in Montes and Rico [28].

### 3.2. Box-Type Digester Assays

Two sets of two experiments in parallel were performed. In the first roll of experiments, which we'll call E1 and E2, only DGSBs were used as substrate. In experiment E1, the box tank for solid substrate was filled with 9.97 kg of feedstock, while 10 L of inoculum were put in the percolate tank. The amounts of feedstock and inoculum for experiment E2 were 10.006 kg and 10 L, respectively. In this experiment, GAC (20 × 40) was added to the percolate tank with the purpose to assess the possible influence of the presence of a conductive material. To avoid the possible interference with the pumping recirculation system and the clogging of pipes and distribution devices, the charcoal was put in a cage-like container, which allowed the flow of inoculum and percolate to pass through and get in contact with the conductive material.

On the first day, the substrates were inoculated by sprinkling 3 L of inoculum (in one run) in each digester and letting them rest, allowing them to reach mesophilic conditions. The same operation was performed on the following two days. On the third day, recirculation was increased to a total of 4.5 L in three runs (of 1.5 L each). And on the fourth day, it was observed that biogas production, which had been increasing both in the box digesters and in the percolate tanks in the previous days, had plummeted. It was detected that Volatile Fatty Acids (VFA's) COD in the percolate tank had experienced a sudden increase (reaching 13.1 and 12.77 g/L respectively in E1 and E2) so inoculum recirculation was stopped.

On the following days, the efforts were focused on the recovery of the inoculums from the failure by acidification. All actions performed (addition of a total of 15% fresh inoculums by volumeand of calcium bicarbonate on several occasions) proved worthless. VFA's COD kept increasing reaching 40.98 and 34.11 g/L in each experiment. Biogas production disappeared in percolate tanks and was residual in box digesters, being observed (but not measured) the presence of $H_2$ in the latter. Finally, after 15 days from the beginning of the experiments, it was decided to stop and reevaluate decisions and the course of actions.

On the following days, two new batches of inoculum were acclimatized by progressively adding 2 L of inactivated inoculum from the previous run of experiment E1 and with a VFA's COD of around 41 g/L to 9 L of fresh inoculum each. Biogas production in the new batches was observed and after 21 days the adaptation seemed satisfactory, so these new batches of 11 L of inoculums each were used in a new run to end the experiments with the partially digested substrate.

The digesters were again inoculated by sprinkling another 3 L of the new inoculums and allowed to rest and reach the mesophilic range of temperatures for the rest of the day. After that, a 1 L recirculation was performed. Biogas production started, but on the following day it was highly reduced, so recirculation was stopped for 10 days while biogas production and VFA's COD in the percolate tanks were controlled. After that and since VFA's COD had been highly reduced, recirculations started again, starting with 0.5 L in two runs (of 0.25 L each) per day. After a week and as the process stabilized, the volume of percolate recirculation was progressively increased, first with another run of 0.25 L (totalling 0.75 L) at which point, the box digesters (which had been inactive) started showing activity and biogas production. Percolate recirculation was increased in the following days as VFA's COD was kept under control and its amount was below 7 g/L. Thus, after 12 days the recirculation volume was 12 L spared into 4 runs along the day. After that, and as the reactor setting allowed the closing of the return line and the accumulation of percolate in the box reactor (acting as a percolating reactor system), recirculation was substituted by daily "substrate floodings". The inundation time was also increased along time, starting from 20 min per day, until reaching 2 h by the end of the experiment and with the feedstock showing depletion.

The experiment was called to an end after 49 days of this second run. The substrate was showing signs of exhaustion and, while still producing biogas, daily yield decreased no matter how aggressive the reinoculation system adopted was. Accumulated methane production along time in this recovery run in both experiments can be seen in Figure 2.

The final specific methane productions in E1 and E2 were 119.14 and 125.50 $l_N$ $CH_4 kgVS^{-1}$, respectively. That amounts to 41.6% and 43.8% of the results attained in the BMP tests. It has to be said that, as the experiments went, a lot of the biomethanogenic feed substances in the substrate might have been lost in the first runs wasted percolate and dissipated or aerobically decomposed during the time the second batch of inoculums was in the adaptation process. Both experiments running in parallel showed very similar behavior, which allowed us to perform the same actions throughout the experimentation process. While the experiment with the conductive material had slightly better results in terms of VFA's COD reduction on the first stages and in final biogas production, the differences in performance in both experiments were negligible.

Two more experiments were performed. In this second run, it was decided to go on the cautious side. Thus, the box reactors were filled with just 8 kg of substrate to keep a lesser substrate/inoculum ratio. The substrates assayed were BSGs in E3 and a mixture consisting of 6.4 kg of BSGs and 1.6 kg of DGSBs (thus, 80% BSGs and 20% DGSBs) in E4. The percolate tanks were filled with 10 L each of the percolate used in the previous experiment and which was now supposed to be perfectly adapted to the substrates after the previous runs.

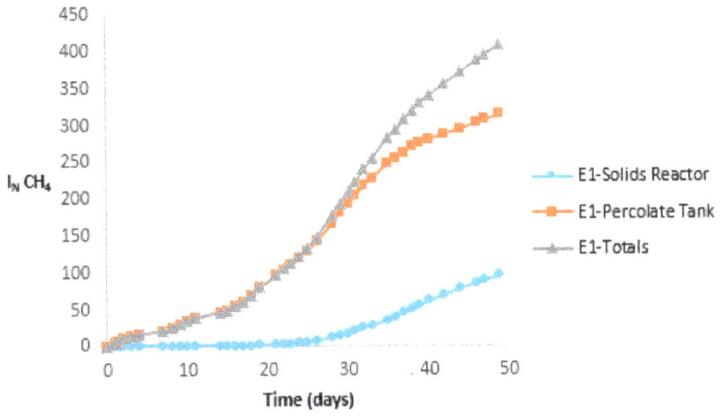


(a)

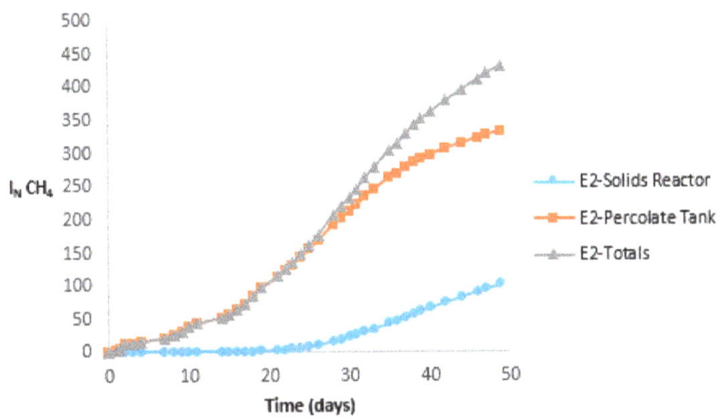


(b)

**Figure 2.** Methane production through time in $l_N$ $CH_4$ in experiments E1 (DGSBs as substrate, subfigure (**a**) and E2 (DGSBs as substrate with GAC used in the percolate tank, subfigure (**b**)) and the second phase of the experiment.

After the first inoculation of the feedstocks with a 3 L sprinkling of inoculum from the percolate tank and leaving the experiments to rest and reach mesophilic conditions for the first day, the adopted recirculation rate was of a total of 0.75 L distributed in 3 runs throughout the day. From then on, recirculation was adjusted according to biogas production and the evolution of VFA's COD in the percolate, which was carefully controlled throughout the experiments. In fact, after the 7th experimentation day, while recirculation was increased in E3 to 1 L distributed in 4 runs, it had to be temporarily stopped in E4 for 2 days and, after that, reduced to a single 0.25 L run per day as VFA's COD showed a steady increasing trend reaching levels around 16 g/L, which we deemed as dangerous. On the 13th day, when conditions were apparently stabilized in E4 and recirculation was increased to 0.5 L distributed in 2 runs, the sudden increase in VFA's COD (which reached 17.64 g/L) forced to stop temporarily recirculations in the experiment and let it rest again for two days and start again with a single recirculation of 0.25 L. At the same time, conditions in E3 had allowed to gradually increase the recirculation rate up to 2.4 L distributed in 6 runs.

After those first two weeks, conditions got permanently stabilized in E4, which allowed a gradual increase in the recirculation rate, performing the same operational way as in E3 but only with a 14-day lag. Thus, at the beginning of the 4th week, while inundation cycles were starting to be performed in E3, the recirculation rate in E4 was of 9 L spared in 6 daily runs.

The experiments were called to an end after 49 days. Accumulated methane production along time in both experiments are shown in Figure 3.

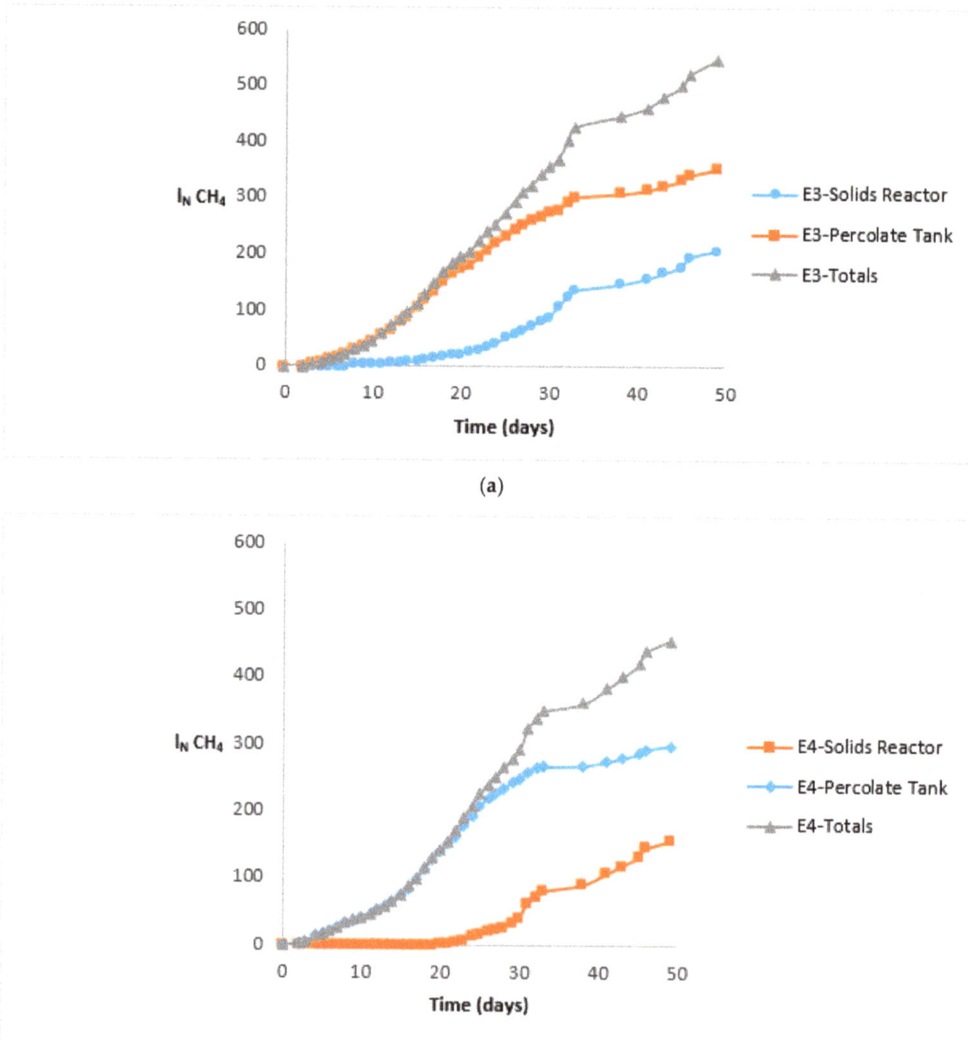


**Figure 3.** Methane production through time in $l_N$ $CH_4$ in experiments E3 (BSGs as substrate, subfigure (**a**)) and E4 (a mixture of 80% BSGs and 20% DGSBs by weight, subfigure (**b**)).

The final specific methane productions in E3 and E4 were 295.06 and 217.57$l_N$ $CH_4$kgVS$^{-1}$ respectively. That amounts to 80.6% and 61.6% of the results attained in the BMP tests and can be considered quite satisfactory in terms of the translation of biomethanogenic potential to work conditions in E3, not so much in E4. As a comment, while biogas production in the percolate tank was the main methane source in all experiments, only in E3 (using 100% BSGs as feedstock) was there a steady production in the box-type solids reactor from the start of the experiment. In all cases, biogas production in the solids reactor increased when the inoculation system was changed from sprinkling to substrate inundation.

When the experiments were finished and the solid digestatecould be observed, it was noticeable that, while BSGs appeared fairly decomposed and its volume had experienced a very noticeable reduction, where DGSBs had been used, some individual distinguishable items could be seen quite unaltered. This was the case with juniper berries, for example.

### 3.3. Calorific Value

Complementary to the previously described tests and assays, it was estimated useful to assess the calorific value of the substrates to complement the studies about their potential value as energy sources. Thus, both substrates were tested for their higher heating value (HHV) using a Parr oxygen bomb calorimeter.

The attained results were of an HHV of 19.95 MJ/kg for DGSBs and 18.51 MJ/kg for BSGs, respectively. These values are typical of materials with a composition high in lignin and cellulose as will be discussed in the discussion section.

### 3.4. Discussion

Several conclusions can be extracted after the experimentation process. First, the importance of having a well-adapted and strong inoculum can never be underestimated.

BSGs have resulted in a good feedstock for anaerobic digestion processes. While the usual way to dispose of both brewers' and distillers' spent grains is to use them as feed for livestock [29,30] (with the limitations for DSGs of copper toxicity, especially for ovine and caprine livestock) or even for human consumption [31], a lot of interest has arisen lately for their use in biogas production [32,33], with a good final biomethane production and, as a by-product, a digestate with a good volumetric reduction and which could still be useful in agriculture as a fertilizer or as an organic amendment for soils [34]. This fact has lead to the appearance of several studies and works focused on its use. In this experimental process, the results attained were fairly good. The only objection could be the long time needed throughout the experiment to attain good results with a good amount of final biogas production. In that sense, it should be commented that, after the first experiences and failures with the other substrate, the experimentation was performed on a very cautious side. The values of VFA's COD accumulated in the percolate tank never reached levels which, after the experience, could be considered dangerous for the process. Thus, the general process could be accelerated using a more intense recirculation in terms of volume and frequency (always bearing in mind it should be a progressive process). Further studies would be needed to adjust and optimize the operational actions in order to speed up the process while keeping it on the safe side to avoid process failure.

On the other side, DGSBs have resulted in a problematic feedstock. This could be the product of the substrate nature in itself. As commented in the introduction, juniper berries are a mandatory ingredient for distilled gin by definition, according to the EU regulations for alcoholic beverages [21]. Juniper galbuli (berries) essential oils have a complex composition with a large number of chemical compounds [35] and have shown anti-microbial and anti-fungal effects [35,36]. While it could be objected that other distilled gin producers could use different recipes and mixes of botanicals, affecting the repeatability of similar experiments, the obligatory nature of the use of a high proportion of juniper berries in the botanicals mixture allows one to think that results similar to ours would be attained, unless an unusually high inoculum-to-substrate ratio (normally not feasible in real-scale dry batch anaerobic digestion reactors) were used. Other frequent distilled gin ingredients, suchas angelica roots and seeds or cinnamon bark, have also shown anti-microbial effects [37,38]. As a first measure to deal with the treatment of these products, procedural changes in the distillation process could be adopted. The use of "infusers" (mesh containers made of an inert material which don't transmit unexpected flavors to the distillate while allowing free contact between botanicals and the liquid phase of water and ethanol) for each ingredient, for example, could make easier to separatethe different species and fractions of botanicals after distillation is completed. This way, the different types of waste could be treated accordingly to their own characteristics, enabling the extraction of valuable chemicals from the separate fractions and using the less problematic products to be used in anaerobic digestion.

After these experiments, it could be supported that DGSBs are not suited for dry batch anaerobic digestion, as it is usually performed in current standard appliances; and its addition as a co-substrate, whileitshould be analyzed on a case-to-case basis, could result in process problems, especially in the initial phases. Some experiences can be found in literature about the problems of dealing with a recalcitrant substrate associated with the alcoholic beverage production industry [39]. In that sense, UASB systems have proven robust and adequate to deal with liquid compounds that could be deleterious in other types of anaerobic processes. That suggests that it could be possible to use hybrid UASB-Dry batch box-type systems like the ones described in Panjičko et al. [40] as the systems of choice which could enable the anaerobic digestion of DGSBs.

On the other hand, both products have shown a good performance as biomass fuels, with a good HHV, very similar to that of wood in the case of DGBs (with net calorific values of 12.5, 14.7, 17 and 19 MJ/kg for wood chips, stacked log wood, wood pellets, and oven dried log wood respectively) [41] and, in general, in line with that of other materials from agri-food production rich in lignin and cellulose (among which we can cite HHV's of 17.1 MJ/kg for wheat straw and for banana waste, 18.17 MJ/kg for sugarcane bagasse, or 19.3 and 20.2 MJ/kg for hazelnut and almond shells respectively) [42]. While extraction of their water content should be necessary for their use as fuels, their nature and appearance would only make necessary a minimum grinding in the case of DGSBs, not even so for BSGs, previously to a pelletization to attain commercial characteristics as a biomass fuel that could be commercialized as it is, or used for heat production in the industrial premises. Though these heating values might appear as modest compared to those of traditional fuels (with HHV's of 46.03 and 45.56 MJ/kg for diesel and gasoline) or those of products originated in the valorization of plastic waste (with HHV's of 44.5, 44.22, 44.63 and 40.17 MJ/kg for pyrolysis generated oils and 99.83, 99.46, 105.04 and 121.18 MJ/kg for pyrolysis generated gas for high-density polyethylene, low-density polyethylene, polypropylene, and polystyrene, respectively) [43,44], it should be taken in account the renewable origin of the biomass fuels and their significantly lower carbon footprint under an environmental point of view.

## 4. Conclusions

This work has dealt with two types of waste from alcoholic beverage production, focusing on their energetic valorization. BSGs have proven to be a good material for this purpose, either as feedstock for anaerobic digestion processes using technologies commonly used in dry batch digestion or by itself as a pelletisable biomass fuel, attaining good heating values similar to those from other natural feedstock rich in lignin and cellulose. This fact opens new ways of dealing with them, other than their traditional use as feed for livestock.

The other material, however, has shown to have less potential. While technically feasible, dry batch anaerobic digestion of DGSBs seems to be a very delicate process, especially in the first phases, where there might be a strong presence of antimicrobial chemical compounds and high accumulations in VFA's COD. Either an effective previous treatment might be required or the adoption of hybrid systems. In any case, it seems to be a difficult to digest substrate which, when used in co-digestion, might hamper and delay the general process. It has good potential as a pelletisable biomass fuel with good calorific power, however. And it is worth noting that those chemical compounds in the substance could be extracted and used, transforming waste into a valuable raw material.

**Author Contributions:** Conceptualization, C.R. and J.A.M.; methodology, C.R. and J.A.M.; formal analysis, C.R.; investigation, C.R. and J.A.M.; data curation, C.R. and J.A.M.; writing—original draft preparation, J.A.M.; writing—review and editing, J.A.M.; supervision, C.R. All authors have read and agreed to the published version of the manuscript.

**Funding:** This research received no external funding.

**Institutional Review Board Statement:** Not applicable.

**Informed Consent Statement:** Not applicable.

**Acknowledgments:** The authors want to thank the management and staff of Destilería Siderit S.L. and of Cervezas Artesanales de Cantabria S.L. (Dougall's) for their support, providing information and samples of waste and by-products used in this article.

**Conflicts of Interest:** The authors declare no conflict of interest.

**Author Contributions:** While I. S. units have been used in this work, some others (as the standard atmosphere atm, liters L, grams per liter g/L and so on) have been kept due to figure representativity and normal use in works dealing with similar matters. The unit $l_N CH_4 kg\ VS^{-1}$, liters of methane in standard conditions per kilogram of volatile solids in the feedstock, has been defined as the unit of choice to express methane yield since Holliger et al. (2016). To translate into SI units ($m^3{}_N CH_4 kg\ VS^{-1}$), the values should be multiplied by $10^{-3}$.

## References


1. Fillaudeau, L.; Blanpain-Avet, P.; Daufin, G. Water, waste water and waste management in brewing industries. *J. Clean. Prod.* **2006**, *14*, 463–471. [CrossRef]
2. Krishnamoorthy, S.; Premalatha, M.; Vijayasekaran, M. Characterization of distillery waste water—An approach to retrofit existing effluent treatment plant operation with phycoremediation. *J. Clean. Prod.* **2017**, *148*, 735–750. [CrossRef]
3. Janhom, T.; Wattanachira, S.; Pavasant, P. Characterization of brewery waste water with spectrofluorometry analysis. *J. Environ. Manag.* **2009**, *90*, 1184–1190. [CrossRef]
4. Cavert, W. The Environmental Policy of Charles I: Coal Smoke and the English Monarchy, 1624–1640. *J. Br. Stud.* **2014**, *53*, 310–333. [CrossRef]
5. Environmental Protection Authority of the Government of South Australia. EPA Guidelines for Wineries and Distilleries [PDF]. Issued January 2004, Revised April 2017. ISBN 1-876562-66-8. Available online: https://www.epa.sa.gov.au/files/4771373_guide_wineries.pdf. (accessed on 10 August 2021).
6. Agriculture and Resource Management Council of Australia and New Zealand and the Australian and New Zealand Environment and Conservation Council. Effluent Management Guidelines for Australian Wineries and Distilleries [PDF]. 1998. Available online: https://www.waterquality.gov.au/sites/default/files/documents/pub6-effluent-wineries-distilleries.pdf. (accessed on 10 August 2021).
7. NetRegs UK. GPP 29 Microbreweries and Distilleries Guidance for Small Scale Breweries and Distilleries (March2019). Available online: https://www.netregs.org.uk/media/1687/gpp29-micro-breweries-and-distilleries.pdf (accessed on 10 August 2021).
8. Olajire, A.A. The brewing industry and environmental challenges. *J. Clean. Prod.* **2020**, *256*, 102817. [CrossRef]
9. Kubule, A.; Zogla, L.; Ikaunieks, J.; Rosa, M. Highlights on energy efficiency improvements: A case of a small brewery. *J. Clean. Prod.* **2016**, *138*, 275–286. [CrossRef]
10. Sturm, B.; Butcher, M.; Wang, Y.; Huang, Y.; Roskilly, T. The feasibility of the sustainable energy supply from bio wastes for a small scale brewery—A case study. *Appl. Therm. Eng.* **2012**, *39*, 45–52. [CrossRef]
11. Muster-Slawitsch, B.; Weiss, W.; Schnitzer, H.; Brunner, C. The green brewery concept—Energy efficiency and the use of renewable energy sources in breweries. *Appl. Therm. Eng.* **2011**, *31*, 2123–2134. [CrossRef]
12. O'Shea, R.; Wall, D.M.; McDonagh, S.; Murphy, J.D. The potential of power to gas to provide green gas utilising existing $CO_2$ sources from industries, distilleries and waste water treatment facilities. *Renew. Energy* **2017**, *114*, 1090–1100. [CrossRef]
13. Skornia, K.; Safferman, S.I.; Rodriguez-Gonzalez, L.; Ergas, S.J. Treatment of winery wastewater using bench-scale columns simulating vertical flow constructed wetlands with adsorption media. *Appl. Sci.* **2020**, *10*, 1063. [CrossRef]
14. Toscano, G.; Riva, G.; Duca, D.; Foppa Pedrretti, E.; Corinaldesi, F.; Rossini, G. Analysis of the characteristics of the residues of the wine production chain finalized to their industrial and energy recovery. *Biomass Bioenergy* **2013**, *55*, 260–267. [CrossRef]
15. Da Ros, C.; Cavinato, C.; Bolzonella, D.; Pavan, P. Renewable energy from thermophilic anaerobic digestion of winery residue: Preliminary evidence from batch and continuous lab-scale trials. *Biomass Bioenergy* **2016**, *91*, 150–159. [CrossRef]
16. Goodwin, J.A.S.; Stuart, J.B. Anaerobic digestion of malt whisky distillery pot ale using upflow anaerobic sludge blanket reactors. *Bioresour. Technol.* **1994**, *49*, 75–81. [CrossRef]
17. Goodwin, J.A.S.; Finlayson, J.M.; Low, E.W. A further study of the anaerobic biotreatment of malt whisky distillery pot ale using an UASB system. *Bioresour. Technol.* **2001**, *78*, 155–160. [CrossRef]

18. Buitrón, G.; Kumar, G.; Martinez-Arce, A.; Moreno, G. Hydrogen and methane production via a two-stage processes (H2-SBR + CH4-UASB) using tequila vinasses. *Int. J. Hydrogen Energy* **2014**, *39*, 19249–19255. [CrossRef]
19. Espinoza-Escalante, F.M.; Pelayo-Ortíz, C.; Navarro-Corona, J.; González-García, Y.; Bories, A.; Gutiérrez-Pulido, H. Anaerobic digestion of the vinasses from the fermentation of *Agave tequilana* Weber to tequila: The effect of pH, temperature and hydraulic retention time on the production of hydrogen and methane. *Biomass Bioenergy* **2009**, *33*, 14–20. [CrossRef]
20. Aylott, R.I. Vodka, Gin and Other Flavored Spirits. In *Fermented Beverage Production*, 2nd ed.; Lea, A.G.H., Piggott, J.R., Eds.; Springer: Boston, MA, USA, 2003; pp. 289–308, ISBN 978-0-306-47706-5. [CrossRef]
21. EUR-LEX, Official Website of European Union Law. Available online: https://eur-lex.europa.eu/legal-content/EN/TXT/?qid=1556781080548&uri=CELEX:32008R0110. (accessed on 10 August 2021).
22. Montes, J.A.; Leivas, R.; Martínez-Prieto, D.; Rico, C. Biogas production from the liquid waste of distilled gin production: Optimization of UASB reactor performance with increasing organic loading rate for co-digestion with swine wastewater. *Bioresour. Technol.* **2019**, *274*, 43–47. [CrossRef]
23. Valero, D.; Alzate-Gaviria, L.; Montes, J.A.; Rico, C. Influence of a Conductive Material and Different Anaerobic Inocula on Biochemical Methane Potential of Substrates from Alcoholic Beverage Production. *Waste Biomass Valoriz.* **2020**, *11*, 5957–5964. [CrossRef]
24. Rico, C.; Montes, J.A.; Lobo, A. Dry batch anaerobic digestion of food waste in a box-type reactor system: Inoculum preparation and reactor performance. *J. Clean. Prod.* **2020**, *251*, 119751. [CrossRef]
25. Valero, D.; Montes, J.A.; Rico, J.L.; Rico, C. Influence of headspace pressure on methane production in Biochemical Methane Potential (BMP) tests. *Waste Manag.* **2016**, *48*, 193–198. [CrossRef]
26. Holliger, C.; Alves, M.; Andrade, D.; Angelidaki, I.; Astals, S.; Baier, U.; Bougrier, C.; Buffiere, P.; Carballa, M.; DeWilde, V.; et al. Towards a standardization of biomethane potential tests. *Water Sci. Technol.* **2016**, *74*, 2515–2522. [CrossRef]
27. Sežun, M.; Grilc, V.; Zupančič, G.D.; Logar, R.M. Anaerobic Digestion of Brewery Spent Grain in a Semi-Continuous Bioreactor: Inhibition by Phenolic Degradation Products. *Acta Chim. Slov.* **2011**, *58*, 158–166. Available online: https://scholar.google.com/scholar_lookup?title=Anaerobic%20digestion%20of%20brewery%20spent%20grain%20in%20a%20semi-continuous%20bioreactor%3A%20inhibition%20by%20phenolic%20degradation%20products&publication_year=2011&author=M.%20Se%C5%BEun&author=V.%20Grilc&author=G.D.%20Zupan%C4%8Di%C4%8D&author=R.M.%20Logar. (accessed on 10 August 2021). [PubMed]
28. Montes, J.A.; Rico, C. Biogas Potential of Wastes and By-Products of the Alcoholic Beverage Production Industries in the Spanish Region of Cantabria. *Appl. Sci.* **2020**, *10*, 7481. [CrossRef]
29. Klopfenstein, T.J.; Erickson, G.E.; Bremer, V.R. BOARD-INVITED REVIEW: Use of distillers by-products in the beef cattle feeding industry. *J. Anim. Sci.* **2008**, *86*, 1223–1231. [CrossRef] [PubMed]
30. Stein, H.H.; Shurson, G.C. Board-invited review: The use and application of distillers dried grains with solubles in swine diets. *J. Anim. Sci.* **2009**, *87*, 1292–1303. [CrossRef] [PubMed]
31. Lynch, K.M.; Steffen, E.J.; Arendt, E.K. Brewers' spent grain: A review with an emphasis on food and health. *J. Inst. Brew.* **2016**, *122*, 553–568. [CrossRef]
32. Gonçalves, I.C.; Fonseca, A.; Morão, A.M.; Pinheiro, H.M.; Duarte, A.P.; Ferra, M.I.A. Evaluation of anaerobic co-digestion of spent brewery grains and an azo dye. *Renew. Energy* **2015**, *74*, 489–496. [CrossRef]
33. Weger, A.; Jung, R.; Stenzel, F.; Hornung, A. Optimized Energetic Usage of Brewers' Spent Grains. *Chem. Eng. Technol.* **2017**, *40*, 306–312. [CrossRef]
34. Wang, T.-T.; Wang, S.-P.; Zhong, X.-Z.; Sun, Z.-Y.; Huang, Y.-L.; Tan, L.; Tang, Y.-Q.; Kida, K. Converting digested residue eluted from dry anaerobic digestion of distilled grain waste into value-added fertilizer by aerobic composting. *J. Clean. Prod.* **2017**, *166*, 530–536. [CrossRef]
35. Angioni, A.; Barra, A.; Russo, M.T.; Coroneo, V.; Dessí, S.; Cabras, P. Chemical composition of the essential oils of Juniperus from ripe and unripe berries and leaves and their antimicrobial activity. *J. Agric. Food Chem.* **2003**, *51*, 3073–3078. [CrossRef]
36. Pepeljnjak, S.; Kosalec, I.; Kalodera, Z.; Blažević, N. Antimicrobial activity of juniper berry essential oil (*Juniperus communis* L., Cupressaceae). *Acta Pharm.* **2005**, *55*, 417–422. [PubMed]
37. Joshi, R.K. Chapter 21—Angelica (Angelica glauca and A. archangelica) oils. In *Essential Oils in Food Preservation, Flavor and Safety*; Preedy, V.R., Ed.; Academic Press: London, UK, 2016; pp. 203–208, ISBN 9780124166417.
38. Hersch-Martínez, P.; Leaños-Miranda, B.E.; Solórzano-Santos, F. Antibacterial effects of commercial essential oils over locally prevalent pathogenic strains in Mexico. *Fitoterapia* **2005**, *76*, 453–457. [CrossRef]
39. Harada, H.; Uemura, S.; Chen, A.; Jayadevan, J. Anaerobic treatment of a recalcitrant distillery wastewater by a thermophilic UASB reactor. *Bioresour. Technol.* **1996**, *55*, 215–221. [CrossRef]
40. Panjičko, M.; Zupančič, G.D.; Fanedl, L.; Marinšek Logar, R.; Tišma, M.; Zelić, B. Biogas production from brewery spent grain as a mono-substrate in a two-stage process composed of solid-state anaerobic digestion and granular biomass reactors. *J. Clean. Prod.* **2017**, *166*, 519–529. [CrossRef]
41. Forest Research UK 2021. Typical Calorific Values of Fuels. Available online: https://www.forestresearch.gov.uk/tools-and-resources/fthr/biomass-energy-resources/reference-biomass/facts-figures/typical-calorific-values-of-fuels/ (accessed on 10 August 2021).

42. Gupta, G.K.; Mondal, M.K. Chapter 15—Bioenergy generation from agricultural wastes and enrichment of end products. In *Refining Biomass Residues for Sustainable Energy and Bioproducts*, 1st ed.; Kumar, R.P., Gnansounou, E., Raman, J.K., Baskar, G., Eds.; Academic Press: London, UK, 2020; pp. 337–356, ISBN 9780128189962. [CrossRef]
43. Constantinescu, M.; Bucura, F.; Ionete, E.I.; Ion-Ebrasu, D.; Sandru, C.; Zaharioiu, A.; Marin, F.; Miricioiu, M.G.; Niculescu, V.C.; Oancea, S.; et al. From Plastic to Fuel—New Challenges. *Mater. Plast.* **2019**, *56*, 721–729. [CrossRef]
44. Constantinescu, M.; Bucura, F.; Ionete, R.; Niculescu, V.; Ionete, E.I.; Zaharioiu, A.; Oancea, S.; Miricioiu, M.G. Comparative Study on Plastic Materials as a New Source of Energy. *Mater. Plast.* **2019**, *56*, 41–46. [CrossRef]

*applied sciences*

MDPI

*Article*

# High Purity of α-Lactalbumin from Binary Protein Mixture by Charged UF Membrane Far from the Isoelectric Point to Limit Fouling

**Rosalinda Mazzei [1,*], Anna Maria Szymczak [1], Enrico Drioli [1], Mohamed Al-Fageeh [2], Mohammed A. Aljohi [2] and Lidietta Giorno [1,*]**

1 Institute on Membrane Technology, CNR-ITM c/o University of Calabria, Via P. Bucci 17C, 87036 Rende, Italy; a.szymczak@itm.cnr.it (A.M.S.); e.drioli@itm.cnr.it (E.D.)
2 National Centre for Biotechnology, King Abdulaziz City of Science and Technology (KACST), P.O. Box 6086, Riyadh 11442, Saudi Arabia; mfageeh@kacst.edu.sa (M.A.-F.); maljohi@kacst.edu.sa (M.A.A.)
* Correspondence: r.mazzei@itm.cnr.it (R.M.); l.giorno@itm.cnr.it (L.G.); Tel.: +39-0984-492076 (R.M.); +39-0984-492076 (L.G.)

**Abstract:** Separation and high recovery factor of proteins similar in molecular mass is a challenging task, and heavily studied in the literature. In this work, a systematic study to separate a binary protein mixture by charged ultrafiltration membranes without affecting membrane performance was carried out. α-lactalbumin (ALA, 14.4 kDa) and β-lactoglobulin (BLG, 18.4 kDa) were used as a binary model system. These two proteins are the main proteins of whey, a very well-known byproduct from the dairy industry. Initially, a systematic characterization of individual proteins was carried out to determine parameters (protein size and aggregation, zeta potential) which could influence their passage through a charged membrane. Then, the influence of operating parameters (such as initial protein concentration, pH, and critical pressure) on the UF process was investigated, so as to identify conditions that limit membrane fouling whilst maximizing protein recovery factor and purity. The study permitted to identify process conditions able to fully separate ALA from BLG, with high purity (95%) and recovery factor (80%), in a single UF step. Compared to studies reported in literature, here, the main approach used was to carry out a charged UF process far from proteins isoelectric point (pI) to limit protein aggregation and membrane fouling.

**Keywords:** charged ultrafiltration membranes; α-lactalbumin; β-lactoglobulin; protein fractionation; protein electrostatic interaction; protein ultrafiltration; fouling

**Citation:** Mazzei, R.; Szymczak, A.M.; Drioli, E.; Al-Fageeh, M.; Aljohi, M.A.; Giorno, L. High Purity of α-Lactalbumin from Binary Protein Mixture by Charged UF Membrane Far from the Isoelectric Point to Limit Fouling. *Appl. Sci.* **2021**, *11*, 9167. https://doi.org/10.3390/app11199167

Academic Editor: Carlos Rico de la Hera

Received: 16 July 2021
Accepted: 24 September 2021
Published: 2 October 2021



## 1. Introduction

Fractionation of proteins by ultrafiltration (UF) is efficient when proteins differ in molecular mass by at least a factor of 10. To this aim, various membrane processes were investigated focusing on (i) tuning pH and ionic strength to maximize differences in the hydrodynamic size of particles in solution, (ii) using electrical charged membranes to retain charged proteins [1–4], (iii) diafiltration process [5], and (iv) selective aggregation of proteins using thermal process [6].

The careful control of pH and ionic strength of bulk proteins' solutions might result in an improved performance of protein fractionation by ultrafiltration.

Different protein mixtures were fractionated by charged UF process [7–15] by using both organic and inorganic membranes, achieving maximum transmission of the protein being at its isoelectric point (pI) and high retention of the one that had the same electrical charge as the membrane.

The overall observation was that large increase in protein transmission through membranes can be reached using charged membranes, but the protein at its isoelectric point can create membrane fouling [16–23]. Membrane fouling, in the mentioned conditions, increases when the protein is at the isoelectric point, since protein aggregation easily occurs [24–26].

A good compromise between high selectivity and low fouling must be found. Strategies to overcome the permeability–selectivity trade-off were published by Arunkumar et al. [4], in which a 300 kDa regenerated cellulose membrane positively charged was used for the fractionation of α-lactalbumin (ALA) from a binary protein mixture. Using a two-stage process, 87% of pure ALA was obtained in the permeate with a flux of 170 $L{\cdot}h^{-1}{\cdot}m^{-2}$ [4]. The same work was carried out [27] using milk serum permeate (MSP) as feed solution, obtaining 87% of ALA purity with a three-stage system. Another research group [15] achieved the complete separation of bovine serum albumin (BSA, 66.5 kDa) in one step, starting from a binary protein mixture with lactoferrin (LF, 78 kDa) by using diafiltration process and charged membranes.

Despite the research efforts and knowledge promoted on the separation of proteins with similar size by charged UF membrane processes, some aspects (such as preventing irreversible fouling) need further insights and improvements.

The novelty of this work, with respect to the current literature, is to promote proteins separation with similar molecular weight, but in conditions in which irreversible fouling can be prevented or limited. Besides, a deep study of protein aggregation state and charge density, when both proteins bore same charge of the membrane, was carried out for the first time, in order to have high purity of the separated protein in a single stage, currently carried out by multistage processes [4,27]. In order to identify operating conditions that might prevent irreversible fouling, as well as guarantee high purification efficiency, we speculated that using binary protein mixture in bulk conditions far from their isoelectric points and having both proteins with the same electrical charge as the membrane would promote the target improvements, since electrical repulsion and then low adsorption between proteins and membrane is promoted. We also aimed at achieving purification in a single step. To easily compare results with literature data, the well-studied β-lactoglobulin and α-lactalbumin binary mixture was used as a model system. Charged regenerated cellulose ultrafiltration membrane was applied.

The work started with a systematic characterization of single protein solutions to determine parameters which could affect their separation (zeta potential, protein size, and tendency to aggregate). The abovementioned characterization at pH around 3 was carried out, since both proteins (ALA IP: 4.4; BLG IP: 5.2–5.4) are positively charged; this limits the proteins/positively charged membrane interaction during UF and then irreversible membrane fouling.

Then, the influence of operation variables (initial binary mixture protein concentration, pH, critical pressure) to limit fouling during charged UF process and to maximize the difference between the two proteins was studied. The obtained results were then used to identify conditions in which to carry out UF process in concentration mode using binary protein mixture.

## 2. Materials and Methods

### 2.1. Chemicals

Phosphoric acid ($H_3PO_4$) (Fluka, Milan, Italy) and sodium phosphate monobasic anhydrous ($NaH_2PO_4$) (Sigma Aldrich, Milan, Italy) were used to prepare buffer solutions; NaCl (Sigma Aldrich) was used to keep constant ionic strength to 0.1 M. Regenerated cellulose flat membranes of 30 kDa nominal molecular weight cut-off (NMWCO) (Millipore) were used. The structure of this kind of membranes is asymmetric. The membrane surface area was $1.25 \times 10^{-3}$ $m^2$. Prior to permeability test, membranes were first washed with ultrapure water (PurelabTM Classic, UF) to remove soluble additives normally used to preserve the membranes. The membrane was mounted in a homemade cross-flow ultrafiltration system (glossy side toward solution) and rinsed by filtering ultrapure water for 10 min at 170 kPa. BLG (cod. L3908) and ALA (cod. L6010) were purchased from Sigma Aldrich (Milan, Italy). To study protein size and to carry out ultrafiltration tests around pH 3, 25 mM sodium phosphate was prepared with phosphoric acid ($H_3PO_4$) (Fluka, Milan, Italy) and sodium phosphate monobasic anhydrous ($NaH_2PO_4$) (Sigma Aldrich, Milan, Italy).

### 2.2. Protein Quantification

The bicinchoninic acid protein assay kit (BCA, QuantiPro™ BCA Assay Kit, Sigma-aldrich, Milan, Italy) used to measure protein concentration (1–20 µg/mL) was purchased from Sigma-Aldrich (Milan, Italy). In solutions in which both ALA and BLG were present, the protein amount was calculated by one-dimensional SDS-PAGE electrophoresis on precast protein gel (NuPAGE®Novex® 4–12% Bis-Tris Gels, 1.0 mm, 1 well, ThermoFisher scientific, Monza, Italy). The gel has a continuous 4 to 12% gradient gel zone. The buffer system used was MES (50 mM MES, 50 mM Tris Base, 0.1% SDS, 1 mM EDTA, pH 7.3). Sample treatment: 8 µL of sample, 5 µL of Nu PAGE LDS sample buffer (4×), and 2 µL of Nu PAGE reducing agent (10×) were added to 5 µL of water to a final volume of 20 µL. Each sample was loaded onto a separate lane of the gel containing 20 µL of sample. The gels were stained with silver staining (Sigma-Aldrich, sensitivity: low nanogram range). In order to evaluate the mass of the protein, gel images were captured by scanner and analyzed by GelQuant Express Analysis Software (Life Technologies, Monza, Italy), which facilitate identification of both molecular weights (MW) and concentration of each band on the gel. The MWs of the proteins of unknown samples were calculated from the logarithm curve fitting, which relate the standard MWs with the relative mobility as pixel position by using calibration kit proteins.

### 2.3. Protein Size and Charge Measurement

Size measurements of protein aggregates and determination of molecular weight, as well as protein charge, was carried out by Zetasizer Nano ZS (Alfatest, Milan, Italy). The Zetasizer system determines the particles size by measuring the Brownian motion of the particles in a sample using dynamic light scattering (DLS). The size range is from 0.3 nm to 10 µm. DLS provides a fast, noninvasive, and sensitive method to determine the size of a protein [28]. The molecular weight was determined by static light scattering (SLS) measuring the sample at different concentrations and applying the Rayleigh equation, which describes the intensity of light scattered from a particle, in static conditions, in solution. The protein charge was measured using a combination of two measurement techniques: electrophoresis and laser Doppler velocimetry. This method measures how fast a particle moves in a liquid when an electrical field is applied. The velocity of particle measured, and the electrical field applied, considering viscosity and dielectrical constant of the solution, work out the zeta potential. A total of 15 different consecutive measurements were carried out for single protein solution at the different pH, and the error reported is the one obtained from the average and standard deviation between the measurements carried out at 25 °C.

### 2.4. Experimental Setup

Functionalized membranes were assembled in a cross-flow ultrafiltration cell system (schematic representation in Figure 1). The system was composed of a stainless-steel module, which contained the membrane, two pressure gauges (Wika, Klingenberg, Germany) to measure inlet and outlet pressure, a feed tank, and a peristaltic pump (Masterflex, Chongqing, China) to feed the protein solution to the membrane. Before ultrafiltration experiments, the membrane was characterized by measuring the pure water permeance ($m \cdot Pa^{-1} \cdot s^{-1}$) (Lp). The permeate flux as function of time at different transmembrane pressure (TMP) values was measured; the steady-state values of flux were then plotted versus TMP, and, from the slope of the straight line obtained, the pure water permeance was calculated from Equation (1).

$$J = Lp \cdot \Delta P \quad (1)$$

where J is the permeate flux ($L \cdot h^{-1} \cdot m^{-2}$), and $\Delta P$ is the TMP (bar). The reason for measuring this parameter pertains to the need to check the initial membrane performance as reference for subsequent use of membrane after protein fractionation experiments. Furthermore, pure water permeance was necessary to evaluate hydraulic resistances, adsorption, and irreversible fouling caused by the different protein solutions with respect to the initial condition. The effect of protein solutions on membrane fouling at different TMP was evaluated by a resistance in series model described by Equation (2):

$$\text{Rtot} = \text{Rm} + \text{Rirr} + \text{Rrev} \tag{2}$$

where Rm is the hydraulic resistance of the membrane itself, Rirr is the hydraulic resistance due to irreversible fouling, Rrev is the hydraulic resistance due to reversible fouling (concentration polarization and reversible deposited material), and Rtot is the total resistance given by the sum of different contributions. The membrane hydraulic resistance Rm can be calculated by the following equation [29,30]:

$$\text{Rm} = \frac{1}{Lp\mu} \tag{3}$$

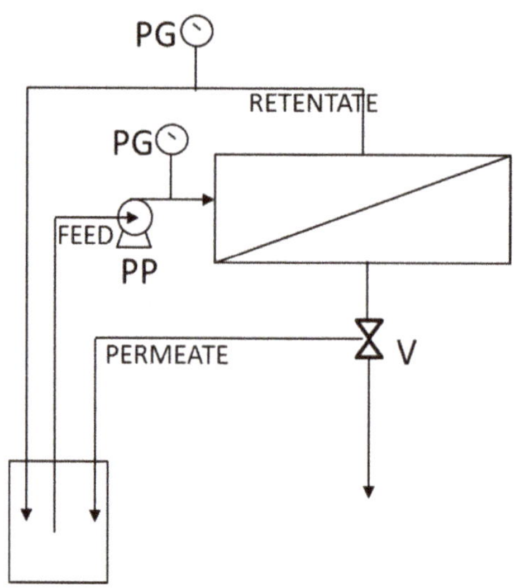


**Figure 1.** Schematic representation of cross-flow ultrafiltration system used. PG: pressure gauges, PP: peristaltic pump, V: valve. The same system was used to work in recycle mode (recirculating the permeate) and in concentration mode.

Here, *Lp* is the pure water permeance ($m \cdot Pa^{-1} \cdot s^{-1}$) of virgin membrane, and $\mu$ (Pa s) the viscosity of the solution. The hydraulic resistance due to irreversible fouling is given by the pure water permeance of the used membrane (that is, membrane after ultrafiltration test and rinsed with water). The total hydraulic resistance is calculated by the membrane permeance using the given working solution and the viscosity of that solution. The reversible hydraulic resistance is calculated by difference from Equation (2). The purity of the protein in the collected fraction is calculated by Equation (4):

$$\text{Purity} = \frac{[C_A]}{[C_A] + [C_B]} \tag{4}$$

where $C_A$ and $C_B$ are the concentration of the protein $A$ and $B$ in a given stream/solution. The recovery factor is calculated by Equation (5):

$$\text{R.F.} = \frac{mg_{AP}}{mg_{AF}} \quad (5)$$

where $mg_{AP}$ is the amount (mg) of the protein $A$ in the permeate, and $mg_{AF}$ is the amount of the same protein in the feed.

### 2.5. Ultrafiltration Experiments

Ultrafiltration experiments aimed at identifying the critical flux at the different pH tested, recirculating the permeate back to the feed tank so as to keep the feed volume and concentration constant, were carried out. The critical flux value allowed the identification of the maximum pressure to be used in order to limit fouling. Beyond this value, no further linear increase of flux would be obtained. The critical flux was measured by using protein solution dissolved in 25 mM phosphate buffer at pH 3 and 3.4. Each experiment was carried out in triplicate by using the same membrane. Between the different experiments, the membrane was washed with water to remove reversible fouling.

When binary protein mixture was used, the UF process was carried out in concentration mode (that is, the permeate was removed and the volume of the retentate correspondingly reduced). The ultrafiltration experiments by using binary protein mixture were monitored by measuring the permeate flux at different TMP. In a first step, the TMP was initially increased and subsequently decreased. Results obtained in terms of critical flux at a given pH were then used to carry out UF in concentration mode. Ultrafiltration of binary protein mixture in concentration mode was carried out by varying the initial protein concentration from 0.5 to 2 $g{\cdot}L^{-1}$. Sieving coefficient and membrane resistance were determined together with recovery factor and protein purity as a function of the volume reduction factor (VRF = ratio of the initial feed volume with respect to the final retentate volume). Each experiment was carried out in triplicate, and membrane cleaning was carried out by using a 0.1 M NaOH. Conductivity was fixed at 1.5 ($\pm$0.2) mS/cm.

### 2.6. Imparting Positive Charge to Regenerated Cellulose Membranes

A number of 30 kDa regenerated cellulose membranes (Millipore), $1.25 \times 10^{-3}$ $m^2$, were functionalized using the method previously described by van Reis et al. [31]. Briefly, membranes were rinsed with 0.1M NaOH recirculating along membrane surface and across the membrane, applying a TMP of 0.5 bar and a cross-flow velocity of about 0.014 m/s. After this passage, the membrane reacted with (3-bromopropyl) trimethylammonium bromide (Sigma-Aldrich, Cat. No. 347604, Milan, Italy) in 0.1 M NaOH for 21 h at room temperature. By a nucleophilic substitution, the alkyl ammonium group was covalently attached to the membrane. Washing steps were then performed by using ultrapure water and followed with 1% of acetic acid solution in phosphoric acid (0.12 M). The hydraulic permeance was measured before and after membrane functionalization.

## 3. Results

The aim of this work is to promote the separation of proteins with similar molecular weight (when present as monomers) and charge by charged UF process far from their isoelectric point, preventing membrane fouling. For this purpose, the ALA and BLG (IP: 4.4 and 5.4, respectively, Supplementary Figure S1) binary protein mixture was used just as the model system, since their separation is already fully developed at an industrial scale.

### 3.1. Properties of Individual Proteins in Bulk Solution

Zeta potential measurements, as well as protein size determination, were carried out on single protein solutions using different initial protein concentration (0.5, 1, 2 $g{\cdot}L^{-1}$). This kind of characterization was carried out in order to find differences between the two proteins, in terms of aggregation state and charge density, in conditions in which they both bore positive charge as the membrane. In particular, pH around 3 was analyzed, since both proteins (IP: 4.4 and 5.4, respectively) are positively charged as the membrane; this will promote low membrane/proteins interaction and then irreversible fouling prevention. In Figure 2, the trend of zeta potential of the two proteins varying the pH around 3 and by using a concentration of 1 g·L$^{-1}$ was reported as an example, since a similar trend for the other two concentration was obtained.

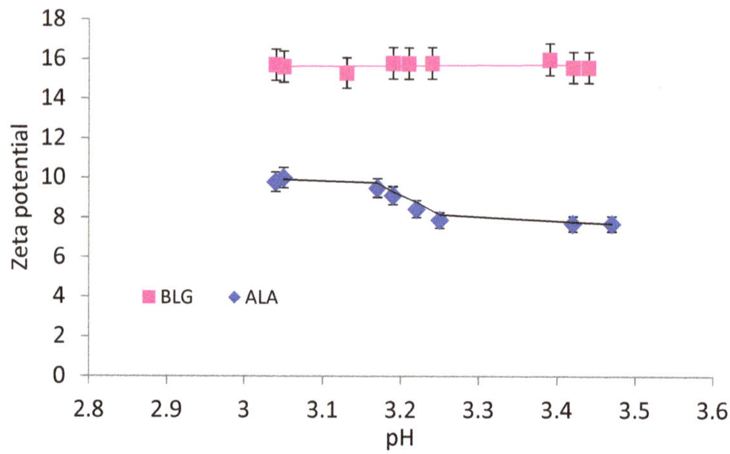


**Figure 2.** Zeta potential measurement of pure BLG and ALA solutions within pH range 3.0–3.5: ionic strength 0.1 M.

BLG is positively charged and did not change its value of zeta potential for all the analyzed pH values (16 mV) and initial protein concentration tested (Figure 2). On the contrary, although ALA bore always positive charge, its zeta potential at pH 3 was 63% lower (10 mV) compared to that for BLG at pH 3 (16 mV), and it dropped further at pH 3.17. A further decrease of ALA zeta potential at around 3.2 was observed, reaching about 50% of BLG value (8 mV) from 3.25–3.50. In Table 1, proteins' size and molecular weight were reported at pH 3.0, 3.2, and 3.4. At these pH values, the difference in zeta potential between the two proteins is most representative. As it is possible to see, ALA is present as a monodisperse monomer at all the pH values analyzed, while BLG is present as monodisperse monomer at pH 3, as a monodisperse monomer and dimer at pH 3.2, and as polydisperse monomer and dimer at pH 3.4. The higher polydispersity in the last case is a clear demonstration of the increase of protein aggregation state, which means a higher presence of dimers [31]. Comparing the results between the two proteins (Figure 2 and Table 1), at pH 3, both proteins are present as monomer and have about 16 and 10 mV of zeta potential, respectively; while at pH 3.4, ALA is still present as monomer, while BLG is present as polydisperse dimer solution. In addition, in this last case, both proteins are positively charged but BLG showed a higher charge density (16 mV) with respect to ALA (8 mV) and to the situation observed at pH 3. So, in order to study the effect of protein aggregation state and charge on the UF separation performance, these two values of pH were considered for further investigation.

**Table 1.** Protein diameter and molecular weight of ALA and BLG, varying pH from 3 to 3.4.

| | pH | Protein Diameter (nm) | Molecular Weight (kDa) | Pd (%) * |
|---|---|---|---|---|
| ALA | 3.0 | 3.62 (±0.60) | 13.5 (±4.5) | 14.7 |
| | 3.2 | 3.62 (±0.48) | 13.5 (±3.9) | 15.2 |
| | 3.4 | 3.62 (±0.36) | 13.5 (±5.3) | 16.0 |
| BLG | 3.0 | 4.19 (±0.71) | 19.0 (±5.4) | 15.4 |
| | 3.2 | 4.19 (±0.99) | 26.7 (±10.1) | 19.5 |
| | 3.4 | 4.89 (±1.36) | 26.7 (±6.5) | 27.1 |

* Polydispersity Pd (%): Pd < 20 = monodisperse; %Pd > 20 = polydisperse.

### 3.2. Determination of Critical Pressure

In this work, both the two analyzed proteins have the same charge as the membrane, and this means that electrical repulsion occurs among them and the membrane. However, during ultrafiltration, a pressure is applied as driving force to promote transport through the membrane.

When the applied pressure overcomes the electrical repulsion, proteins approach the membrane surface and may gelify on it, if the local concentration in the boundary layer reaches the gelling conditions, creating membrane fouling. In order to limit fouling phenomena during ultrafiltration process in concentration mode, the critical pressure at the two selected pH values and at protein concentration from 0.5 to 2 $g \cdot L^{-1}$ was investigated. The critical pressure was 0.2 bar when the initial protein concentration was 0.5 or 1 $g \cdot L^{-1}$, and it decreased down to 0.1 bar when the initial concentration was increased up to 2 $g \cdot L^{-1}$ (Table 2). However, in all the analyzed cases, on the basis of hydraulic resistance measurements (Table 2), no significant irreversible fouling was caused.

**Table 2.** Critical pressure, hydraulic resistance of membrane, and fouling components obtained by ultrafiltration of binary protein mixture at different initial concentration and pH.

| Protein Mixture ($g \cdot L^{-1}$) | pH | Critical Pressure (bar) | Critical Flux ($L \cdot h^{-1} \cdot m^{-2}$) | $R_{tot}$ ($m^{-1}$) | $R_m$ ($m^{-1}$) | $R_{frev}$ ($m^{-1}$) | $R_{firr}$ ($m^{-1}$) |
|---|---|---|---|---|---|---|---|
| 0.5 | 3.0 | 0.2 | 68 (±5) | $1.00 \times 10^{12}$ ($\pm 3.00 \times 10^{10}$) | $9.67 \times 10^{11}$ ($\pm 5.80 \times 10^{10}$) | $3.34 \times 10^{10}$ ($\pm 2.34 \times 10^{9}$) | 0 |
| | 3.4 | 0.2 | 64 (±5) | $1.24 \times 10^{12}$ ($\pm 7.44 \times 10^{10}$) | $1.01 \times 10^{12}$ ($\pm 5.00 \times 10^{10}$) | $1.64 \times 10^{11}$ ($\pm 2.40 \times 10^{10}$) | $2.20 \times 10^{10}$ * |
| 1 | 3 | 0.2 | 70 (±3) | $1.18 \times 10^{12}$ ($\pm 9.44 \times 10^{10}$) | $8.29 \times 10^{11}$ ($\pm 5.80 \times 10^{10}$) | $3.50 \times 10^{11}$ ($\pm 2.45 \times 10^{10}$) | 0 |
| | 3.4 | 0.2 | 68 (±3) | $1.68 \times 10^{12}$ ($\pm 1.01 \times 10^{10}$) | $8.7 \times 10^{11}$ ($\pm 5.22 \times 10^{10}$) | $8.10 \times 10^{11}$ ($\pm 4.86 \times 10^{10}$) | 0 |
| 2 | 3.0 | 0.1 | 35 (±5) | $1.16 \times 10^{12}$ ($\pm 6.96 \times 10^{10}$) | $9.20 \times 10^{11}$ ($\pm 4.60 \times 10^{10}$) | $2.15 \times 10^{11}$ ($\pm 2.60 \times 10^{10}$) | $2.56 \times 10^{10}$ * |
| | 3.4 | 0.1 | 25 (±5) | $1.68 \times 10^{12}$ ($\pm 8.40 \times 10^{10}$) | $9.73 \times 10^{11}$ ($\pm 5.84 \times 10^{10}$) | $6.79 \times 10^{11}$ ($\pm 4.75 \times 10^{10}$) | $2.62 \times 10^{10}$ * |

$R_m$ = hydraulic resistance due to the membrane; $R_{tot}$ = hydraulic resistance due to membrane and fouling; $R_{firr}$ = hydraulic resistance due to irreversible fouling; $R_{frev}$ = hydraulic resistance due to reversible fouling. * It is worth underlining that the $R_{firr}$ is within the error range of $R_m$ and $R_{tot}$; this confirms its negligible contribution to $R_{tot}$.

*3.3. Binary Protein Mixture Ultrafiltration at pH 3 in Concentration Mode*

As already mentioned in the Materials and Methods section, the ultrafiltration of binary protein mixtures was carried out in concentration mode. A constant flux as a function of time was observed when the ultrafiltration process was carried out using initial protein concentrations of 0.5 or 2 $g \cdot L^{-1}$ (Figure 3). In particular, using an initial protein concentration of 0.5 $g \cdot L^{-1}$ and applying a TMP of 0.2 bar, a steady-state flux of 68 (±5) $L \cdot h^{-1} \cdot m^{-2}$ was obtained, while using an initial protein concentration of 2 $g \cdot L^{-1}$ and applying a TMP of 0.1 bar, a steady-state flux of 30 (±2) $L \cdot h^{-1} \cdot m^{-2}$ was obtained. The similar flux obtained operating in concentration mode (Figure 3) or at constant feed volume (Table 2) is a further confirmation that no significant fouling is observed, since just reversible fouling is obtained, which can be easily removed by washing steps. The TMP values were selected according to the results of the critical pressure study. Furthermore, in this series of experiments, the initial pure water permeance ($6.70 \times 10^{-8}$ ($\pm 1.68 \times 10^{-9}$) $mPa^{-1} \cdot s^{-1}$ was also completely restored after UF with protein solutions ($6.65 \times 10^{-8}$ ($\pm 2.52 \times 10^{-9}$) $mPa^{-1} \cdot s^{-1}$). In Figure 4, the electrophoretic profile of samples analyzed as a function of ultrafiltration time were reported together with the MW standards of used proteins (Figure 4a). Analyzing the electrophoresis of permeates collected as a function of time (Figure 4b,c), it is possible to note that both proteins can pass through the membrane; however, ALA (that at this pH value has a lower charge density) is less rejected by the positively charged membrane. This is clear evidence that at pH 3 the membrane cannot be used to purify one protein with respect to the other, since both can pass through the membrane. Similar results were obtained for both initial protein concentrations used (Figure 4b,c).

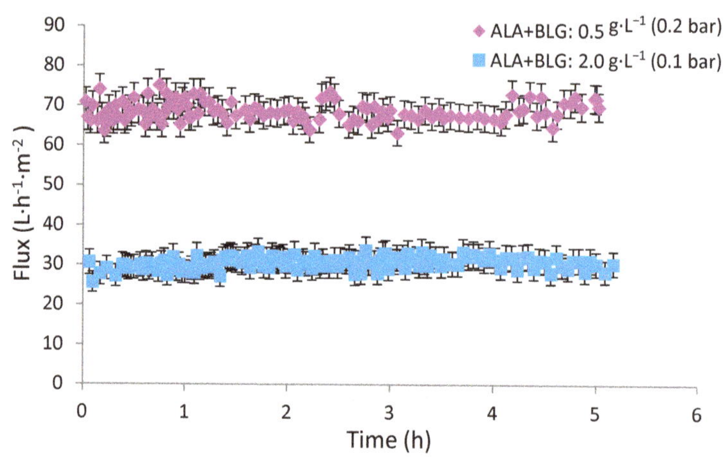


**Figure 3.** Time course of the UF carried out in concentration mode through charged membrane, by using different initial binary protein mixture concentration; pH: 3, T: 25 °C, ionic strength: 0.1 M.

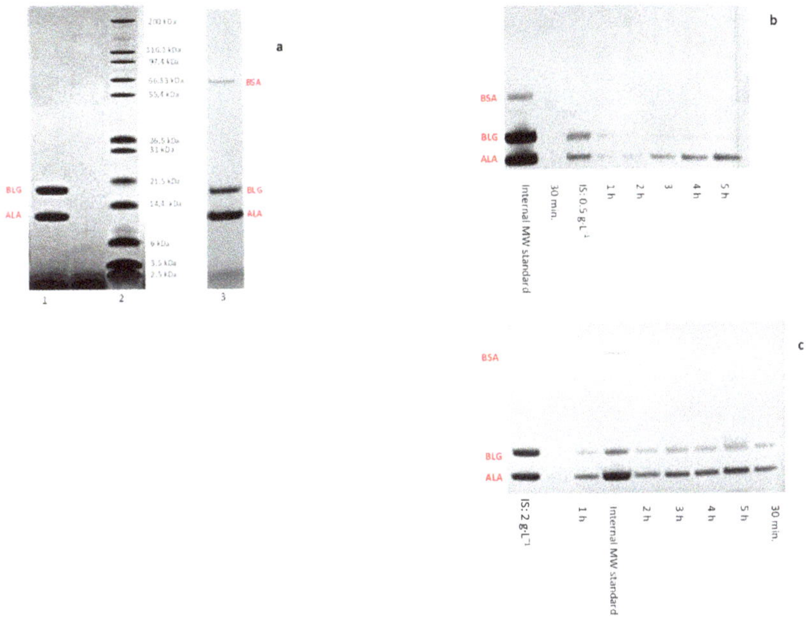


**Figure 4.** SDS-PAGE carried out on (**a**) standard solutions used to carry out molecular weight (MW) determination and quantification of proteins in binary protein mixture. 1: solution containing both ALA and BLG (1 g·L$^{-1}$); 2: wide range molecular weight marker (1:20 dilution); 3: internal molecular weight standard (formed by ALA, BLG, and BSA); (**b**,**c**) on permeates collected as a function of time, when UF process was carried out at pH 3. IS: initial solution concentration. The dilution of permeates samples and IS of (**c**) is: 1:2.

### 3.4. Binary Protein Mixture Ultrafiltration at pH 3.4

In Figure 5, the fluxes obtained during the ultrafiltration of binary protein mixture (0.5, 1.0, 2.0 g·L$^{-1}$) at pH 3.4 are reported. When the initial protein concentration was 0.5 g·L$^{-1}$, a constant flux was observed for about three hours (65 L·h·m$^{-2}$); after that, the flux starts to decrease, reaching a value of 50 L·h·m$^{-2}$ after 5 h of continuous UF process in concentration mode. However, after washing the membrane with the buffer, the initial pure water permeance (6.70 × 10$^{-8}$ (±1.68 × 10$^{-9}$) m·Pa$^{-1}$·s$^{-1}$) was restored (6.68 × 10$^{-8}$ (±1.60 × 10$^{-9}$) m·Pa$^{-1}$·s$^{-1}$), demonstrating that in this case also, no irreversible fouling did occur. If the initial protein concentration was doubled to 1 g·L$^{-1}$, a constant flux was observed (Figure 5) (64 L·h$^{-1}$·m$^{-2}$) for about two hours; after that it started to decrease, reaching a value of about 10 L·h$^{-1}$·m$^{-2}$ after 4.3 h of UF process. If the initial protein mixture concentration was further increased to 2 g·L$^{-1}$, a constant flux was observed for about 1.8 h; after that, a severe flux decrease was observed (Figure 5).

Although a flux decrease was observed for both the highest concentrations used, the initial pure water permeance, also in this case, was completely recovered. The absence of irreversible fouling can be attributed to proteins and membrane electrostatic repulsion which avoids a possible chemical interaction between membrane and protein. The flux decrease was mainly due to reversible fouling, caused by the accumulation of the most retained protein on the retentate side of the membrane. In Figure 6, the electrophoretic profile related to permeates collected as a function of time for all the three initial protein concentrations tested is reported. When 0.5 g·L$^{-1}$ was used (Figure 6a), just ALA can pass through the membrane, increasing its concentration as a function of time. Until a VRF of about 1.5, 95% of pure ALA was obtained with a recovery factor of 33% (Table 3).

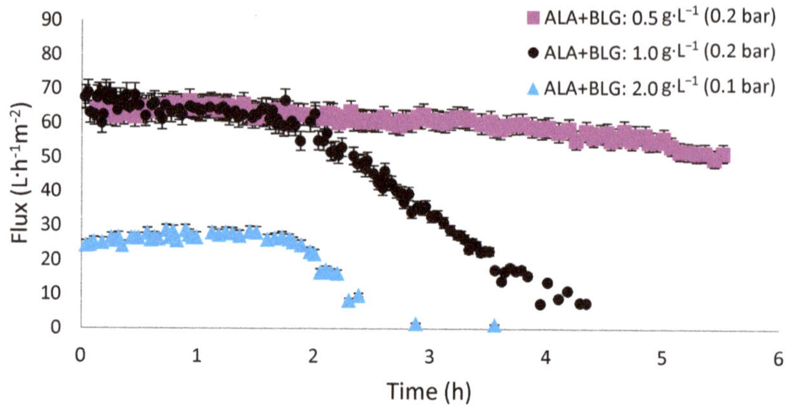


**Figure 5.** Time course of the UF flux of binary protein mixture (ALA + BLG), carried out in concentration mode through charged membrane by using different initial mixture concentration; pH: 3.4, T: 25 °C, ionic strength: 0.1 M.

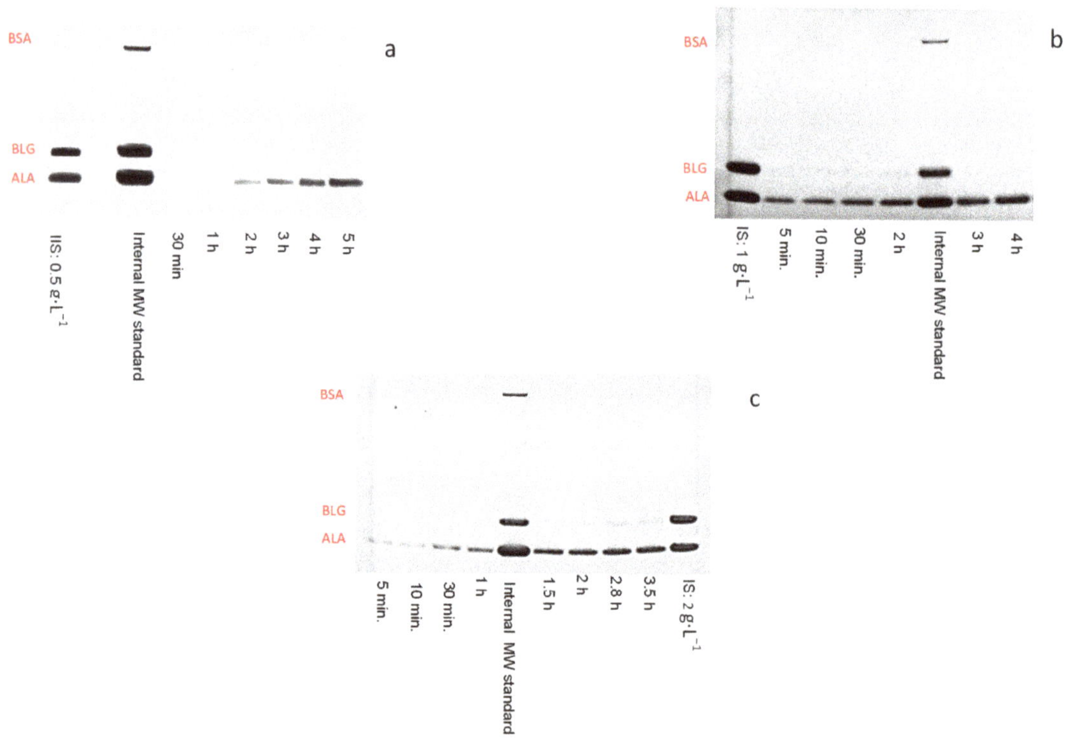


**Figure 6.** SDS-PAGE carried on permeates collected as a function of time when the UF process was carried out at pH 3.4. IS: initial solution concentration. (**a**) IS: 0.5 g·L$^{-1}$; (**b**) IS: 1 g·L$^{-1}$; (**c**) IS: 2 g·L$^{-1}$ (IS and samples dilution: 1:2).

**Table 3.** ALA recovery factor and purity in the permeate as a function of initial protein mixture concentration and constant flux during UF carried out at pH 3.4.

| Initial Binary Protein Mixture Concentration (mg/mL) | ALA Recovery Factor (%) | ALA Purity (%) |
|---|---|---|
| 0.5 | 33 | 95 |
| 1.0 | 32 | |
| 2.0 | 33 | |

Using 1 g·L$^{-1}$ as initial protein concentration, in this case just ALA was also present in all the collected permeates (Figure 6b), and its concentration increased as a function of time. Despite the abovementioned flux decrease after two hours of UF, no BLG was present in the permeate, but it was retained on the retentate side of the membrane, causing reversible fouling and flux decrease. Nevertheless, the membrane did not change its selectivity throughout the whole process, reaching 80% recovery factor of ALA in the permeate with 95% purity after 4.4 h. On the contrary, BLG is concentrated in the retentate (see Supplementary Figure S2, and due to the presence of concentrated ALA, it cannot be purified by charged UF process similar to ALA, but a diafiltration process is needed. When the charged UF process was carried out using 2 g·L$^{-1}$, during the timeframe that the flux was constant, only ALA was present in the collected permeates (Figure 6c) with a purity of 95%, and a recovery factor of 33% at a VRF of 1.4. However, as soon as the flux started to decrease, BLG appeared in the permeate also. In case the UF process was carried out at pH 3.4, even though proteins were positively charged, ALA could pass through the membrane because it had a lower charge density (8 mV) compared to BLG, and a lower size (~14 kDa) compared to the membrane pore size (30 kDa), as it was present as monomeric form. BLG was completely rejected because it had a higher density of positive charge and larger molecular size, BLG being prevalently present in dimeric form. Considering BLG dimer size (~36 kDa) and charge density (16 mV), compared to the pore size (30 kDa) and surface positive charge of the membrane, it is reasonable to expect a high retention due to electrostatic repulsion and size sieving mechanisms. Furthermore, BLG, being prevalently present in dimeric form, is also retained from the membrane, but when the initial protein concentration is further increased, the reversible fouling, due to the accumulation of the most rejected protein, also increases, causing the passage of BLG monomer. The higher passage of ALA through the membrane, before BLG concentration increase in the retentate, was also given by an associative affect between the two positively charged proteins due to the Donnan effect, as already reported in the UF of different binary protein mixtures [2,32,33]. In this particular condition, the largely retained BLG (prevalently in dimer form) tends to push the more transmittable ALA (present as monomer, and less charged) through the membrane to permit charge balance. The positively charged membrane used repulsed the positively charged ALA and did not interact with it, ensuring low fouling, but thanks to the convection flow through the membrane promoted by the applied pressure, the repulsive interaction between the membrane and ALA was overcome, permitting the selective passage of the protein through the membrane. Overall, results demonstrate that the initial protein concentration of 2 g·L$^{-1}$ is not suitable to achieve high recovery factor of purified proteins.

## 4. Conclusions

In this work, the possibility to separate ALA from a binary protein mixture of proteins having similar MW (when present as monomers) and charge by charged UF process was demonstrated. Compared to current literature, here, the UF process was carried out far from the isoelectric point of both proteins and in conditions where they both bore the same charge as the membrane. This was performed in order to limit fouling and to achieve ALA separation on the basis of electrostatic repulsion, Donnan exclusion, and size exclusion.

The work identified the operating conditions leading to high recovery factor and protein purity in the absence of irreversible fouling. When the ultrafiltration of the binary protein mixture was carried out at pH 3.4, and with initial protein concentration of 1 g·L$^{-1}$, a fractionation of ALA with respect to BLG was obtained with a recovery factor of about 80% and a purity of 95%. The pure water permeance of the membrane was fully recovered after protein separation by simply rinsing the membrane with buffer solution. Measurements of hydraulic resistance confirmed that reduction of flux with increasing of VRF was due to reversible fouling. Even though, in the presence of real whey, antifouling properties might change due to the presence of proteins with different charge, this study provides a solution tuned for protein pairs after their easy separation from larger/smaller proteins that differ by at least a factor of 10 in terms of molecular weight.

**Supplementary Materials:** The following are available online at https://www.mdpi.com/article/10.3390/app11199167/s1, Figure S1: Zeta potential measurement as a function of pH of ALA and BLG (1 g·L$^{-1}$). Figure S2: SDS-page carried out on final retentate (4.4. h) after UF by using binary protein mixture (1 g·L$^{-1}$) pH 3.4 and charged regenerated cellulose membrane. 1: IS (1 g·L$^{-1}$); 2: internal MW standard; 3: retentate obatained n the UF process after 4.4 h.

**Author Contributions:** R.M.: conceptualization, performing experiments, data curation, writing, review and editing; A.M.S.: performing experiments, data curation; E.D., M.A.-F. and M.A.A.: funding review and editing; L.G.: funding, writing, review, and editing. All authors have read and agreed to the published version of the manuscript.

**Funding:** This work was financially supported by the project "Membrane systems in regenerative medicine, tissue engineering, and biotechnology" AGREEMENT No. KACST-ITM-CNR/03.

**Institutional Review Board Statement:** Not applicable.

**Informed Consent Statement:** Not applicable.

**Data Availability Statement:** Not applicable.

**Conflicts of Interest:** The authors declare no conflict of interest.

## References

1. Bhushan, S.; Etzel, M.R. Charged ultrafiltration membranes increase the selectivity of whey protein separations. *J. Food Sci.* **2009**, *74*, E131–E139. [CrossRef]
2. Saxena, A.; Tripathi, B.P.; Kumar, M.; Shahi, V.K. Membrane-based techniques for the separation and purifi cation of proteins: An overview. *Adv. Colloid Interface Sci.* **2009**, *145*, 1–22. [CrossRef] [PubMed]
3. Rohani, M.M.; Mehta, A.; Zydney, A.L. Development of high performance charged ligands to control protein transport through charge-modified ultrafiltration membranes. *J. Membr. Sci.* **2010**, *362*, 434–443. [CrossRef]
4. Arunkumar, A.; Etzel, M.R. Fractionation of α-lactalbumin from β-lactoglobulin using positively charged tangential flow ultrafiltration membranes. *Sep. Purif. Technol.* **2013**, *105*, 121–128. [CrossRef]
5. Cheang, B.; Zydney, A.L. Separation of α-lactalbumin and β-lactoglobulin using membrane ultrafiltration. *Biotechnol. Bioeng.* **2003**, *83*, 201–209. [CrossRef]
6. Toro-Sierra, J.; Schumann, J.; Kulozik, U. Impact of spray-drying conditions on the particle size of microparticulated whey protein fractions. *Dairy Sci. Technol.* **2013**, *93*, 487–503. [CrossRef]
7. Ratnaningsih, E.; Reynard, R.; Khoiruddin, K.; Wenten, I.G.; Boopathy, R. Recent Advancements of UF-Based Separation for Selective Enrichment of Proteins and Bioactive Peptides—A Review. *Appl. Sci.* **2021**, *11*, 1078. [CrossRef]
8. van Reis, R.; Brake, J.; Charkoudian, J.; Burns, D.; Zydney, A. High-performance tangential flow filtration using charged membranes. *J. Membr. Sci.* **1999**, *159*, 133–142. [CrossRef]
9. Rabiller-Baudry, M.; Chaufer, B.; Lucas, D.; Michel, F. Ultrafiltration of mixed protein solutions of lysozyme and lactoferrin: Role of modified inorganic membranes and ionic strength on the selectivity. *J. Membr. Sci.* **2001**, *184*, 137–148. [CrossRef]
10. Chun, K.-Y.; Stroeve, P. Protein transport in nanoporous membranes modified with self-assembled monolayers of functionalized thiols. *Langmuir* **2002**, *18*, 4653–4658. [CrossRef]
11. Ebersold, M.F.; Zydney, A.L. Separation of protein charge variants by ultrafiltration. *Biotechnol. Prog.* **2008**, *20*, 543–549. [CrossRef] [PubMed]
12. Striemer, C.C.; Gaborski, T.R.; McGrath, J.L.; Fauchet, P.M. Charge- and size-based separation of macromolecules using ultrathin silicon membranes. *Nature* **2006**, *445*, 749–753. [CrossRef] [PubMed]
13. Mahlicli, F.Y.; Altinkaya, S.A.; Yurekli, Y. Preparation and cha racterization of polyacrylonitrile membranes modified with polyelec trolytedepositionfor separatin g simil ar sized proteins. *J. Membr. Sci.* **2012**, *415–416*, 383–390. [CrossRef]

14. Sorci, M.; Gu, M.; Heldt, C.L.; Grafeld, E.; Belfort, G. A multi-dimensional approach for fractionating proteins using charged membranes. *Biotechnol. Bioeng.* **2013**, *110*, 1704–1713. [CrossRef] [PubMed]
15. Valiño, V.; San Román, M.F.; Ibañez, R.; Ortiz, I. Improved separation of bovine serum albumin and lactoferrin mixtures using charged ultrafiltration membranes. *Sep. Purif. Technol.* **2014**, *125*, 163–169. [CrossRef]
16. Fane, A.G.; Fell, C.J.D.; Suki, A. The effect of the pH and ionic environment on the ultrafiltration of protein solutions with retentive membranes. *J. Membr. Sci.* **1983**, *16*, 195–210. [CrossRef]
17. Fane, A.G.; Fell, C.J.D.; Waters, A.G. Ultrafiltration of protein solutions through partially permeable membranes—The effect of adsorption and solution environment. *J. Membr. Sci.* **1983**, *16*, 211–224. [CrossRef]
18. Kelly, S.T.; Zydney, A.L. Mechanisms for BSA fouling during microfiltration. *J. Membr. Sci.* **1995**, *107*, 115–124. [CrossRef]
19. Ricq, L.; Narçon, S.; Reggiani, J.-C.; Pagetti, J. Streaming potential and protein transmission ultrafiltration of single proteins and proteins in mixture: B-lactoglobulin and lysozyme. *J. Membr. Sci.* **1999**, *156*, 81–96. [CrossRef]
20. Martín, A.; Martinez, F.; Calvo, J.I.; Prádanos, P.; Palacio, L.; Hernández, A. Protein adsorption onto na inorganic microfiltration membrane solute-solid interactions and surface coverage. *J. Membr. Sci.* **2002**, *207*, 199–207. [CrossRef]
21. Kujundzic, E.; Greenberg, A.R.; Fong, R.; Hernandez, M. Monitoring Protein Fouling on Polymeric Membranes Using Ultrasonic Frequency-Domain Reflectometry. *Membranes* **2011**, *1*, 195–216. [CrossRef]
22. Portugal, C.A.M. Functional membranes for control of protein-surface interactions and transport properties. *Curr. Org. Chem.* **2017**, *21*, 1725–1739. [CrossRef]
23. Aguero, R.; Bringas, E.; Román, M.S.; Ortiz, I.; Ibañez, R. Membrane processes for whey proteins separation and purification. A review. *Curr. Org. Chem.* **2017**, *21*, 1740–1752. [CrossRef]
24. Bramaud, C.; Aimar, P.; Daufin, G. Whey protein fractionation: Isoelectric precipitation of a-lactalbumin under gentle heat treatment. *Biotechnol. Bioeng.* **1997**, *56*, 391–397. [CrossRef]
25. Salgin, S.; Salgin, U.; Bahadir, S. Zeta potentials and isoelectric points of biomolecules: The effect of ion types and ionic strengths. *Int. J. Electrochem. Sci.* **2012**, *7*, 12404–12414.
26. Yan, Y.; Seeman, D.; Zheng, B.; Kizilay, E.; Xu, Y.; Dubin, P.L. pH-dependent aggregation and disaggregation of native β-lactoglobulin in low salt. *Langmuir* **2013**, *29*, 4584–4593. [CrossRef]
27. Arunkumar, A.; Etzel, M.R. Fractionation of α-lactalbumin and β-lactoglobulin from bovine milk serum using staged, positively charged, tangential flow ultrafiltration membranes. *J. Membr. Sci.* **2014**, *454*, 488–495. [CrossRef]
28. Chayen, N.; Dieckmann, M.; Dierks, K.; Fromme, P. Size and shape determination of proteins in solution by a noninvasive depolarized dynamic light scattering instrument. *Ann. N. Y. Acad. Sci.* **2004**, *1027*, 20–27. [CrossRef] [PubMed]
29. Mazzei, R.; Drioli, E.; Giorno, L. Biocatalytic membrane reactor and membrane emulsification concepts combined in a single unit to assist production and separation of water unstable reaction products. *J. Membr. Sci.* **2010**, *352*, 166–172. [CrossRef]
30. Mazzei, R.; Drioli, E.; Giorno, L. Comprehensive Membrane Science and Engineering. *Biocatal. Membr. Membr. Bioreact. Compr. Membr. Sci. Eng.* **2010**, *3*, 195–212.
31. van Reis, R.D. Charged Filtration Membranes and Uses Therefor. U.S. Patent 7,001,550B2, 26 December 2006. Available online: https://app.dimensions.ai/details/patent/US-7153426-B2 (accessed on 21 September 2021).
32. Filipe, C.D.; Ghosh, R. Effects of protein–protein interaction in ultrafiltration based fractionation processes. *Biotechnol. Bioeng.* **2005**, *91*, 678–687. [CrossRef] [PubMed]
33. Holland, B.; Kackmar, J.; Corredig, M. Short communication: Isolation of a whey fraction rich in α-lactalbumin from skim milk using tangential flow ultrafiltration. *J. Dairy Sci.* **2012**, *95*, 5604–5607. [CrossRef] [PubMed]

applied sciences

MDPI

Article

# Effect of Bioaugmentation with Anaerobic Fungi Isolated from Ruminants on the Hydrolysis of Corn Silage and *Phragmites australis*

**Bhargavi Ravi [1,2], Valentine Nkongndem Nkemka [2], Xiying Hao [2], Jay Yanke [2], Tim A. McAllister [2], Hung Lee [3], Chitraichamy Veluchamy [1] and Brandon H. Gilroyed [1,*]**

1 School of Environmental Sciences, University of Guelph Ridgetown Campus, Ridgetown, ON N0P 2C0, Canada; bhargaviravi20@gmail.com (B.R.); chitraiv@uoguelph.ca (C.V.)
2 Agriculture and Agri-Food Canada, Lethbridge Research Station, Lethbridge, AB T1J 4P4, Canada; vallynkemka@gmail.com (V.N.N.); xiying.hao@canada.ca (X.H.); jay.yanke@canada.ca (J.Y.); tim.mcallister@canada.ca (T.A.M.)
3 School of Environmental Sciences, University of Guelph, Guelph, ON N1G 2W1, Canada; hlee@uoguelph.ca
* Correspondence: bgilroye@uoguelph.ca; Tel.: +1-519-674-1500 (ext. 63605)

**Abstract:** Anaerobic fungi produce extracellular hydrolytic enzymes that facilitate degradation of cellulose and hemicellulose in ruminants. The purpose of this work was to study the impact of three different anaerobic fungal species (*Anaeromyces mucronatus* YE505, *Neocallimastix frontalis* 27, and *Piromyces rhizinflatus* YM600) on hydrolysis of two different lignocellulosic substrates, corn (*Zea mays* L.) silage and reed (*Phragmites australis* (Cav.) Trin. ex Steud.). Biomass from each plant species was incubated anaerobically for 11 days either in the presence of live fungal inoculum or with heat-inactivated (control) inoculum. Headspace gas composition, dry matter loss, soluble chemical oxygen demand, concentration of volatile fatty acids, and chemical composition were measured before and after hydrolysis. While some microbial activity was observed, inoculation with anaerobic fungi did not result in any significant difference in the degradation of either type of plant biomass tested, likely due to low fungal activity or survival under the experimental conditions tested. While the premise of utilizing the unique biological activities of anaerobic fungi for biotechnology applications remains promising, further research on optimizing culturing and process conditions is necessary.

**Keywords:** *Anaeromyces mucronatus*; lignocellulose; *Neocallimastix frontalis*; *Piromyces rhizinflatus*; pretreatment; hydrogen; biomass

**Citation:** Ravi, B.; Nkongndem Nkemka, V.; Hao, X.; Yanke, J.; McAllister, T.A.; Lee, H.; Veluchamy, C.; Gilroyed, B.H. Effect of Bioaugmentation with Anaerobic Fungi Isolated from Ruminants on the Hydrolysis of Corn Silage and *Phragmites australis*. *Appl. Sci.* **2021**, *11*, 9123. https://doi.org/10.3390/app11199123

Academic Editor: Ramaraj Boopathy

Received: 9 August 2021
Accepted: 24 September 2021
Published: 30 September 2021

**Publisher's Note:** MDPI stays neutral with regard to jurisdictional claims in published maps and institutional affiliations.



## 1. Introduction

In Europe, more than 13,638 biogas plants (72%) utilize the agricultural feedstocks out of 18,943 biogas plants [1], among which corn (*Zea mays* L.) silage is the major feedstock [2]. While corn silage is the most widely used energy crop for biogas production, there is interest in using other sources of plant biomass for environmental, economic, and societal reasons [3]. For example, *Phragmites australis* (Cav.) Trin. ex Steud. (common reed) is a perennial invasive wetland plant species in North America that produces substantial quantities of biomass of up to 30 t $ha^{-1}$ $y^{-1}$ [4]. While all parts of the common reed can be used for both biogas and biofuel production [5], the estimated biogas yields reported in the literature are only 150 L $kg^{-1}$ volatile solids (VS) of fresh material compared to grass and pig manure that yield more than 280 and 340 L $kg^{-1}$ VS, respectively [4,6]. The issues of low degradability and poor conversion to biogas are also applicable to other potential energy crops, such as *Miscanthus* and *Arundo donax* L. [7,8].

In nature, one of the most efficient systems for unlocking the energy found in lignocellulosic substrates is the rumen animals such as cattle and sheep. While the stepwise fermentation process (hydrolysis, acidification, acetogenesis, methanogenesis) that occurs during anaerobic digestion (AD) is crudely similar to the digestive process in the rumen, it is far less efficient [9]. One reason for the reduced efficiency of AD compared to the rumen likely lies in differences in the microbial populations between these two environments. The anaerobic digestive system of the rumen has been extensively studied, and anaerobic fungi (AF) are known to be involved in the digestion of the most recalcitrant lignocellulose within the rumen [10]. Anaerobic fungi use rhizoids to physically penetrate and disrupt the lignin layer of lignocellulose, while also enzymatically degrading plant cell walls using a diverse suite of extracellular hydrolytic enzymes, including cellulases, hemicellulases, pectinases, and phenolic acid esterases [11]. Some of the extracellular hydrolytic enzymes produced by these organisms are freely released into the milieu; others are bound to the cellular surface as components of multienzyme cellulosomes [12]. Using feruloyl esterase activity, AF cleaves the bonds between hemicellulose and lignin, increasing the access of microbial enzyme to hemicelluloses. Although AF are known to degrade lignin, they do not utilize the lignin themselves [13]. While AF are known to play an essential role within the rumen, their presence, abundance, and activity level in AD is not well understood.

Bioaugmentation involves adding specific microorganisms into a system or process in order to improve its efficiency [14]. Several studies have been conducted using bioaugmentation with bacteria or fungi as a pretreatment for the hydrolysis of lignocellulosic substrates prior to AD [12]. In one study, the addition of thermophilic *Geobacillus* sp. strain AT1 to a biogas reactor using sewage sludge as substrate resulted in a 210% increase in biogas production due to the protease activity of the microbe [15]. In another study, 22 isolates of white rot fungi were used individually to pretreat wheat straw, with the greatest lignin degradation and subsequent increase in biogas yield (from 0.293 L $g^{-1}$ to 0.343 L $g^{-1}$) obtained from an isolate of *Pleurotus florida* [16].

Recently, studies utilizing AF to improve biogas production and speed up substrate degradation have been reported [17–19]. To date, isolates of the genera *Anaeromyces*, *Neocallimastix*, and *Piromyces* have been added to AD systems in an effort to improve lignocellulose degradation and ultimately improve methane yield [18,19]. A previous study [18] demonstrated increased biogas yields from different substrates, such as maize silage, anaerobic sludge, and microcrystalline cellulose, with bioaugmentation of AF in fed batch semicontinuous digesters. In that study, addition of 8 mg dry mycelium of *Anaeromyces* sp. (strains KF8 or JF1) or mixed cultures of 1.9 mg dry mycelium of *Anaeromyces* sp. KF8 and *Piromyces* sp. KF9 increased biogas yield by up to 22%. Although the study demonstrated an increase in biogas yield with AF, the researchers did not determine if the increase in biogas occurred as a result of the addition of AF or the anaerobic microbes that were already present in the sludge. Another study [18] explored bioaugmentation of a two-stage reactor with *Piromyces rhizinflata*, using corn silage and cattail as substrates, which resulted in an initial increase of $H_2$ and $CH_4$ production but with no overall increase in biogas production. They proposed that this response occurred as a result of rapid wash out of AF from the anaerobic digester systems. There may also have been additional challenges with integration of AF into the microbial populations within the AD. A recent study [20] surveyed 10 agricultural biogas plants for the presence and transcriptional activity of AF, concluding that survival and activity were impeded by the process conditions prevalent in commercial scale biogas systems.

Based on the seeming lack of activity from AF in commercial biogas systems [20] and poor survival of AF bioaugmented into lab-scale AD systems [18], this study was designed to evaluate the efficacy of AF as a hydrolytic pretreatment for lignocellulosic biomass. We evaluated the effect of three different fungal species (*Anaeromyces mucronatus* YE505, *Neocallimastix frontalis* 27, and *Piromyces rhizinflatus* YM600), which were previously isolated from ruminants and known to possess hydrolytic activity against lignocellulosic substrates, on microbial hydrolysis of corn silage and common reed.

## 2. Materials and Methods

### 2.1. Feedstock

Corn silage (*Zea mays* L.) and common reed (*Phragmites australis* (Cav.) Trin. ex Steud.) were used as substrates for fungal hydrolysis. Corn silage was obtained from a commercial beef cattle feedlot in Lethbridge County, Alberta, Canada. Common reed, harvested in July, was obtained from Ridgetown, ON, Canada.

### 2.2. Anaerobic Fungal Strains, Media, and Culturing Conditions

Pure cultures of three AF were obtained from the microbial collection lab at the Agriculture and Agri-Food Canada Lethbridge Research and Development Centre: *Anaeromyces mucronatus* YE505 (elk isolate), *Neocallimastix frontalis* 27 (cow isolate), and *Piromyces rhizinflatus* YM600 (moose isolate). Inocula of the fungal cultures were maintained anaerobically at 39 °C in modified semi-defined Lowe's medium B [21] with barley straw (ground <1 mm) as the sole carbon source. The ground barley comprised 5% of the mass (0.05 g) of the anaerobic media (about 5 mL) in the test tube and was then autoclaved for 20 min at 120 °C with 103.4 kPa pressure. After autoclaving, the media was cooled down and fungal cultivation was carried out using the Hungate technique [22]; tubes were inoculated by transferring fungal biomass from already existing culture tubes using a Pasteur pipet under anaerobic conditions. After inoculation, tubes were incubated at 39 °C in an incubator for 4 days to allow for fungal growth, and then the AF with spent medium was transferred to Erlenmeyer flasks at the start of the hydrolysis experiment.

### 2.3. Hydrolysis Experiment

Hydrolysis of plant biomass was evaluated in 0.5 L Erlenmeyer flasks. The total solids (TS) content of all flasks was set at 7.9% (*w*/*w*). A single lot of anaerobic sludge was obtained from a commercial scale biogas facility (Lethbridge Biogas LP) that co-digests livestock manures with industrial food processing waste. Anaerobic sludge was autoclaved for 20 min at 120 °C with 103.4 kPa pressure to inactivate background microbial activity and then used as a buffering solution in each flask. Triplicate samples of autoclaved sludge were analyzed and used to determine the chemical and physical properties. The sludge had a pH of 7.88, total bicarbonate alkalinity of 16.66 g $L^{-1}$ and TS of 1.66%.

A total of 36 flasks were used for this hydrolysis experiment. Flasks containing either corn silage or common reed were individually inoculated with each of the three AF in triplicate. Each corn silage flask contained 200 mL of anaerobic sludge, 80 mL fungal inoculum (comprising 20% of the total working volume), 92.8 g of corn silage, and 100 mL of distilled water. Each common reed flask contained 200 mL of anaerobic sludge, 80 mL fungal inoculum, 57.6 g common reed, and 140 mL of distilled water. Control flasks were also set up in triplicate in a manner identical to those described above, except that the fungal inocula were first killed by autoclaving prior to addition to the flasks. Inoculated flasks were then flushed with nitrogen for 1–2 min to ensure anaerobic conditions and sealed with butyl rubber stoppers connected to aluminum gas tight bags (Multilayer Transofoil, Flextrus AD, Sweden) as described in [23]. Flasks were equipped with sampling ports for gas and liquid sample extraction. The experiment was conducted under mesophilic conditions (40 ± 1 °C) by placing flasks in a water bath (2870; Thermo Fisher Scientific, Waltham, MA, USA) and manually agitated at least three times a day.

### 2.4. Analytical Methods

#### 2.4.1. Gas Analysis

Gas samples (10 mL) were taken daily from the headspace of each flask and transferred to 5.9 mL evacuated glass vials (Exetainer; Labco Limited, Lampeter, UK) prior to analysis using gas chromatography (GC). Gas samples were analyzed for $CO_2$ and $CH_4$ concentrations using a two-channel micro-GC (Varian 4900, Palo Alto, CA, USA) equipped with a thermal conductivity detector [24]. Operational parameters of the GC were as follows: channel A ($H_2$ analysis) injector 110 °C, column oven 40 °C, argon carrier gas at

150 kPa; channel B ($CH_4$, $CO_2$ analysis) injector 80 °C, column oven 40 °C, helium carrier gas at 100 kPa. Total gas volume from each flask was captured in individual gas-tight bags and quantified using a 0.1 L glass syringe (Perfektum™ Jumbo Glass Syringes, Cadence Science™, Cranston, RI, USA). Gas volumes reported were normalized to 0 °C and 1 atm.

### 2.4.2. Liquid Analysis

Liquid samples were extracted from a sampling port on each flask every 48 h using a 10 mL syringe and divided into aliquots for further analysis as described below. The TS and VS of liquid samples were measured following a standard protocol [25]. To estimate the extent of lignocellulose hydrolysis and the amount of remaining dissolved organic matter, soluble chemical oxygen demand (COD) was determined according to the manufacturer's protocol (Dr. Lange test kit HR mercury free, 20–1500 mg $L^{-1}$, Mississauga, ON, Canada). Samples used for COD analysis were first syringe filtered through 0.45 μm nylon filter (Chromatographic Specialties Inc., Brockville, ON, Canada) and then digested using a digital reactor block (HACH DRB200, Loveland, CO, USA) at 150 °C for 2 h. After digestion, absorbance of the sample was measured using a spectrophotometer (DR900, HACH, Mississauga, ON, Canada).

The pH and total bicarbonate alkalinity were measured using a BIOGAS titration Manager (R41T114, HACH, Vésenaz, Switzerland). Liquid samples were also analyzed for volatile fatty acids (VFA; acetate acid, n-butyrate, iso-butyrate, propionate, n-valerate, iso-valerate, and caproate) by GC (Agilent 6890 N, Agilent, Mississauga, ON, Canada). The samples were prepared by first filtering using 0.45 μm nylon filter (Chromatographic Specialties Inc., Brockville, ON, Canada), then 25% meta phosphoric acid was added to the filtered sample in the ratio of 5:1 sample to acid. The gas chromatograph was equipped with a flame ionization detector maintained at 250 °C, and a fused silica capillary column (ZB-FFAP, 30 m × 0.32 mm × 1.0 μm: Phenomenex, Torrance, LA, USA). The equipment was set at split mode and the split injection ratio was 5:1. Helium was used as the carrier gas and the analytical steps were performed according to the procedures outlined in Gilroyed et al. [26].

Concentration of soluble ions ($NH_4^+$, $Na^+$, $K^+$, $Ca^{2+}$, $Mg^{2+}$) were determined after filtration through 0.45 μm filter paper using ion chromatography (ICS-1000 and DX-600, Dionex, Sunnyvale, CA, USA). The concentration of free (unionized) $NH_3$ was calculated as previously reported [27]. To determine the ratio of total carbon to total nitrogen in samples, a subsample (5 mg) was freeze dried for 1 week and ground to a size < 0.15 mm using a Cyclone sample mill (UDY Corporation, Fort Collins, CO, USA), and then analyzed using a CNS analyzer (NA-1500, Carlo Erba, Rodano, Italy) linked via a continuous flow interface to an Optima isotope ratio mass spectrometer (Micromass, Manchester, UK).

### 2.4.3. Fiber Analysis and C:N

Fiber analysis was performed to characterize the composition (cellulose, hemicellulose, and lignin content) of corn silage and common reed before hydrolysis. Before taking samples for analysis, corn silage and common reed were thoroughly mixed in the containers that they stored to obtain a uniform and unbiased sample for analysis. Triplicate samples of each feedstock were air dried for 1 week and then ground through a screen of 1 mm mesh size in a tabletop mill grinder (Wiley mill standard model 4; Arthur H. Thomas Co., Philadelphia, PA, USA). The contents of lignin, hemicellulose, and cellulose were analyzed according to a modified method of [28] with thermal stable amylase (Termamyl® 120, Sigma-Aldrich Co. LLC., St. Louis, MO, USA) and sodium sulfite (S430-3 sodium sulfite anhydrous, Fisher Scientific Int., Inc., Pittsburgh, PA, USA) included in the NDF procedure [29]. Total carbon and total nitrogen concentrations were determined from freeze-dried, finely ground samples using a Model 1500 Nitrogen/Carbon analyzer (Carlo Erba Instruments, Milan, Italy).

### 2.5. Statistical Analysis

One-way ANOVA and repeated measures ANOVA tests were performed for statistical analysis using IBM SPSS version 24.0. The different treatments (*Anaeromyces mucronatus* YE505, *Neocallimastix frontalis* 27, and *Piromyces rhizinflatus* YM600) were kept as independent variables, and the different analytical tests, such as average cumulative hydrogen and $CO_2$ gas production, and changes in COD, pH, and VFA, were considered as dependent variables.

## 3. Results and Discussion

### 3.1. Feedstock Characteristics

The TS of corn silage and common reed were 33.5% ± 0.6 and 54.7% ± 0.8, of which 96.3% ± 0.6 and 94.9% ± 0.4 were VS, respectively (Table 1). The C:N ratio of corn silage and common reed was 31.2% ± 0.1 and 26.3% ± 0.7, respectively; both values were almost within the optimum range of 20 to 30 for AD [30]. In terms of fiber composition, corn silage had about two times lower ($p < 0.05$) concentrations of cellulose, hemicellulose, and lignin than common reed (Table 1).

**Table 1.** The physicochemical characteristics of corn silage and common reed.

| Parameters | Corn Silage | Common Reed |
|---|---|---|
| TS [1] (%) | 33.5 ± 0.6 | 54.7 ± 0.8 |
| VS [2] (% TS) | 96.3 ± 0.6 | 94.9 ± 0.4 |
| VS added (g) | 30 | 30 |
| Moisture content (%) | 66 ± 0.6 | 44.3 ± 0.8 |
| Total carbon: Total nitrogen ratio | 31.2 ± 0.1 | 26.3 ± 0.7 |
| Hemicellulose (% of TS) | 12 ± 4.3 | 28.7 ± 0.4 |
| Cellulose (% of TS) | 17.2 ± 1.8 | 38.7 ± 0.4 |
| ADL [3] (% TS) | 2.7 ± 0.0 | 7.9 ± 0.3 |

[1] Total solids. [2] Volatile solid. [3] Acid detergent lignin.

### 3.2. Gas Production and Composition

Cumulative methane production was <1 mL $g^{-1}$ VS in all treatments for both corn silage and common reed substrates. Since only hydrolysis was conducted in this study, minimal methane volume was expected. Anaerobic fungi are known to produce $H_2$ and $CO_2$ during substrate hydrolysis [31]. Nkemka and Gilroyed [18] demonstrated that bioaugmentation with anaerobic AF into a two-stage AD can increase $H_2$ production within the system in the days following inoculation. In our study, an initial increase in hydrogen (Figure 1) and $CO_2$ (Figure 2) production were observed with all fungal species that were added to digesters containing corn silage. Over the course of the 11-day hydrolysis experiment, all three fungal treatments produced similar cumulative volumes of $H_2$, in the range of 46–60 mL $g^{-1}$ VS ($p > 0.05$) (Figure 1a). Similar trends were observed for $CO_2$ gas production, with cumulative $CO_2$ production for all three treatments of corn silage in the range of 78–93 mL $g^{-1}$ VS ($p > 0.05$) (Figure 2a). The initial increase in gas production observed for corn silage may have been due to the fact that the material had already undergone ensiling. Additionally, background microbes such as $H_2$-producing *Clostridia* are known to be present in corn silage and were likely actively contributing to the gas production observed [32].

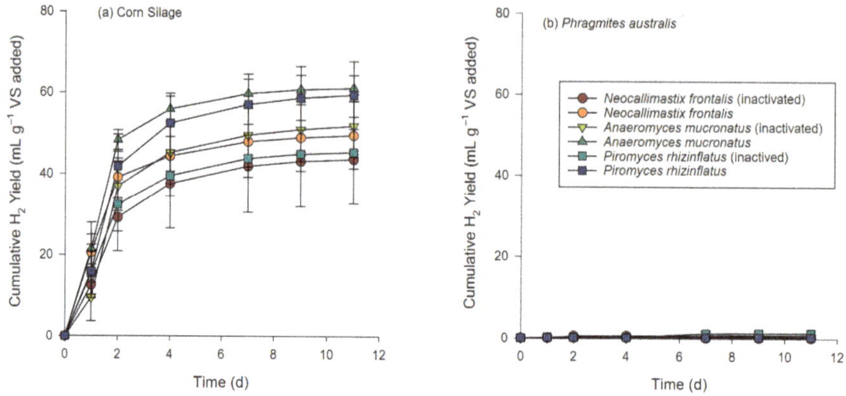


**Figure 1.** Cumulative hydrogen yield in the bioaugmentation of three different anaerobic fungal species: (**a**) corn silage (*Zea mays* L.); (**b**) common reed (*Phragmites australis*). Error bars show standard deviation.

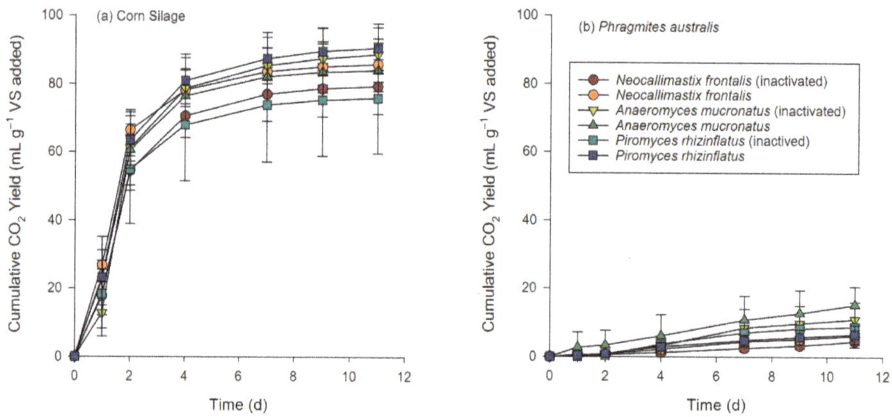


**Figure 2.** Cumulative carbon dioxide yield in the bioaugmentation of three different anaerobic fungal species: (**a**) corn silage (*Zea mays* L.); (**b**) common reed (*Phragmites australis*). Error bars show standard deviation.

During hydrolysis of common reed, smaller volumes of $H_2$ (<1 mL $g^{-1}$ VS) (Figure 1b) and $CO_2$ (<15.5 mL $g^{-1}$ VS) (Figure 2b) gas were evolved from all three treatments than was observed with corn silage ($p > 0.05$). Further studies are required to either eliminate the contribution of background microflora present on the substrate, or to account for the magnitude of their activity within the overall microbial consortia present during substrate hydrolysis.

### 3.3. Chemical Changes during Hydrolysis

Chemical oxygen demand was measured over the course of the hydrolysis experiment to examine the amount of soluble COD released due to hydrolysis and for further AD (Figure 3). Soluble COD concentrations for corn silage treatments trended upwards over time but did not significantly differ pre- and post-hydrolysis (Figure 3a). Similarly, soluble COD concentration did not increase in common reed (Figure 3b), regardless of treatment ($p > 0.05$), and was approximately half the value compared to corn silage. The higher initial COD concentration in corn silage compared to common reed was likely due to conversion of some corn biomass to soluble fermentation products during the ensiling process.

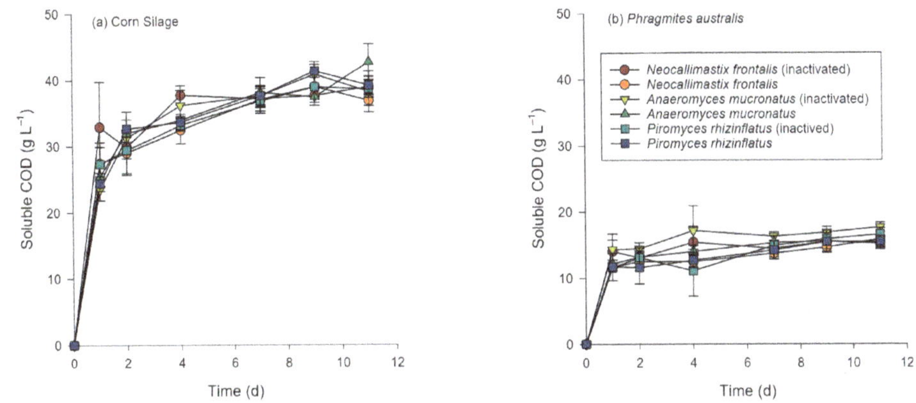


**Figure 3.** Soluble COD production in the bioaugmentation of three different anaerobic fungal species: (**a**) corn silage (*Zea mays* L.); (**b**) common reed (*Phragmites australis*). Error bars show standard deviation.

The main VFA produced during hydrolysis of corn silage were acetic, propionic, and butyric acids. Total VFA concentration trended upward over the course of the hydrolysis experiment for corn silage in all treatments; however, these increases were not statistically significant (Figure 4). In comparison, VFA production during hydrolysis of common reed was limited, with no significant difference in concentration observed over the course of the experiment for any treatments. It is unlikely that the lack of VFA accumulation could have been attributed to microbial conversion, as minimal gas volume was produced, suggesting an overall lack of microbial activity. The absence of a functional methanogenic phase in the experimental system could have been inhibitory. The inhibitory concentration of VFA for the specific fungal species investigated here is unknown, but it is possible that the concentrations observed were detrimental to continued fungal growth.

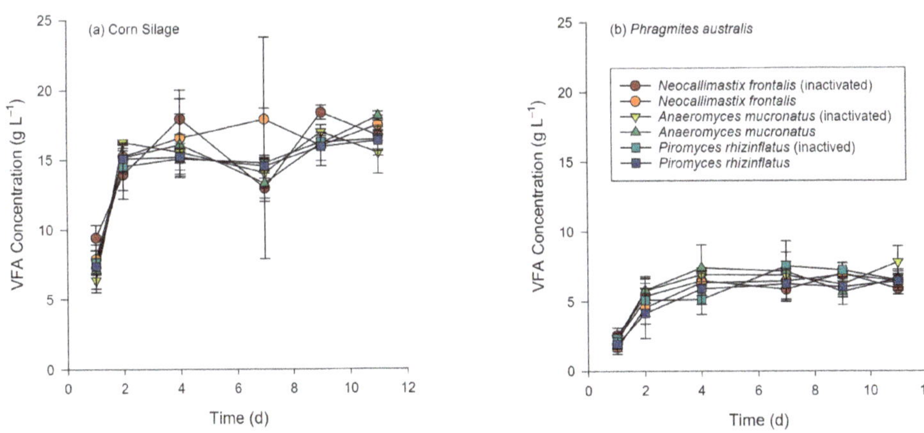


**Figure 4.** Volatile fatty acids (VFA) production in the bioaugmentation of three different anaerobic fungal species with and without activation: (**a**) corn silage (*Zea mays* L.); (**b**) common reed (*Phragmites australis*). Error bars show standard deviation.

The pH of the digestate in corn silage flasks decreased from approximately 7.0 on day 1 to 5.6–5.8 on day 11 for all treatments (Figure 5a). The pH of digestate in common reed flasks ranged between 7.2–8.0 on day 1 (Figure 5b), which was not statistically different from corn silage ($p > 0.05$). At the end of hydrolysis, the pH values of common reed decreased to 6.1–6.9, which again was not statistically different from corn silage ($p > 0.05$).

The optimum pH range for the growth of AF is between 6.0 and 7.0 [33], so reduction of pH to <6.5 due to VFA accumulation may have contributed to conditions unfavorable for anaerobic fungal activity. In the rumen, there is both a constant supply of buffering capacity as well as organic acid removal through the production of saliva and the symbiotic activities of the host animal and microbial consortium, respectively. The complexity of the rumen system is difficult to mimic in vitro in the laboratory, but a better approximation of the conditions which are favorable for AF to survive and be active will be essential for future success in this area of research.

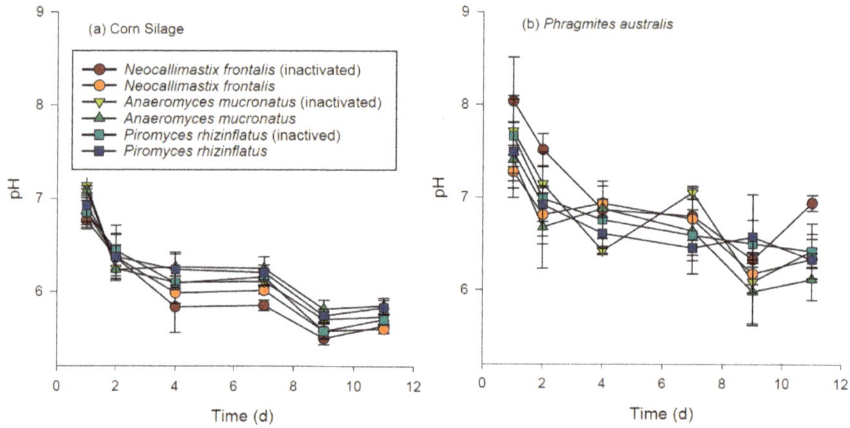


**Figure 5.** Changes in pH in the bioaugmentation of three different anaerobic fungal species with and without activation: (**a**) corn silage (*Zea mays* L.); (**b**) common reed (*Phragmites australis*). Error bars show standard deviation.

Ammonia is known to inhibit hydrolysis during AD at >200 mg $L^{-1}$ [34], but that threshold was not exceeded in our study (Table 2). Similarly, metals can inhibit biological hydrolysis processes when present in sufficient concentration. Alkaline metals, such as $Na^+$, $Mg^{2+}$, $K^+$, and $Ca^{2+}$, up to a range of 400 mg $L^{-1}$ help maintain alkalinity and pH in AD. However, higher concentrations would cause toxicity and inhibit AD processes [34]. In our study, $Na^+$ and $Mg^{2+}$ concentrations were very low (<1 mg $L^{-1}$) in all treatments and well below reported inhibitory levels, i.e., <750 mg $L^{-1}$ for $Mg^{2+}$ [35] and 3500–5500 mg $L^{-1}$ for $Na^+$ [36]. Concentration of $K^+$ was 1–3 mg $L^{-1}$ for all treatments, again below the inhibitory concentration of 400 mg $L^{-1}$ [34]. Similarly, for $Ca^{2+}$ the values were <1 mg $L^{-1}$ for all treatments and below inhibitory concentrations (>7000 mg $L^{-1}$). Based on this, the low degree of hydrolysis observed in all treatments was not likely caused by inhibition from ammonia or metals.

When considering our overall results, we can conclude (1) that there was limited hydrolytic activity in any of the reactors, regardless of fungal species or substrate type, and (2) the activity that was present was likely due to background microflora, including the bacteria that are present on the feedstock and not the AF. The most likely explanation for these results is that the AF were unable to survive, or at least be active, in the environment provided in this study. Low activity and survival of anaerobic fungi when applied to non-rumen environments has been reported by others [17,18,20].

**Table 2.** Effect of ammonia and alkaline metals in the bioagumentation of three different anaerobic fungal species.

| Substrate | Fungal Species | Active/Inactivated | $NH_3$ mg $L^{-1}$ | $Na^{+}$ mg $L^{-1}$ | $K^{+}$ mg $L^{-1}$ | $Mg^{2+}$ mg $L^{-1}$ | $Ca^{2+}$ mg $L^{-1}$ |
|---|---|---|---|---|---|---|---|
| Corn Silage | *Neocallimastix frontalis* | Inactivated | N/A | 0.87 ± 0.18 | 2.14 ± 0.13 | 0.14 ± 0.00 | 0.17 ± 0.00 |
| | | Active | N/A | 0.80 ± 0.08 | 2.20 ± 0.08 | 0.17 ± 0.02 | 0.23 ± 0.02 |
| | *Anaeromyces mucronatus* | Inactivated | N/A | 0.79 ± 0.08 | 2.19 ± 0.17 | 0.15 ± 0.02 | 0.21 ± 0.03 |
| | | Active | N/A | 0.75 ± 0.02 | 2.07 ± 0.02 | 0.14 ± 0.01 | 0.20 ± 0.02 |
| | *Piromyces rhizinflata* | Inactivated | N/A | 0.72 ± 0.03 | 2.00 ± 0.11 | 0.12 ± 0.00 | 0.18 ± 0.01 |
| | | Active | N/A | 0.72 ± 0.03 | 2.08 ± 0.09 | 0.13 ± 0.01 | 0.20 ± 0.01 |
| *Phragmites australis* | *Neocallimastix frontalis* | Inactivated | 78.45 ± 5.62 | 0.57 ± 0.03 | 1.63 ± 0.02 | 0.12 ± 0.01 | 0.19 ± 0.02 |
| | | Active | 14.71 ± 0.09 | 0.61 ± 0.02 | 1.64 ± 0.03 | 0.12 ± 0.00 | 0.22 ± 0.01 |
| | *Anaeromyces mucronatus* | Inactivated | 41.10 ± 0.42 | 0.53 ± 0.04 | 1.59 ± 0.02 | 0.10 ± 0.00 | 0.21 ± 0.00 |
| | | Active | 37.66 ± 14.56 | 0.64 ± 0.02 | 1.58 ± 0.03 | 0.08 ± 0.00 | 0.16 ± 0.01 |
| | *Piromyces rhizinflata* | Inactivated | 44.11 ± 2.05 | 0.59 ± 0.03 | 1.63 ± 0.07 | 0.08 ± 0.00 | 0.16 ± 0.00 |
| | | Active | 25.81 ± 0.36 | 0.58 ± 0.03 | 1.58 ± 0.08 | 0.10 ± 0.00 | 0.21 ± 0.01 |

Anaerobic fungi are known for having a close symbiotic activity and interspecies $H_2$ transfer with other microbes in the rumen [37,38]. The absence of these relationships, or the lack of time for such relationships to develop using the experimental design of this study, may also account for the poor hydrolytic activity observed. Coculturing AF with other rumen hydrolytic bacteria, such as *Fibrobacter succinogenes*, could be a potential approach to take in the future to increase viability [39]. Joblin et al. [40] inoculated *F. succinogenes* together with methanogenic cocultures of *Caecomyces/M. smithii* grown on rye grass. They found that there was an increase in stem degradation and attributed this to complementary fibrolytic activities between the two species. By comparison, our study utilized only AF to help degrade the substrate. The heat sterilization used to eliminate background microflora from anaerobic digestate may have limited potential for symbiotic relationships to develop between AF and bacteria. Conversely, antibiosis has been reported between ruminal bacteria and AF in laboratory studies [41], which highlights our current poor understanding of AF ecology.

Yıldırım et al. [42] recently reported up to a 60% increase in biogas yield from animal manures bioaugmented anaerobic fungi. In that study, the authors used an undefined mixture of AF isolated from a cow's rumen, resulting in an AF community composed of >6 groups (including *Anaeromyces* spp., *Neocallimastix* spp., and *Piromyces* spp. used in our study) [43]. It is possible that the mixed AF culture approach is a better strategy for ensuring AF activity and survival when used in bioaugmentation compared to single species inoculations, as we have described here. The benefit of a mixed AF culture could be due simply to higher diversity increasing the chances for survival under artificial conditions, and/or could be due to interactions between the different community members. The combination of our results and those by [42] strongly suggest that successful outcomes from the addition of AF to hydrolysis and/or anaerobic digestion may require use of mixed complex communities.

## 4. Conclusions

Hydrolysis of corn silage and common reed were not improved by bioaugmentation with three different species of AF, as evidenced by a lack of significant $H_2$ production or substrate degradation compared to controls. The most likely explanation for these results is that AF had low activity and/or survival in the anaerobic fermentation systems used in this study. More research is required to better understand survival of AF in anaerobic digestion processes to determine the feasibility of exploiting these organisms for lignocellulosic degradation.

**Author Contributions:** Conceptualization, B.R. and B.H.G.; methodology, B.R., V.N.N. and J.Y.; software, B.R. and B.H.G.; validation, B.R., V.N.N. and B.H.G.; formal analysis, B.R.; investigation, B.R. and V.N.N.; resources, X.H., J.Y., T.A.M. and B.H.G.; data curation, B.R. and V.N.N.; writing—original draft preparation, B.R.; writing—review and editing, V.N.N., C.V., X.H., J.Y., T.A.M., H.L. and B.H.G.; visualization, B.R. and B.H.G.; supervision, X.H., H.L. and B.H.G.; project administration,

B.H.G.; funding acquisition, X.H., T.A.M. and B.H.G. All authors have read and agreed to the published version of the manuscript.

**Funding:** This work was supported by the Natural Resources Canada Program of Energy Research and Development (grant number Inc001), Agriculture and Agri-Food Canada (grant number T.2002.36), and the Natural Sciences and Engineering Research Council of Canada (grant number 400806).

**Institutional Review Board Statement:** Not applicable.

**Informed Consent Statement:** Not applicable.

**Data Availability Statement:** Not applicable.

**Acknowledgments:** The authors wish to thank Kurtis Baute, Kimberley Van Overloop, Greg Travis, Douglas Marchbank, Stephanie Montienth, Winnie Lao, and Raene Barber for their technical assistance on this research project.

**Conflicts of Interest:** The authors declare no conflict of interest.

## References


1. EBA. *EBA Stastical Report 2020*; European Biogas Association: Brussels, Belgium, 2020.
2. Hutňan, M. Maize Silage as Substrate for Biogas Production. In *Advances in Silage Production and Utilization*; Silva, T.D., Ed.; IntechOpen: Rijeka, Croatia, 2016.
3. Lask, J.; Guajardo, A.M.; Weik, J.; von Cossel, M.; Lewandowski, I.; Wagner, M. Comparative environmental and economic life cycle assessment of biogas production from perennial wild plant mixtures and maize (*Zea mays* L.) in southwest Germany. *GCB Bioenergy* **2020**, *12*, 571–585. [CrossRef]
4. Baute, K.; Van Eerd, L.L.; Robinson, D.E.; Sikkema, P.H.; Mushtaq, M.; Gilroyed, B.H. Comparing the Biomass Yield and Biogas Potential of *Phragmites australis* with *Miscanthus* x *giganteus* and *Panicum virgatum* Grown in Canada. *Energies* **2018**, *11*, 2198. [CrossRef]
5. Vaičekonytė, R.; Kiviat, E.; Nsenga, F.; Ostfeld, A. An exploration of common reed (Phragmites australis) bioenergy potential in North America. *Mires Peat* **2014**, *13*, 1–9.
6. Akula, V.R. Wetland Biomass-Suitable for Biogas Production? Master's Thesis, Halmsted University, Halmstad, Sweden, 2013.
7. Wang, C.; Kong, Y.; Hu, R.; Zhou, G. Miscanthus: A fast-growing crop for environmental remediation and biofuel production. *GCB Bioenergy* **2021**, *13*, 58–69. [CrossRef]
8. e Silva, C.F.L.; Schirmer, M.A.; Maeda, R.N.; Barcelos, C.A.; Pereira, N. Potential of giant reed (*Arundo donax* L.) for second generation ethanol production. *Electron. J. Biotechnol.* **2015**, *18*, 10–15. [CrossRef]
9. Bayané, A.; Guiot, S. Animal digestive strategies versus anaerobic digestion bioprocesses for biogas production from lignocellulosic biomass. *Rev. Environ. Sci. Bio/Technol.* **2011**, *10*, 43–62. [CrossRef]
10. Dollhofer, V.; Podmirseg, S.M.; Callaghan, T.M.; Griffith, G.W.; Fliegerová, K. Anaerobic Fungi and Their Potential for Biogas Production. *Adv. Biochem. Eng. Biotechnol.* **2015**, *151*, 41–61.
11. Wei, Y.Q.; Yang, H.J.; Luan, Y.; Long, R.J.; Wu, Y.J.; Wang, Z.Y. Isolation, identification and fibrolytic characteristics of rumen fungi grown with indigenous methanogen from yaks (Bos grunniens) grazing on the Qinghai-Tibetan Plateau. *J. Appl. Microbiol.* **2016**, *120*, 571–587. [CrossRef]
12. Dollhofer, V.; Podmirseg, S.M.; Callaghan, T.M.; Griffith, G.W.; Fliegerová, K. Anaerobic Fungi and Their Potential for Biogas Production. In *Biogas Science and Technology*; Guebitz, G.M., Bauer, A., Bochmann, G., Gronauer, A., Weiss, S., Eds.; Springer International Publishing: Cham, Switzerland, 2015; pp. 41–61.
13. Janusz, G.; Pawlik, A.; Sulej, J.; Swiderska-Burek, U.; Jarosz-Wilkolazka, A.; Paszczynski, A. Lignin degradation: Microorganisms, enzymes involved, genomes analysis and evolution. *FEMS Microbiol. Rev.* **2017**, *41*, 941–962. [CrossRef]
14. Vogel, T.M. Bioaugmentation as a soil bioremediation approach. *Curr. Opin. Biotechnol.* **1996**, *7*, 311–316. [CrossRef]
15. Miah, M.S.; Tada, C.; Sawayama, S. Enhancement of Biogas Production from Sewage Sludge with the Addition of Geobacillus sp. Strain AT1 Culture. *Jpn. J. Water Treat. Biol.* **2004**, *40*, 97–104. [CrossRef]
16. Müller, H.; Trösch, W. Screening of white-rot fungi for biological pretreatment of wheat straw for biogas production. *Appl. Microbiol. Biotechnol.* **1986**, *24*, 180–185. [CrossRef]
17. Saye, L.M.; Navaratna, T.A.; Chong, J.P.; O'Malley, M.A.; Theodorou, M.K.; Reilly, M. The anaerobic fungi: Challenges and opportunities for industrial lignocellulosic biofuel production. *Microorganisms* **2021**, *9*, 694. [CrossRef] [PubMed]
18. Nkemka, V.N.; Gilroyed, B.H.; Yanke, J.; Gruninger, R.; Vedres, D.; McAllister, T.; Hao, X. Bioaugmentation with an anaerobic fungus in a two-stage process for biohydrogen and biogas production using corn silage and cattail. *Bioresour. Technol.* **2015**, *185*, 79–88. [CrossRef] [PubMed]
19. Liang, J.; Zhang, H.; Zhang, P.; Zhang, G.; Cai, Y.; Wang, Q.; Zhou, Z.; Ding, Y.; Zubair, M. Effect of substrate load on anaerobic fermentation of rice straw with rumen liquid as inoculum: Hydrolysis and acidogenesis efficiency, enzymatic activities and rumen bacterial community structure. *Waste Manag.* **2021**, *124*, 235–243. [CrossRef] [PubMed]

20. Dollhofer, V.; Callaghan, T.M.; Griffith, G.W.; Lebuhn, M.; Bauer, J. Presence and transcriptional activity of anaerobic fungi in agricultural biogas plants. *Bioresour. Technol.* **2017**, *235*, 131–139. [CrossRef]
21. Lowe, S.E.; Theodorou, M.K.; Trinci, A.P.J.; Hespell, R.B. Growth of anaerobic rumen fungi on defined and semi-difined media lacking rumen fluid. *J. Gen. Microbiol.* **1985**, *131*, 2225–2229.
22. Hungate, R.E. The anaerobic mesophilic cellulolytic bacteria. *Bacteriol. Rev.* **1950**, *14*, 1–49. [CrossRef]
23. Nkemka, V.N.; Murto, M. Two-stage anaerobic dry digestion of blue mussel and reed. *Renew. Energy* **2013**, *50*, 359–364. [CrossRef]
24. Gilroyed, B.H.; Li, C.; Hao, X.; Chu, A.; McAllister, T.A. Biohydrogen production from specified risk materials co-digested with cattle manure. *Int. J. Hydrogen Energy* **2010**, *35*, 1099–1105. [CrossRef]
25. APHA. *Standard Methods for the Examination of Water and Wastewater*; APHA-AWWA-WEF: Washington, DC, USA, 2005.
26. Gilroyed, B.H.; Li, C.; Reuter, T.; Beauchemin, K.A.; Hao, X.; McAllister, T. Influence of distiller's grains and condensed tannins in the diet of feedlot cattle on biohydrogen production from cattle manure. *Int. J. Hydrogen Energy* **2015**, *40*, 6050–6058. [CrossRef]
27. Gilroyed, B.H.; Reuter, T.; Chu, A.; Hao, X.; Xu, W.; McAllister, T.A. Anaerobic digestion of specified risk materials with cattle manure for biogas production. *Bioresour. Technol.* **2010**, *101*, 5780–5785. [CrossRef] [PubMed]
28. Van Soest, P.J.; Robertson, J.B.; Lewis, B.A. Methods for dietary fiber, neutral detergent fiber, and nonstarch polysaccharides in relation to animal nutrition. *J. Dairy Sci.* **1991**, *74*, 3583–3597. [CrossRef]
29. Mertens, D.R.; Allen, M.; Carmany, J.; Clegg, J.; Davidowicz, A.; Drouches, M.; Frank, K.; Gambin, D.; Garkie, M.; Gildemeister, B.; et al. Gravimetric determination of amylase-treated neutral detergent fiber in feeds with refluxing in beakers or crucibles: Collaborative study. *J. AOAC Int.* **2002**, *85*, 1217–1240.
30. Rabii, A.; Aldin, S.; Dahman, Y.; Elbeshbishy, E. A review on anaerobic co-digestion with a focus on the microbial populations and the effect of multi-stage digester configuration. *Energies* **2019**, *12*, 1106. [CrossRef]
31. Gruninger, R.J.; Puniya, A.K.; Callaghan, T.M.; Edwards, J.E.; Youssef, N.; Dagar, S.S.; Fliegerova, K.; Griffith, G.W.; Forster, R.; Tsang, A.; et al. Anaerobic fungi (phylum *Neocallimastigomycota*): Advances in understanding their taxonomy, life cycle, ecology, role and biotechnological potential. *Fems. Microbiol. Ecol.* **2014**, *90*, 1–17. [CrossRef]
32. Borreani, G.I.; Tabacco, E.R.; Schmidt, R.J.; Holmes, B.J.; Muck, R.E. Silage review: Factors affecting dry matter and quality losses in silages. *J. Dairy Sci.* **2018**, *101*, 3952–3979. [CrossRef]
33. Uwineza, C.; Mahboubi, A.; Atmowidjojo, A.C.; Ramadhani, A.N.; Wainaina, S.; Millati, R.; Wikandari, R.; Niklasson, C.; Taherzadeh, M.J. Cultivation of edible filamentous fungus Aspergillus oryzae on volatile fatty acids derived from anaerobic digestion of food waste and cow manure. *Bioresour. Technol.* **2021**, *337*, 125410. [CrossRef]
34. Chen, Y.; Cheng, J.J.; Creamer, K.S. Inhibition of anaerobic digestion process: A review. *Bioresour. Technol.* **2008**, *99*, 4044–4064. [CrossRef]
35. Romero-Güiza, M.S.; Mata-Alvarez, J.; Chimenos, J.M.; Astals, S. The effect of magnesium as activator and inhibitor of anaerobic digestion. *Waste Manag.* **2016**, *56*, 137–142. [CrossRef]
36. Veluchamy, C.; Kalamdhad, A.S. Influence of pretreatment techniques on anaerobic digestion of pulp and paper mill sludge: A review. *Bioresour. Technol.* **2017**, *245*, 1206–1219. [CrossRef] [PubMed]
37. Ivarsson, M.; Schnürer, A.; Bengtson, S.; Neubeck, A. Anaerobic fungi: A potential source of biological H2 in the oceanic crust. *Front. Microbiol.* **2016**, *7*, 674. [CrossRef] [PubMed]
38. Hess, M.; Paul, S.S.; Puniya, A.K.; van der Giezen, M.; Shaw, C.; Edwards, J.E.; Fliegerová, K. Anaerobic Fungi: Past, Present, and Future. *Front. Microbiol.* **2020**, *11*, 584893. [CrossRef]
39. Kobayashi, Y.; Shinkai, T.; Koike, S. Ecological and physiological characterization shows that Fibrobacter succinogenes is important in rumen fiber digestion. *Folia Microbiol.* **2008**, *53*, 195–200. [CrossRef] [PubMed]
40. Joblin, K.N.; Matsui, H.; Naylor, G.E.; Ushida, K. Degradation of Fresh Ryegrass by Methanogenic Co-Cultures of Ruminal Fungi Grown in the Presence or Absence of Fibrobacter succinogenes. *Curr. Microbiol.* **2002**, *45*, 46–53. [CrossRef]
41. Dehority, B.A.; Tirabasso, P.A. Antibiosis between Ruminal Bacteria and Ruminal Fungi. *Appl. Environ. Microb.* **2000**, *66*, 2921–2927. [CrossRef]
42. Yıldırım, E.; Ince, O.; Aydin, S.; Ince, B. Improvement of biogas potential of anaerobic digesters using rumen fungi. *Renew. Energy* **2017**, *109*, 346–353. [CrossRef]
43. Aydin, S.; Yıldırım, E.; Ince, O.; Ince, B. Rumen anaerobic fungi create new opportunities for enhanced methane production from microalgae biomass. *Algal Res.* **2017**, *23*, 150–160. [CrossRef]

*applied sciences*

MDPI

*Article*

# Diagnosis and Monitoring of Volatile Fatty Acids Production from Raw Cheese Whey by Multiscale Time-Series Analysis

**Antonio Lara-Musule [1,2], Ervin Alvarez-Sanchez [1], Gloria Trejo-Aguilar [3], Laura Acosta-Dominguez [2], Hector Puebla [4] and Eliseo Hernandez-Martinez [2,*]**

1 Doctorado en Ingeniería, Facultad de Ingeniería Mecánica-Eléctrica, Universidad Veracruzana, Xalapa-Enríquez 91000, Mexico; antolara@uv.mx (A.L.-M.); eralvarez@uv.mx (E.A.-S.)
2 Facultad de Ciencias Químicas, Universidad Veracruzana, Xalapa-Enríquez 91000, Mexico; lacosta@uv.mx
3 Departamento de Biotecnología, Universidad Autónoma Metropolitana-Iztapalapa, Mexico City 09340, Mexico; gmta@xanum.uam.mx
4 Departamento de Energía, Universidad Autónoma Metropolitana-Azcapotzalco, Mexico City 02200, Mexico; hpuebla@azc.uam.mx
* Correspondence: elisehernandez@uv.mx

**Abstract:** Anaerobic treatment is a viable alternative for the treatment of agro-industrial waste. Anaerobic digestion reduces organic load and produces volatile fatty acids (VFA), which are precursors of value-added products such as methane-rich biogas, biohydrogen, and biopolymers. Nowadays, there are no low-cost diagnosis and monitoring systems that analyze the dynamic behavior of key variables in real time, representing a significant limitation for its practical implementation. In this work, the feasibility of using the multiscale analysis to diagnose and monitor the key variables in VFA production by anaerobic treatment of raw cheese whey is presented. First, experiments were carried out to evaluate the performance of the proposed methodology under different operating conditions. Then, experimental pH time series were analyzed using rescaled range ($R/S$) techniques. Time-series analysis shows that the anaerobic VFA production exhibits a multiscale behavior, identifying three characteristic regions (i.e., three values of Hurst exponent). In addition, the dynamic Hurst exponents show satisfactory correlations with the chemical oxygen demand (COD) consumption and VFA production. The multiscale analysis of pH time series is easy to implement and inexpensive. Hence, it could be used as a diagnosis and indirect monitoring system of key variables in the anaerobic treatment of raw cheese whey.

**Keywords:** anaerobic treatment; raw cheese whey; multiscale analysis; pH time series

**Citation:** Lara-Musule, A.; Alvarez-Sanchez, E.; Trejo-Aguilar, G.; Acosta-Dominguez, L.; Puebla, H.; Hernandez-Martinez, E. Diagnosis and Monitoring of Volatile Fatty Acids Production from Raw Cheese Whey by Multiscale Time-Series Analysis. *Appl. Sci.* **2021**, *5803*, 915. https://doi.org/10.3390/app11135803

Academic Editor: Carlos Rico de la Hera

Received: 17 May 2021
Accepted: 11 June 2021
Published: 23 June 2021



## 1. Introduction

Cheese whey is a waste generated in cheese production, which is characterized by its high organic load [1]. Different biological and physicochemical treatments have been proposed to recover compounds with added value, such as proteins and lactose. However, a significant number of small and medium industries do not have the necessary economic resources to implement technologies for the treatment of whey, which leads to the inadequate disposal of the waste, causing serious problems of environmental pollution [2]. Anaerobic treatment is an alternative for the valorization of this effluent. The anaerobic digestion of cheese whey reduces the organic load with simultaneous recovery of value-added by-products such as VFA, methane-rich biogas, biohydrogen, and biopolymers [3,4].

Anaerobic digestion (AD) is a complex and intrinsically unstable process. Indeed, the anaerobic treatment is formed by consecutive stages conducted by a specific group of microorganisms with different metabolic characteristics which are not fully understood. Moreover, variations of the input variables (hydraulic flow rate, influent organic load) and the effects of inhibitory agents may easily lead to the wash-out condition [5].

Thus, proper diagnosis and control of AD are of utmost relevance to avoid serious breakdowns and guarantee operational stability [6,7]. To this end, it is necessary to monitor the variables that provide information on the performance of each stage. The key variables of each stage can be defined by their relationship with the performance of the associated microbial consortium. Indeed, carbohydrate and protein degradation are related to hydrolysis, VFA production with acidogenesis-acetogenesis, and biogas production with acetogenesis-methanogenesis [8]. However, at the industrial level, in situ and in-line measurements in AD plants are limited to pH, temperature, water flow, biogas flow, level, and pressure [9–11].

Although pH and biogas production can be used as indicators of the overall status of the AD process, they fail to indicate the stress level of the reactor in real time [12]. On the one hand, biogas production can show the overall process performance. However, a decrease in biogas production often occurred after the process was severely inhibited. On the other hand, pH is not a suitable parameter for a well-buffered bioreactor since pH has low sensitivity. VFA concentration has been suggested as an ideal indicator [12]. Indeed, VFA accumulation is usually interpreted as the result of methanogenesis inhibition or organic overloading, leading to a risk of process breakdown. Thus, for AD process diagnosis, several authors have proposed different thresholds based on VFA measurements as early warning indicators to judge the reactor status [5,13–16].

Hence, for the AD diagnosis, different model-based and data-driven estimation approaches have been proposed to determine key variables, including VFA concentration, using the available information in real time (e.g., temperature, pH, conductivity measurements). Model-based estimation methods are based on the reconstruction of unmeasured variables from the measurements available through the mathematical model of the process [11,17]. Model-based estimation approaches applied to AD processes include state-observers [18–22], Bayesian methods [23], and filtering techniques [24,25]. The most significant limitation of model-based approaches is that they depend on the accuracy of the process model, i.e., the lower the precision of the model, the greater the estimation error. Moreover, model uncertainties degrade model predictions, and the measurements are affected by noise and disturbances.

Data-driven methods are based on statistical models trained to learn trends from collected historical [26]. Data-driven estimation approaches are gaining broad interest at the industrial level because of their practicability, robustness, and flexibility to be developed and applied to a wide range of processes, in addition to their independence from a mathematical process model [17,27]. Examples of data-driven methods used in AD processes include artificial neural networks (ANN) [28,29], regression models [30], machine learning [16,31], and fuzzy inference systems [32]. Data-driven methods based on multiscale approaches have been introduced due to the superior robustness and accuracy concerning single-scale analysis [33]. Indeed, the motivation for using multiscale methods for feature extraction is that the data from almost all practical processes are of a multiscale nature. For instance, the events can occur at different locations and with different localization in time and frequency. Moreover, conventional statistical methods based on a single scale are not ideally suited for separating the deterministic component from the normal process variation. Lee et al. [34] applied a multiscale approach with principal component analysis and wavelets to a full-scale biological anaerobic process. Data-driven methods have two main disadvantages: (i) High implementation costs in terms of design effort (architecture selection, training stage, extensive multivariable data sets) and computational cost. (ii) The lack of phenomenological interpretability since they are based on black-box models to predict the desired estimated states.

In the last few years, multiscale data-driven methods based on fractal analysis have shown great potential to infer the key variables of the process. Méndez-Acosta et al. [35] proposed the rescaled range ($R/S$) analysis of pH time series obtained from an up-flow reactor for the anaerobic digestion of tequila vinasses, where high correlations between the multiscale indices and critical variables of the process (i.e., COD, VFA, and biogas production) were determined. Similar results were found by Hernandez-Martinez et al. [36] applying multiscale analysis in pH time series obtained from anaerobic treatment of tequila vinasses in continuous digesters (FBR and CSTR). They concluded that the $R/S$ analysis could qualitatively estimate essential variables for the process evolution. Finally, Sanchez-Garcia et al. [37] studied the anaerobic digestion of cheese whey, determining that multiscale analysis can be used for monitoring the anaerobic treatment of complex substrates. Hence, the above works have shown that through multiscale analysis of time series, it is possible to dynamically characterize the key variables of the anaerobic digestion process using different substrates. Multiscale fractal analysis has two nice features for its practical applicability. The multiscale analysis allows a better understanding of the underlying process phenomena occurring at different scales. Furthermore, a reduction of the implementational costs is obtained (in terms of both design effort and computational cost) since it is based on using a single easy-to-measure and low-cost variables, such as pH and electrical conductivity.

In this paper, the application of the multiscale fractal analysis is presented for diagnosis purposes of the anaerobic treatment of raw cheese whey. It should be noted that previous contributions on the multiscale fractal analysis of anaerobic digestion have been applied to stable operating conditions favoring methane production. However, multiscale analysis in situations where the process is affected by inhibition or problems that do not favor methane production has not been evaluated. An efficient monitoring system must assess the dynamic behavior of the variables in a broad region of operation, especially in the region where the process tends to destabilize since, in these conditions, it is when action is required to control the process. Indeed, the need for a proper diagnosis and monitoring system is more significant when the process is in critical conditions. The multiscale fractal analysis is performed on pH time series of two different experimental conditions of pH and temperature. The time series are analyzed using $R/S$ multiscale analysis. The results indicate that it is possible to track the key variables using dynamically calculated multiscale indices, which open the possibility of developing reliable diagnosis and monitoring systems of the anaerobic processes that contribute to the operational improvement of the process.

## 2. Materials and Methods

In this section, the experimental conditions used for the anaerobic treatment of raw cheese whey are presented. The operating conditions, as well as the offline and online measurements performed during the experimental runs, are also described.

### 2.1. Experimental Setup

The anaerobic fermentation of raw cheese whey was carried out in an anaerobic sequencing batch reactor (AnSBR) with an effective volume of 3.5 L. The sludge used as inoculum was obtained from a laboratory-scale anaerobic digester fed with raw cheese whey OLR = 3.6 $g_{COD}$ $L^{-1}$ $d^{-1}$ and operated at 33 °C, pH = 8.0 and HRT = 30 d. Experimental runs were performed using fresh raw cheese whey as a substrate collected from a community dairy located in Acajete Veracruz, México. The cheese whey was dried, milled, and screened in a 1 mm mesh. After that, it was stored at 4 °C. Table 1 shows the values obtained of the characterization of cheese whey and inoculum.

In order to evaluate the effect of initial pH and temperature on VFA production, four experimental essays to different conditions (pH = 5.5-T = 35 °C, pH = 5.5-T = 40 °C, pH = 7.5-T = 35 °C, and pH = 7.5-T = 40 °C) were carried. In experimental trials performed, the pH control was only established to avoid acidification of the system, i.e., the pH is not less than the initial pH. The pH control for the pH increase was not established, causing pH to increase, so the initial pH value was not maintained.

### 2.2. Analytical Methods

The raw cheese whey and inoculum were characterized by determining the concentration of fat, protein, lactose, total carbohydrates (TCH), VFA, total solids (TS), volatile suspended solids (VSS), as well as the COD and alkalinity. Standard techniques were used to determine COD [38] and VSS [39]. Total carbohydrates and protein concentration were determined according to Dubois et al. [40] and Bradford et al. [41], respectively. An Agilent gas chromatograph (model 7820 A) equipped with a capillary column and a flame ionization detector was used to determine the concentrations of short-chain organic acids. Conversion factors were used to pass VFA, TCH, VSS, and protein concentration to COD equivalent concentration. The biogas production was measured with a Gow Mac 580 series gas chromatograph equipped with an isothermal column oven. Helium and Nitrogen were used as carrier gas.

**Table 1.** Characterization of whey and inoculum.

| Parameters | Whey | Inoculum |
|---|---|---|
| COD (g $L^{-1}$) | 74.24 ± 1.35 | 32.52 ± 4.27 |
| Carbohydrates (CH) ($g_{glucose}$ $L^{-1}$) | 35.04 ± 1.56 | 0.94 ± 0.07 |
| Total Solids (g $L^{-1}$) | 51.21 ± 1.364 | 49.87 ± 0.95 |
| Volatile Solids (g $L^{-1}$) | 36.68 ± 4.68 | 22.46 ± 4.30 |
| Volatile Sedimented solids (g $L^{-1}$) | 1.20 ± 0.24 | 2.15 ± 0.41 |
| VFA ($g_{COD}$ $L^{-1}$) | 0.57 ± 0.08 | 28.56 ± 1.90 |
| Protein (mg $L^{-1}$) | 27.32 ± 2.17 | 7.22 ± 1.09 |
| pH | 4.72 ± 0.04 | 6.85 ± 0.08 |

A NI cRIO-9074 system from National Instrument and an M300 pH/ORP transmitter from Mettler-Toledo were used to acquire pH time series. The pH meter has an accuracy of ±0.02, capturing 1 datapoint per second in each experiment. The pH time series shows temporal variations, reflecting the dynamic changes of the interactions between the characteristic phenomena involved in the process (Figure 1). For all experiments, fluctuations are observed around the initial pH, followed by an increase in pH to a maximum value, to reach an apparent steady state, where the pH stabilizes finally. It can be observed that the pH dynamics depend on the process operating conditions. In the experiments at initial pH = 5.5, the system showed an increase in pH at a time greater than 35 h, while for pH = 7.5, such increase occurred at 80 h.

### 2.3. Multiscale Time Series Analysis

In the literature, several methodologies are available for multiscale time-series analysis. However, due to its simplicity and easy implementation, rescaled range ($R/S$) [42] is widely applied for time-series analysis obtained from physical, chemical, and biological processes.

The $R/S$ statistical measures the range of the deviations of the partial sums in a time series about its average, rescaled by the series standard deviation. For a time series $Z_N = (Zi)$ with length $N$, it is considered a subsequence $X_{NS} = (Xi)$ of length $N_S$, where $N_S < N$. The $R/S$ statistic for $X_{NS}$ is calculated as

$$(R/S) = \frac{1}{\sigma_S(N_S)}\left\{\max_{1\le i\le M}\sum_{k=1}^{i}(X_k - \overline{X}_{Ns}) - \min_{1\le i\le M}\sum_{k=1}^{i}(X_k - \overline{X}_{Ns})\right\} \tag{1}$$

where $\overline{X}_{Ns}$ is the subsequence mean and $\sigma_S(N_S)$ is the sample standard deviation, which are defined as

$$\overline{X}_{Ns} = \frac{1}{N_S}\sum_{i=1}^{Ns} X_i \text{ and } \sigma(N_S) = \sqrt{\frac{1}{N_S}\sum_{k=1}^{N_S}(X_k - \overline{X}_{Ns})^2} \tag{2}$$

The $R/S$ statistic follows a power law, $(R/S) = aN_S^H$ where $a$ is a constant and $H$ is the Hurst exponent. A log-log plot of $(R/S)$ as a function of $N_S \in (N_{S,\min}, N_{S,\max})$, gives a straight line with slope $H$. If the series data is independent (e.g., white-noise process), the plot is roughly a straight line with slope $H = 0.5$. If $H > 0.5$, the time series is persistent, indicating the presence of long-term autocorrelations. Conversely, if $H < 0.5$ autocorrelations in the signal are antipersistent [43,44].

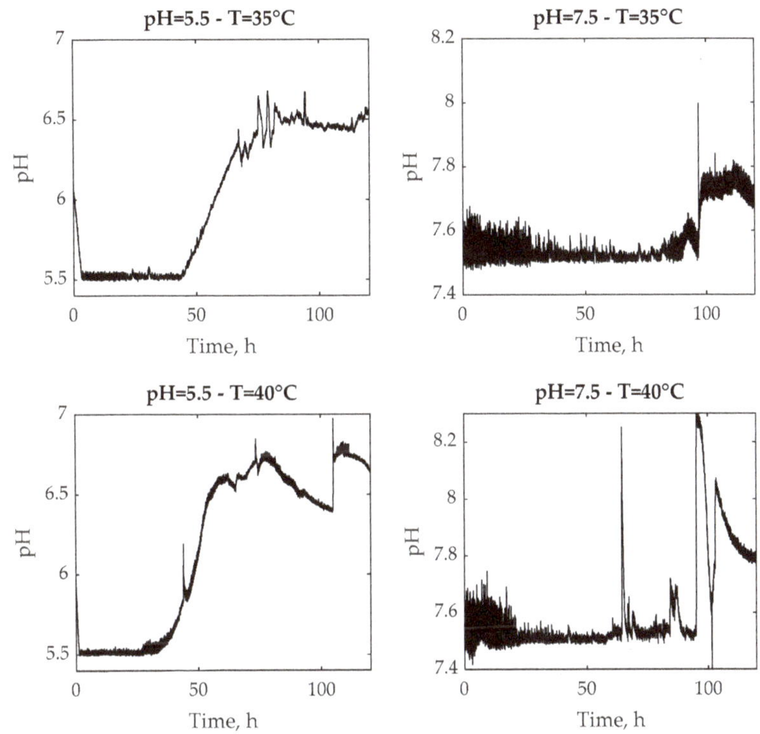


**Figure 1.** pH time series collected in the anaerobic treatment of raw cheese whey at different conditions of initial pH and temperature.

Multifractal Analysis

Multifractality is a valuable tool for explaining many patterns seen in nature. Particularly, multifractal analysis allows for the investigation of a mixture of fractal dimensions characterizing the inherent complexity in some data series. Based on the thoughts of Barabasi and Vicsek [45] and Katsuragi and Honjo [46], the multifractal analysis can be done through the calculation of the rescaled range through the q-norm of σs, that is,

$$\sigma_q(N_S) = \left[ \frac{1}{N_S} \sum_{k=1}^{Ns} (X_k - \overline{X}_{Ns})^q \right]^{\frac{1}{q}} \quad (3)$$

In this case, the $R/S$ statistical is given by $(R/S)_q = (R/S)\sigma(N_S)/\sigma_q(N_S)$, and the average range is expected to follow the scaling behavior $(R/S)_q = aN_S^{2H_q}$, where $H_q$ is the q-th Hurst exponent. If $H_q$ is constant for all $q$ then the underlying phenomena are monofractal. A nontrivial dependence of $H_q$ vs. $q$ indicates that the process is multifractal. It should be recalled that a multifractal system is a generalization of a fractal system in which a single exponent (the fractal dimension) is not enough to describe its dynamics. In general, multifractality also indicates the nonlinear nature of the mechanisms that generated the series [47,48]. To describe the degree of multifractality, we propose a multifractal index ($I_M$) defined as,

$$I_M = \max(H(q)) - \min(H(q)) \quad (4)$$

where $H(q)$ is the $H$ value as a function of $q$.

## 3. Results and Discussion

In this section, the results obtained from applying the multiscale analysis on the pH time series collected in the VFA production process by the anaerobic treatment of raw cheese whey are presented. First, the effect of pH and temperature on the overall performance of the process (i.e., VFA production and COD consumption) is described. Subsequently, the correlations obtained between the multiscale analysis and the experimental data of the key variables of the process are shown. Finally, the application of multifractal analysis to identify dynamic changes in the process associated with the different stages of the process is shown.

### 3.1. Experimental Tests

Whey is considered a complex substrate that is constituted mainly by carbohydrates, proteins, and lipids. At the conditions evaluated in this work, it is observed that total carbohydrates are consumed up to 95% in the first 35 h of operation (Figure 2).

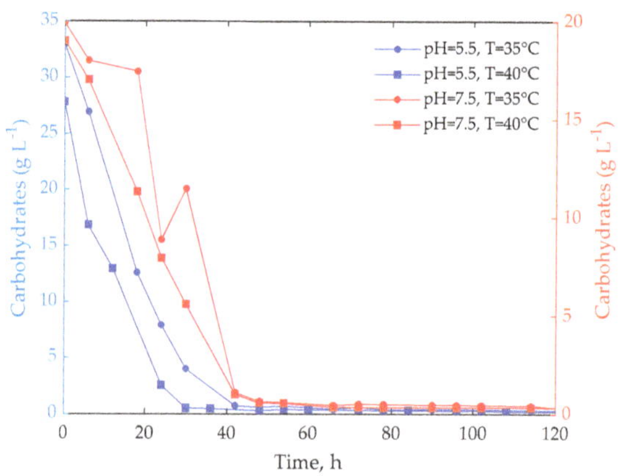


**Figure 2.** The dynamic profile of total carbohydrate consumption.

However, the COD consumption only reaches a consumption between 30 and 50%, which suggests that the consumption of lipids and proteins was less significant. At pH = 7.5 and T = 40 °C, 50% of the COD consumption is reached, while at pH = 5.5 and T = 35 °C the minimum consumption is obtained with 30% (Figure 3). These results correspond to those obtained by Perna et al. [49], who achieved COD removal percentages between 14% and 32% using diluted whey.

Regarding VFA production, the amount and types of VFA produced mainly depend on the initial pH. In Figure 3, the production of VFA is shown, noting that at initial pH = 5.5, butyric, valeric, and caproic acids are favored. In contrast, at initial pH = 7.5, the production of acetic and propionic acids is favored. These results agree with that reported by different authors, who indicate that at pH < 6, butyric acids are favored, and at pH > 6 propionic and acetic acids are favored [50–53].

The pH time series exhibit dynamic changes that correspond to the observed changes in key process variables. For the first 35 h, where the carbohydrate consumption rate is high, the pH series shows strong fluctuations, which could indicate the high activity of hydrolytic microorganisms. Subsequently, after 40 h, the experiments at pH 5.5 present an alkalinization stage, observing an increase in pH values of 6.8–7. This behavior can be associated with the production of biogas. Indeed, the presence of $CO_2$ promotes the generation of bicarbonates which causes an increase in pH. For the experiments at pH = 7.5, the alkalinization stage begins at 80 h, showing a slight increase in pH, reaching a maximum value of pH = 8.3. Finally, when the pH stabilizes, the highest VFA production rate is observed, which indicates the high activity of the acidogenic-acetogenic microorganisms.

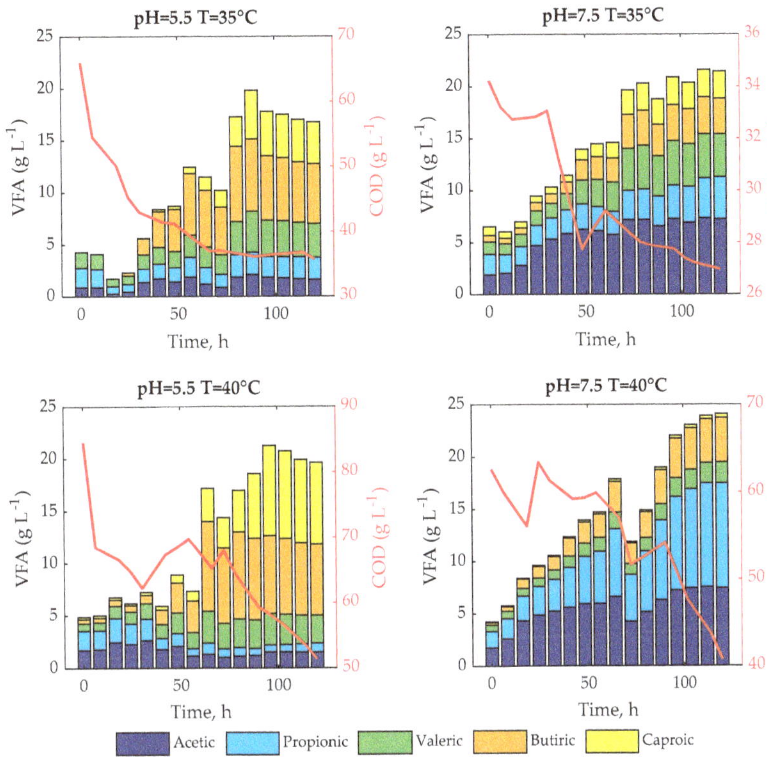


**Figure 3.** The dynamic profile of COD consumption and distribution of VFA produced.

### 3.2. R/S Analysis

$R/S$ analysis is applied to the pH series of the four experimental conditions. The time scale interval is given by $N_{S,\min}$ = 20 data (20 s) and $N_{S,\max}$ = 3600 data (1 h), which according to Sanchez-Garcia et al. [37], is the time scale where the characteristic phenomena corresponding to each of the stages of the anaerobic digestion of crude whey can be identified. Figure 4 shows the statistic $(R/S)2$ as a function of the time scale Ns, observing that the power-law ($(R/S) = aNs^H$) exhibits three changes in slope (i.e., three different values of the Hurst exponent, $H_1$, $H_2$, and $H_3$), named as Zones 1, 2, and 3. The presence of different values of the Hurst exponent suggests that the $R/S$ analysis can identify three characteristic phenomena immersed in the fluctuations of the pH series. These zones correspond to those reported in the anaerobic digestion of whey [37] and tequila vinasse [35,36], where each region was attributed to a different anaerobic digestion stage. Based on the above observations, Zone 1 is related to hydrolysis, Zone 2 corresponds with the acidogenesis and acetogenesis stages, and Zone 3 is related to methanogenesis. It should be noted that in conventional anaerobic digestion processes, the experimental conditions favor methane-rich biogas production. In contrast, in this study, the operating conditions favor the VFA production, so the methanogenic stage is not significant.

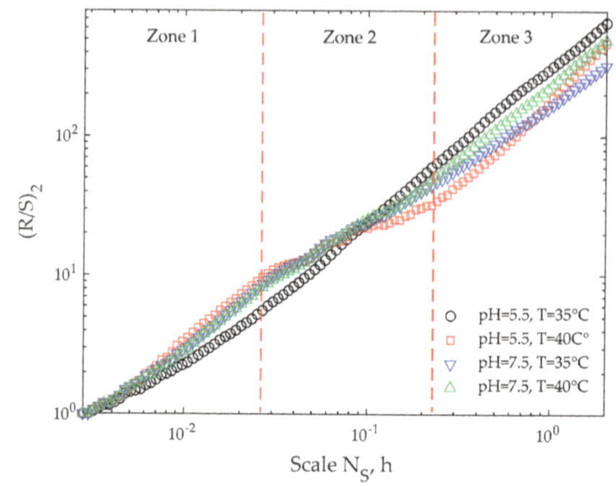


**Figure 4.** $R/S$ statistic vs. time scale $N_S$.

Table 2 shows the values of the Hurst exponent for each zone, where it is observed that the $H_1$ values are around 1.0, indicating that in this time scale, there are characteristic phenomena with similar behavior for all the operating conditions evaluated. On the other hand, the Hurst exponent values in Zones 2 and 3 show more significant variability between operating conditions, which indicates that the phenomena on these characteristic scales are more susceptible to changes in pH and temperature.

**Table 2.** Hurst exponent by zone for the four experimental tests.

| Zone/Test | T = 35 °C pH = 5.5 | T = 35 °C pH = 7.5 | T = 40 °C pH = 5.5 | T = 40 °C pH = 7.5 |
|---|---|---|---|---|
| Zone 1 | 0.85819 | 0.99983 | 1.01371 | 0.9804 |
| Zone 2 | 1.12514 | 0.73615 | 0.51536 | 0.79128 |
| Zone 3 | 1.05433 | 0.89266 | 1.28973 | 1.05855 |

According to the above, Zone 1 could be correlated with the hydrolytic stage. In all experiments, the COD consumption is mainly to carbohydrate degradation, which is consumed up to 95% in the first two days. Zones 2 and 3 could be correlated with the acidogenic-acetogenic stages, which show the formation of different fatty acids depending on the operating conditions evaluated. For example, at pH = 5.5, higher production of butyric and caproic acids is obtained, while at pH = 7.5, higher production of acetic and propionic acids is obtained. Note that the $H_3$ value is higher when there is a more significant formation of butyric and caproic acids.

In order to identify correlations between the key variables and the Hurst exponents, the $R/S$ analysis is implemented considering 6 h windows, which are shifted every hour, covering the whole series. This allows one to obtain the Hurst exponent values as a function of scale and time. For the four experimental conditions, Figure 5 shows the dynamic $R/S$ analysis, where the three zones described in Figure 4 can be identified, noting that the H values show temporal variations in each zone, which can be correlated with those of the process stages.

In order to avoid different interpretations of the dynamic Hurst exponent, the average is carried out in each zone, which leads to an H dynamic profile for each zone (i.e., $Hd_1$, $Hd_2$, and $Hd_3$). These profiles are compared with the experimental measurements of COD and VFA.

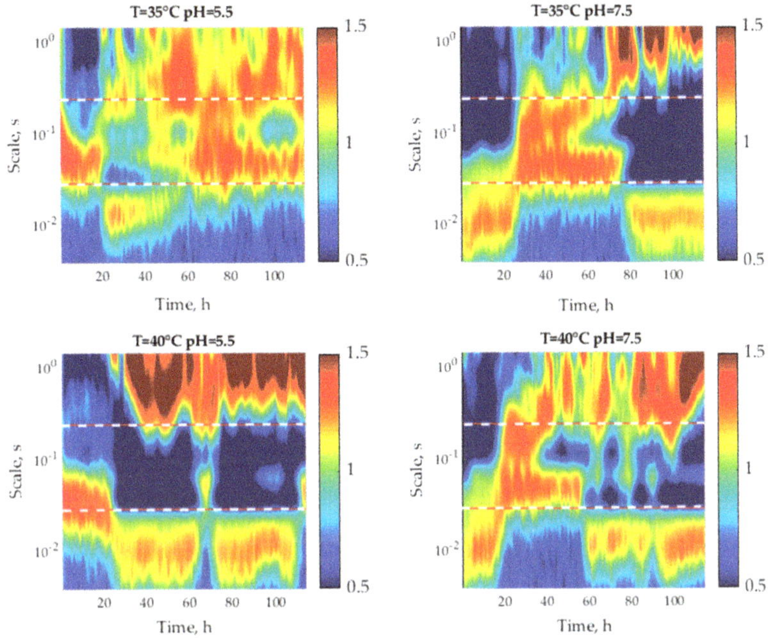


**Figure 5.** Scale-temporal profile of the Hurst exponent considering a 6 h moving window.

Considering the four experimental conditions, Figure 6 shows the comparison between the COD measurements with $Hd_1$, where it is observed that the $Hd_1$ dynamic exhibits the same trend observed with the COD consumption, which confirms that the time scale of Zone 1 corresponds to the hydrolysis stage in the anaerobic treatment of whey.

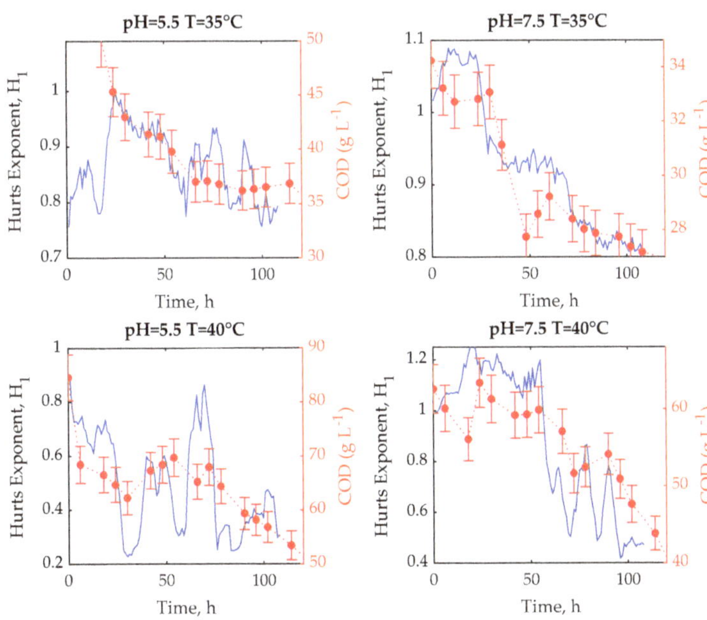


**Figure 6.** Correlations between $Hd_1$ and COD consumption measurements.

Figure 7 shows the comparison between the total VFA measurements produced with the $Hd_3$ dynamic profile, finding that both profiles exhibit the same behavior, suggesting that the time scale of Zone 3 corresponds to the acidogenic-acetogenic stage. $Hd_2$ profile shows behavior that cannot be correlated with any of the process variables.

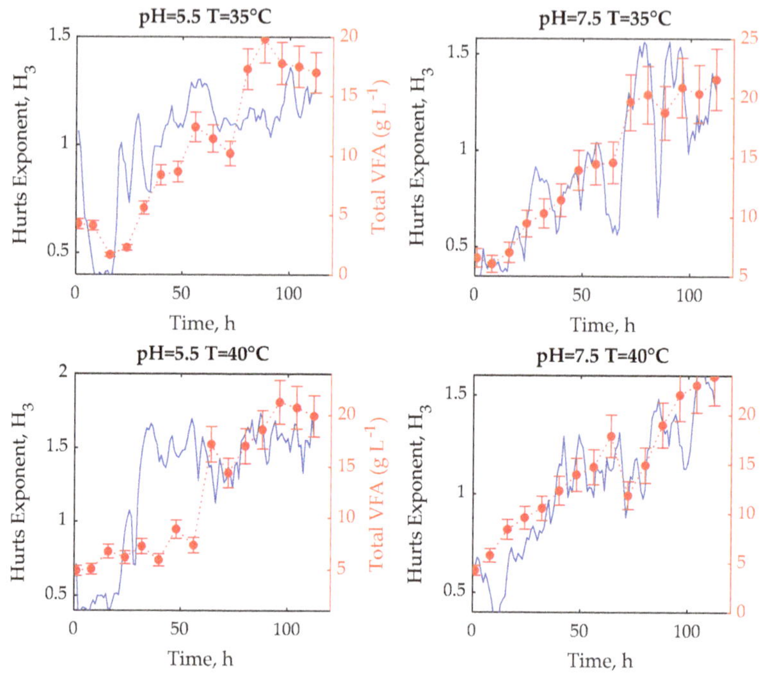


**Figure 7.** Correlations between $Hd_3$ and VFA produced measurements.

Moreover, due to the operating conditions evaluated, there is no significant methane production, so the methanogenic stage is not considered as an active stage of the process. Then, Hurst exponents dynamic could be used as qualitative indexes for monitoring the COD and VFA for different operating conditions.

### 3.3. Multifractal Analysis

The multifractal analysis is a useful tool to identify information at different time scales associated with the inherent complexity in bioprocesses. The anaerobic treatment of raw cheese whey is a highly complex process involving transport phenomena and bioreactions, affecting overall process stability. This interaction depends on the evaluated operating conditions and is reflected in the type and quantity of products generated.

In Figure 8, the application of the multifractal $R/S$ analysis is presented, considering $q \in [0.2, 20]$. For the experimental test at T = 40 °C and initial pH = 5.5, Figure 8a shows the statistic $(R/S)q$, noting that the slopes ($H_q$) of the three zones show variations at different values of q-norm. Figure 8b shows that the $H_q$ values of the three zones present nonlinear variations with respect to the q-norm that exhibit significant differences in each zone, suggesting the presence of a multifractal behavior in the global process, as well as in each of the stages that make up the process.

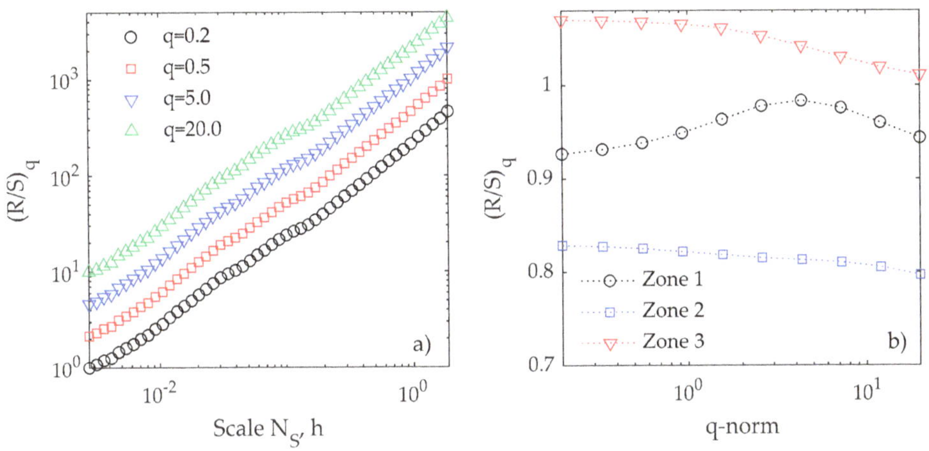


**Figure 8.** Multifractal analysis for pH time series to pH = 7.5 and T = 40 °C, (**a**) $R/S$ analysis applied to different q-norm and (**b**) $H_q$ as a function of q-norm.

The degree of multifractality of each process stage is evaluated with the multifractal index (IM). The higher the IM value, the greater the multifractal behavior. Table 3 shows the multifractal index of the three zones for the four experimental tests. The lowest values of the multifractal index correspond to Zone 1, associated with the hydrolytic stage. Regardless of the operating conditions evaluated, a similar behavior is observed in all experiments. Whereas Zone 3, associated with the acidic stage, shows the highest values of the multifractal index, which correspond to the generation of products obtained in this stage. It should be noted that different types of VFA are produced and the quantity produced depends on the operating conditions.

**Table 3.** Multifractal index by zone for the four experimental tests.

| Zone/Test | T = 35 °C pH = 5.5 | T = 35 °C pH = 7.5 | T = 40 °C pH = 5.5 | T = 40 °C pH = 7.5 |
|---|---|---|---|---|
| Zone 1 | 0.0725 | 0.0919 | 0.0469 | 0.1107 |
| Zone 2 | 0.028 | 0.1929 | 0.0679 | 0.1414 |
| Zone 3 | 0.0115 | 0.1691 | 0.1756 | 0.1039 |

Globally, the multifractal index identifies which stage of the process presents more significant interactions between the underlying phenomena. In order to determine the changes in multifractality as the process is carried out, the multifractal index is calculated considering 6 h moving windows. Figure 9 shows the scale-temporal changes of the multifractal index, where it is observed that the dynamic changes of the IM correspond to the variations observed in the consumption/production rates of the key variables. In Zone 1, it is observed that the most important changes in IM occur in the first 30 h of the process, where the maximum consumption of carbohydrates is carried out. In Zone 3, more significant dynamic changes are observed associated with the changes in the VFA production rate. Increases in the IM correspond to the increases in the VFA production rate.

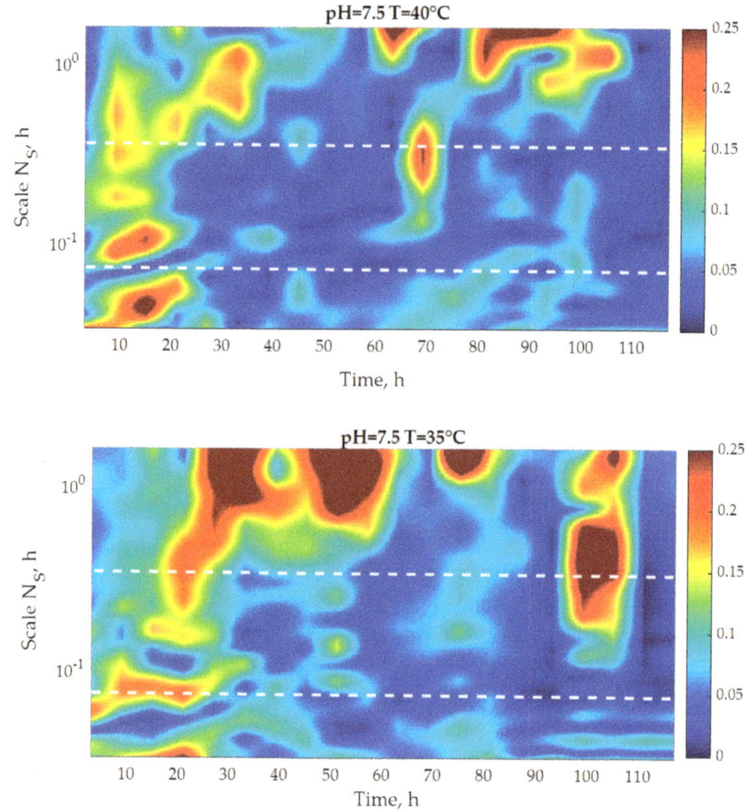


**Figure 9.** Dynamical multifractal index for two experimental tests.

## 4. Conclusions

This work proposed the application of fractal analysis to pH time series collected from the anaerobic treatment of raw cheese whey in an anaerobic batch digester for four different operating conditions variating the pH and temperature. The rescaled range analysis identifies three regions that suggest three underlying phenomena in the anaerobic digestion treatment for raw cheese whey. The dynamic $R/S$ analysis has permitted a qualitative description of two key variables for the four experimental sets. Thus, the $R/S$ fractal methodology can be used as a tool for indirect monitoring of COD consumption and VFA production to different operating conditions. The fact that key variables can be monitored by pH measurements under different operating conditions is a promising advance for developing an economic diagnosis and monitoring and control system that can indicate process performance. Indeed, results in this paper can be used in conjunction with available diagnosis methods using thresholds based on VFA trends. Besides, the proposed multifractal index provides information on dynamic changes in consumption/production rates of key process variables, which are directly associated with the inherent complexity of the process.

**Author Contributions:** Conceptualization E.H.-M.; methodology, A.L.-M., L.A.-D., G.T.-A. and E.H.-M.; software, A.L.-M. and E.H.-M.; validation, A.L.-M.; investigation, A.L.-M. and E.H.-M.; writing—original draft preparation, E.H.-M., H.P.; writing—review and editing, A.L.-M., E.A.-S., L.A.-D. and H.P. All authors have read and agreed to the published version of the manuscript.

**Funding:** This research received no external funding, and the APC was partially funded through the graduate program of the doctorate in engineering at the Universidad Veracruzana.

**Institutional Review Board Statement:** Not applicable.

**Informed Consent Statement:** Not applicable.

**Data Availability Statement:** Not applicable.

**Conflicts of Interest:** The authors declare that they have no known competing financial interests or personal relationships that could appear to have influenced the work reported in this paper.

## Abbreviations

| | | |
|---|---|---|
| VFA | $g\,L^{-1}$ | volatile fatty acids |
| R/S | - | rescaled range |
| COD | $g\,L^{-1}$ | chemical oxygen demand |
| AD | - | Anaerobic digestion |
| FRB | - | fluidized bed reactor |
| CSTR | - | continuous stirred tank reactor |
| AnSBR | - | anaerobic sequencing batch reactor |
| OLR | $g_{COD}\,L^{-1}\,d^{-1}$ | organic loading rate |
| HRT | d | hydraulic retention time |
| VSS | $g\,L^{-1}$ | volatile suspended solids |
| TCH | $g_{glucose}\,L^{-1}$ | total carbohydrates |
| TS | $g\,L^{-1}$ | total solids |
| ORP | mV | oxide reduction potential |
| IM | - | multifractal Index |

## References

1. Mockaitis, G.; Ratusznei, S.M.; Rodrigues, J.A.; Zaiat, M.; Foresti, E. Anaerobic whey treatment by a stirred sequencing batch reactor (ASBR): Effects of organic loading and supplemented alkalinity. *J. Environ. Manag.* **2006**, *79*, 198–206. [CrossRef]
2. Prazeres, A.R.; Carvalho, F.; Rivas, J. Cheese whey management: A review. *J. Environ. Manag.* **2012**, *110*, 48–68. [CrossRef]
3. Zhang, Q.; Hu, J.; Lee, D.-J. Biogas from anaerobic digestion processes: Research updates. *Renew. Energy* **2016**, *98*, 108–119. [CrossRef]
4. Ward, A.J.; Hobbs, P.J.; Holliman, P.J.; Jones, D.L. Optimisation of the anaerobic digestion of agricultural resources. *Bioresour. Technol.* **2008**, *99*, 7928–7940. [CrossRef] [PubMed]
5. Falk, H.M.; Reichling, P.; Andersen, C.; Benz, R. Online monitoring of concentration and dynamics of volatile fatty acids in anaerobic digestion processes with mid-infrared spectroscopy. *Bioprocess Biosyst. Eng.* **2014**, *38*, 237–249. [CrossRef] [PubMed]
6. Pind, P.F.; Angelidaki, I.; Ahring, B.K.; Stamatelatou, K.; Lyberatos, G. Monitoring and Control of Anaerobic Reactors. *Blue Biotechnol.* **2003**, *82*, 135–182. [CrossRef]
7. Wu, D.; Li, L.; Zhao, X.; Peng, Y.; Yang, P.; Peng, X. Anaerobic digestion: A review on process monitoring. *Renew. Sustain. Energy Rev.* **2019**, *103*, 1–12. [CrossRef]
8. Noike, T.; Endo, G.; Chang, J.-E.; Yaguchi, J.-I.; Matsumoto, J.-I. Characteristics of carbohydrate degradation and the rate-limiting step in anaerobic digestion. *Biotechnol. Bioeng.* **1985**, *27*, 1482–1489. [CrossRef]
9. Spanjers, H.; Van Lier, J. Instrumentation in anaerobic treatment—Research and practice. *Water Sci. Technol.* **2006**, *53*, 63–76. [CrossRef]
10. Madsen, M.; Holm-Nielsen, J.B.; Esbensen, K.H. Monitoring of anaerobic digestion processes: A review perspective. *Renew. Sustain. Energy Rev.* **2011**, *15*, 3141–3155. [CrossRef]
11. Jimenez, J.; Latrille, E.; Harmand, J.; Robles, Á.; Ferrer, J.; Gaida, D.; Wolf, C.; Mairet, F.; Bernard, O.; Alcaraz-Gonzalez, V.; et al. Instrumentation and control of anaerobic digestion processes: A review and some research challenges. *Rev. Environ. Sci. Bio/Technol.* **2015**, *14*, 615–648. [CrossRef]
12. Boe, K.; Batstone, D.J.; Steyer, J.-P.; Angelidaki, I. State indicators for monitoring the anaerobic digestion process. *Water Res.* **2010**, *44*, 5973–5980. [CrossRef]
13. Ahring, B.K.; Sandberg, M.; Angelidaki, I.J.A.M. Volatile fatty acids as indicators of process imbalance in anaerobic di-gestors. *Appl. Microbiol. Biotechnol.* **1995**, *43*, 559–565. [CrossRef]
14. Nielsen, H.B.; Uellendahl, H.; Ahring, B.K. Regulation and optimization of the biogas process: Propionate as a key pa-rameter. *Biomass Bioenergy* **2007**, *31*, 820–830. [CrossRef]
15. Molina, F.; Castellano, M.; García, C.; Roca, E.; Lema, J.M. Selection of variables for on-line monitoring, diagnosis, and control of anaerobic digestion processes. *Water Sci. Technol.* **2009**, *60*, 615–622. [CrossRef]

16. Gaida, D.; Wolf, C.; Meyer, C.; Stuhlsatz, A.; Lippel, J.; Bäck, T.; Bongards, M.; McLoone, S. State estimation for anaero-bic digesters using the ADM1. *Water Sci. Technol.* **2012**, *66*, 1088–1095. [CrossRef] [PubMed]
17. Tidriri, K.; Chatti, N.; Verron, S.; Tiplica, T. Bridging data-driven and model-based approaches for process fault diagnosis and health monitoring: A review of researches and future challenges. *Annu. Rev. Control.* **2016**, *42*, 63–81. [CrossRef]
18. Montiel-Escobar, J.L.; Alcaraz-González, V.; Méndez-Acosta, H.O.; González-Álvarez, V. ADM1-Based Robust Interval Observer for Anaerobic Digestion Processes. *Clean Soil Air Water* **2012**, *40*, 933–940. [CrossRef]
19. Lara-Cisneros, G.; Dochain, D. Software Sensor for Online Estimation of the VFA's Concentration in Anaerobic Digestion Processes via a High-Order Sliding Mode Observer. *Ind. Eng. Chem. Res.* **2018**, *57*, 14173–14181. [CrossRef]
20. Dewasme, L.; Sbarciog, M.; Rocha-Cózatl, E.; Haugen, F.; Wouwer, A.V. State and unknown input estimation of an an-aerobic digestion reactor with experimental validation. *Control Eng. Pract.* **2019**, *85*, 280–289. [CrossRef]
21. Draa, K.C.; Zemouche, A.; Alma, M.; Voos, H.; Darouach, M. A Nonlinear observer-based trajectory tracking method applied to an anaerobic digestion process. *J. Process. Control.* **2019**, *75*, 120–135. [CrossRef]
22. Flores-Mejia, H.; Lara-Musule, A.; Hernández-Martínez, E.; Aguilar-López, R.; Puebla, H. Indirect Monitoring of Anaero-bic Digestion for Cheese Whey Treatment. *Processes* **2021**, *9*, 539. [CrossRef]
23. Lee, J.-W.; Hong, Y.-S.T.; Suh, C.; Shin, H.-S. Online nonlinear sequential Bayesian estimation of a biological wastewater treatment process. *Bioprocess Biosyst. Eng.* **2011**, *35*, 359–369. [CrossRef]
24. Jones, R.M.; MacGregor, J.F.; Murphy, K.L. State estimation in wastewater engineering: Application to an anaerobic process. *Environ. Monit. Assess.* **1989**, *13*, 271–282. [CrossRef] [PubMed]
25. Das, L.; Kumar, G.; Rani, M.D.; Srinivasan, B. A novel approach to evaluate state estimation approaches for anaerobic digester units under modeling uncertainties: Application to an industrial dairy unit. *J. Environ. Chem. Eng.* **2017**, *5*, 4004–4013. [CrossRef]
26. MacGregor, J.; Cinar, A. Monitoring, fault diagnosis, fault-tolerant control and optimization: Data driven methods. *Comput. Chem. Eng.* **2012**, *47*, 111–120. [CrossRef]
27. Yin, S.; Ding, S.X.; Xie, X.; Luo, H. A Review on Basic Data-Driven Approaches for Industrial Process Monitoring. *IEEE Trans. Ind. Electron.* **2014**, *61*, 6418–6428. [CrossRef]
28. Nair, V.V.; Dhar, H.; Kumar, S.; Thalla, A.K.; Mukherjee, S.; Wong, J.W. Artificial neural network based modeling to evaluate methane yield from biogas in a laboratory-scale anaerobic bioreactor. *Bioresour. Technol.* **2016**, *217*, 90–99. [CrossRef]
29. Tufaner, F.; Demirci, Y. Prediction of biogas production rate from anaerobic hybrid reactor by artificial neural network and nonlinear regressions models. *Clean Technol. Environ. Policy* **2020**, *22*, 713–724. [CrossRef]
30. Newhart, K.B.; Holloway, R.W.; Hering, A.S.; Cath, T.Y. Data-driven performance analyses of wastewater treatment plants: A review. *Water Res.* **2019**, *157*, 498–513. [CrossRef]
31. Kazemi, P.; Bengoa, C.; Steyer, J.-P.; Giralt, J. Data-driven techniques for fault detection in anaerobic digestion process. *Process. Saf. Environ. Prot.* **2021**, *146*, 905–915. [CrossRef]
32. Asadi, M.; Guo, H.; McPhedran, K. Biogas production estimation using data-driven approaches for cold region municipal wastewater anaerobic digestion. *J. Environ. Manag.* **2020**, *253*, 109708. [CrossRef]
33. Qin, S.J. Survey on data-driven industrial process monitoring and diagnosis. *Annu. Rev. Control.* **2012**, *36*, 220–234. [CrossRef]
34. Lee, H.W.; Lee, M.W.; Park, J.M. Multi-scale extension of PLS algorithm for advanced on-line process monitoring. *Chemom. Intell. Lab. Syst.* **2009**, *98*, 201–212. [CrossRef]
35. Méndez-Acosta, H.O.; Hernandez-Martinez, E.; Jáuregui-Jáuregui, J.A.; Alvarez-Ramirez, J.; Puebla, H. Monitoring anaerobic sequential batch reactors via fractal analysis of pH time series. *Biotechnol. Bioeng.* **2013**, *110*, 2131–2139. [CrossRef] [PubMed]
36. Hernandez-Martinez, E.; Puebla, H.; Mendez-Acosta, H.; Alvarez-Ramirez, J. Fractality in pH time series of continuous anaerobic bioreactors for tequila vinasses treatment. *Chem. Eng. Sci.* **2014**, *109*, 17–25. [CrossRef]
37. Sánchez-García, D.; Hernández-García, H.; Mendez-Acosta, H.O.; Hernández-Aguirre, A.; Puebla, H.; Hernández-Martínez, E. Fractal Analysis of pH Time-Series of an Anaerobic Digester for Cheese Whey Treatment. *Int. J. Chem. React. Eng.* **2018**. [CrossRef]
38. Jirka, A.M.; Carter, M.J. Micro semiautomated analysis of surface and waste waters for chemical oxygen demand. *Anal. Chem.* **1975**, *47*, 1397–1402. [CrossRef] [PubMed]
39. Rice, E.W.; Baird, R.B. *Standard Methods for the Examination of Water and Wastewater*; American Water Works Association: Denver, CO, USA, 1995.
40. Dubois, M.; Gilles, K.A.; Hamilton, J.K.; Rebers, P.A.; Smith, F. Colorimetric Method for Determination of Sugars and Related Substances. *Anal. Chem.* **1956**, *28*, 350–356. [CrossRef]
41. Bradford, M.M. A rapid and sensitive method for the quantitation of microgram quantities of protein utilizing the principle of protein-dye binding. *Anal. Biochem.* **1976**, *72*, 248–254. [CrossRef]
42. Hurst, H.E. Long-Term Storage Capacity of Reservoirs. *Trans. Am. Soc. Civ. Eng.* **1951**, *116*, 770–799. [CrossRef]
43. Kantelhardt, J.W.; Zschiegner, S.A.; Koscielny-Bunde, E.; Havlin, S.; Bunde, A.; Stanley, H. Multifractal detrended fluctuation analysis of nonstationary time series. *Phys. A Stat. Mech. Its Appl.* **2002**, *316*, 87–114. [CrossRef]
44. Mandelbrot, B.B.; Wallis, J.R. Computer Experiments with Fractional Gaussian Noises: Part 2, Rescaled Ranges and Spectra. *Water Resour. Res.* **1969**, *5*, 242–259. [CrossRef]
45. Barabási, A.-L.; Vicsek, T. Multifractality of self-affine fractals. *Phys. Rev. A* **1991**, *44*, 2730–2733. [CrossRef]
46. Katsuragi, H.; Honjo, H. Multiaffinity and entropy spectrum of self-affine fractal profiles. *Phys. Rev. E* **1999**, *59*, 254–262. [CrossRef]

47. Xu, Y.; Qian, C.; Pan, L.; Wang, B.; Lou, C. Comparing Monofractal and Multifractal Analysis of Corrosion Damage Evolution in Reinforcing Bars. *PLoS ONE* **2012**, *7*, e29956. [CrossRef]
48. Sanchez-Ortiz, W.; Andrade-Gómez, C.; Hernandez-Martinez, E.; Puebla, H. Multifractal Hurst analysis for identification of corrosion type in AISI 304 stainless steel. *Int. J. Electrochem. Sci.* **2015**, *10*, 1054–1064.
49. Perna, V.; Castelló, E.; Wenzel, J.; Zampol, C.; Fontes-Lima, D.; Borzacconi, L.; Varesche, M.; Zaiat, M.; Etchebehere, C. Hydrogen production in an upflow anaerobic packed bed reactor used to treatcheese whey. *Int. J. Hydrogen Energy* **2013**, *38*, 54–62. [CrossRef]
50. Calero, R.R.; Lagoa-Costa, B.; Fernandez-Feal, M.M.D.C.; Kennes, C.; Veiga, M.C. Volatile fatty acids produc-tion from cheese whey: Influence of pH, solid retention time and organic loading rate. *J. Chem. Technol. Biotechnol. Technol.* **2018**, *93*, 1742–1747. [CrossRef]
51. Wang, K.; Yin, J.; Shen, D.; Li, N. Anaerobic digestion of food waste for volatile fatty acids (AGVs) production with different types of inoculum: Effect of pH. *Bioresour. Technol.* **2014**, *161*, 395–401. [CrossRef]
52. Lee, C.; Kim, J.; Shin, S.G.; O'Flaherty, V.; Hwang, S. Quantitative and qualitative transitions of methanogen community structure during the batch anaerobic digestion of cheese-processing wastewater. *Appl. Microbiol. Biotechnol.* **2010**, *87*, 1963–1973. [CrossRef] [PubMed]
53. Bengtsson, S.; Hallquist, J.; Werker, A.; Welander, T. Acidogenic fermentation of industrial wastewaters: Effects of chemo-stat retention time and pH on AGV production. *Biochem. Eng. J.* **2008**, *40*, 492–499. [CrossRef]

applied sciences

Article

# Olive Mill and Olive Pomace Evaporation Pond's By-Products: Toxic Level Determination and Role of Indigenous Microbiota in Toxicity Alleviation

Houda Ben Slama [1], Ali Chenari Bouket [2,†], Faizah N. Alenezi [3], Ameur Khardani [1,†], Lenka Luptakova [4], Armelle Vallat [5], Tomasz Oszako [6], Mostafa E. Rateb [7] and Lassaad Belbahri [8,*]

1 NextBiotech, 98 Rue Ali Belhouane, Agareb 3030, Tunisia; benslamahouda92@gmail.com (H.B.S.); ameurkhardani@gmail.com (A.K.)
2 Plant Protection Research Department, East Azarbaijan Agricultural and Natural Resources Research and Education Center, Agricultural Research, Education and Extension Organization (AREEO), Tabriz 5355179854, Iran; a.chenari@areeo.ac.ir
3 The Public Authority for Applied Education and Training, Adailiyah 00965, Kuwait; Fn.alenazi@paaet.edu.kw
4 Department of Biology and Genetics, Institute of Biology, Zoology and Radiobiology, University of Veterinary Medicine and Pharmacy in Košice, 04181 Kosice, Slovakia; lenka.luptakova@uvlf.sk
5 Neuchatel Platform of Analytical Chemistry, Institute of Chemistry, University of Neuchâtel, 2000 Neuchâtel, Switzerland; armelle.vallat@unine.ch
6 Department of Forest Protection of the Forest Research Institute in Sekocin Stary, 05-090 Raszyn, Poland; T.Oszako@ibles.waw.pl
7 School of Computing, Engineering & Physical Sciences, University of the West of Scotland, Paisley PA1 2BE, UK; Mostafa.Rateb@uws.ac.uk
8 Laboratory of Soil Biology, University of Neuchatel, 2000 Neuchatel, Switzerland
* Correspondence: lassaad.belbahri@unine.ch
† Both coauthors equally cooperated in this paper.

**Citation:** Slama, H.B.; Chenari Bouket, A.; Alenezi, F.N.; Khardani, A.; Luptakova, L.; Vallat, A.; Oszako, T.; Rateb, M.E.; Belbahri, L. Olive Mill and Olive Pomace Evaporation Pond's By-Products: Toxic Level Determination and Role of Indigenous Microbiota in Toxicity Alleviation. *Appl. Sci.* **2021**, *11*, 5131. https://doi.org/10.3390/app11115131

Academic Editor: Carlos Rico de la Hera

Received: 30 April 2021
Accepted: 29 May 2021
Published: 31 May 2021

**Publisher's Note:** MDPI stays neutral with regard to jurisdictional claims in published maps and institutional affiliations.



**Abstract:** Diverse vegetable oils are extracted from oleagenic trees and plants all over the world. In particular, olive oil represents a strategic socio-economic branch in the Mediterranean countries. These countries use either two or three-phase olive oil extraction systems. In this work, we focus on the by-products from three-phase olive oil extraction, which are the liquid olive mill wastewater (OMW) and the solid olive mill pomace (OMP) rejected in evaporative ponds. The disposal of this recalcitrant waste poses environmental problems such as the death of different species of insects and animals. In-depth ICP-OES analysis of the heavy metal composition of OMW and OMP revealed the presence of many metals ranging from non-toxic to highly toxic. The LC-HRMS characterization of these by-products indicated the presence of several secondary metabolites harmful to humans or to the environment. Thus, we aimed to identify OMW and OMP indigenous microbiota through metagenomics. The bacterial population was dominated by the *Acetobacter* (49.7%), *Gluconobacter* (17.3%), *Gortzia* (13.7%) and *Nardonalla* (5.3%) genera. The most abundant fungal genera were *Nakazawaea*, *Saccharomyces*, *Lachancea* and *Candida*. These microbial genera are responsible for OMW, OMP and soil toxicity alleviation.

**Keywords:** OMW; OMP; evaporation pond; heavy metals; LC-HRMS; metagenomics; indigenous microbiota; toxicity alleviation

## 1. Introduction

Vegetable oils are a staple food for all countries worldwide. They are extracted from multiple sources such as olive oil, palm oil, coconut oil, sunflower oil, canola oil, soybean oil, peanut oil, cottonseed oil, corn oil, argan oil and many more. Particularly, olive oil is considered one of the most versatile and healthy oils in terms of consumption as a food product. Olive oil extraction occurs in either two- or three-phase systems. The two-phase extraction process generates the olive oil and olive mill pomace (OMP), also called 'alperujo' in Spanish. This method helps in saving water discharge [1,2]. The three-phase extraction process produces the olive oil and two by-products; a liquid olive mill wastewater (OMW) and a solid olive mill pomace (OMP) [3,4].

Olive oil production is centered in the Mediterranean region including Spain, Italy, Greece and Tunisia, which are the major olive oil producers worldwide, accounting for 97% of the global market [5,6]. Particularly, Tunisia provides 6% of the world's olive oil production, it occupies the 4th position worldwide and the 1st position in the north African countries with 60% production [6]. El-Bassi et al. [7] mentioned that Tunisia produces 770.000 $m^3$/year OMW and 550,000 tons/year of OMP by itself. OMW is a liquid discharge fraction containing water (≥90%) and multiple organic and inorganic compounds. OMP is the solid discharge fraction which is mainly composed of olive pulp, seed and skin [8]. These by-products constitute a potential danger to the environment, plants and living organisms in all olive oil-producing countries [7,9,10].

The Mediterranean olive oil industries generate huge amounts of OMW and OMP attaining about 30 million $m^3$ discharges per year [9,11], with an average of 0.5 to 1.5 $m^3$/ton of olive [12,13]. OMW characteristics vary according to the geographic and climatic change, the olive cultivar, the fruit quality, the extraction process and the use of chemical fertilizers [13,14]. Overall, OMW features acidic pH, high-sodium content, blackish-purple color, and distinct smell due to its high polyphenols, polyalcohols, and carbohydrates content [15–18].

Olive mill effluents are frequently discharged in the environment into uncontrolled evaporation ponds. Several studies have proved their high concentrations in toxic waste matters (heavy metals, volatile acids, polyphenols and long-chain fatty acids) leading to foul odors release and several harmful effects to the ecosystem such as soil, water, plants, animals, insects and microorganisms [19–21]. The long term disposal of OMW in the open evaporation ponds cause the transformation of the disposed by-products from a liquid to a solid more toxic form. Recalcitrant compounds either concentrate in a dry sludge or infiltrate inside the soil [22]. For that reason, several treatments have been implemented such as the biological, physicochemical, thermochemical methods or the integration of the above-mentioned methods to ensure better results [18,23,24]. The role of the indigenous microbial communities in the bioremediation of OMW has been extensively studied [25,26]. These microbes could alleviate, remove or bio-transform toxic compounds from the OMW by-products [25,27].

This paper reports the environmental threat of uncontrolled OMW and OMP disposal in evaporation ponds, as well as an in-depth study of the OMW and OMP heavy metals and secondary metabolites composition. Lastly, the indigenous microbial diversity in the OMW was extensively monitored to investigate its role in alleviating the toxicity of the evaporation ponds.

## 2. Materials and Methods

### 2.1. Sampling Site Description

OMW and OMP by-products originate from three-phase olive oil extraction mills. Samples were collected in 2018 from two evaporation ponds located in Agareb (Figure 1a) and Cherarda (Figure 1d), belonging to the governorates of Sfax and Kairouan in Tunisia, respectively (Figure 1, Table 1).

This study involves a total of fourteen samples (Table 1) divided into: thirteen OMW samples comprising five samples collected from the evaporation pond of Cherarda, seven samples from the evaporation pond of Agareb, and one sample collected from soil mixed with dried OMW and deposited in a huge pile near to the evaporation pond of Agareb (MA). It was further used for OMW metagenomic analysis of the indigenous microbiota (Table 1, Figure 1b). The 14th olive mill pomace (OMP) sample was collected from an olive oil industry and used for further tests.

**Figure 1.** (**a**,**d**) Evaporation ponds situated in Agareb and Cherarda, respectively. (**b**) Pile of soil mixed with dried OMW in Agareb. (**e**) Newly established evaporation pond in Cherarda. (**c**,**f**) *Atriplex* plants growing in the vicinity of the evaporation ponds of Agareb and Cherarda, respectively.

**Table 1.** Description of the sampling sites of the evaporation ponds of Agareb and Cherarda.

| Samples Codes | Description | Metagenomique |
|---|---|---|
| A1 | Sample collected from contaminated area around OMW evaporation pond in Agareb | - |
| A2 | Sample collected from the vicinity of evaporation pond of Agareb with lower level of contamination | - |
| A3 | Sample collected outside of the evaporation pond of Agareb with no sign of contamination (control) | - |
| A4 | Sample collected from the center of dried evaporation pond of Agareb | - |
| A5 | OMW sample collected from the center of dried evaporation pond of Agareb but not fully dried still wet | - |
| MA | Sample collected from soil with dried OMW and deposited in huge piles in Agareb | + |
| C1 | Sample collected from contaminated area around OMW evaporation pond in Cherarda | - |
| C2 | Sample collected from the vicinity of evaporation pond of Cherarda with lower level of contamination | - |
| C3 | Sample collected outside of the evaporation pond of Cherarda with no sign of contamination (control) | - |
| C4 | Sample collected from the center of a completely dried out evaporation pond of Cherarda | - |
| C5 | Sample collected from the center of a completely dried out evaporation pond of Cherarda after removal of the dried out residue | - |

**Table 1.** *Cont.*

| Samples Codes | Description | Metagenomique |
|---|---|---|
| C6 | Sample collected from the center of a completely dried out evaporation pond of Cherarda after digging 20 cm deep | - |
| C7 | Sample collected from the center of a completely dried out evaporation pond of Cherarda after digging 50 cm deep | - |
| OMP | Olive Mill Pomace sample collected from an olive oil industry | + |

Briefly, A3 and C3 were the control OMW samples of soil collected from the outside of the evaporation ponds. These samples have no sign of contamination. The remaining samples were taken from contrasting spots situated within (A4, A5, C4 and C5), in the vicinity (A2 and C2) or around (A1 and C1) the 2 evaporation ponds, or even after digging 20 cm (C6) and 50 cm (C7) deep from the center of a completely dried evaporation pond as clearly shown in Table 1 and Figure 2. One additional sample of OMP was collected from an olive oil industry in Sfax (Table 1). Lastly, we have perceived two plants belonging to the *Atriplex* genus (Figure 1c,f) being able to grow in the vicinity of the two evaporation ponds.

**Figure 2.** Sampling process of OMW from several spots within, in the vicinity, around, or after digging in the evaporation ponds of Agareb and Cherarda.

### 2.2. Heavy Metal Analysis of OMW Samples Using Inductively Coupled Plasma Optical Emission Spectrometry (ICP-OES)

Eleven OMW samples (A1, A2, A3, A4, A5, C1, C2, C3, C5, C6 and C7) were analyzed using the ICP-OES method to determine their composition in heavy metals. Samples preparation started with a drying step at a temperature of 60 °C followed by a mashing step until obtaining a smooth texture. Then, elements digestion/extraction was obtained using 65% $HNO_3$ (Suprapur®, Merck KGaA, Germany) for 48 h. Lastly, samples were diluted 100 times by weighing. All samples were analyzed using an ICP-OES (OPTIMA 2100 DV) from Perkin Elmer with AS-93 plus autosampler. The calibration solution contained 2% $HNO_3$_MiliQ water (18 Ω). The detailed protocol was described by Mefteh et al. [28].

### 2.3. Liquid Chromatography-High-Resolution Mass Spectrometry (LC-HRMS) Analysis of OMW and OMP Extracted Samples

The thirteen OMW samples in addition to the OMP sample (10 mL each) were extracted by shaking with 10 mL ethyl acetate (X2), and the combined ethyl acetate extract for each sample was evaporated under reduced pressure to obtain the total organic extract. Each extract was then re-dissolved in 50% aqueous methanol to produce 1 mg/mL final concentration which was then analyzed using LC-HRMS, data were collected on a Thermo Instruments ESI–MS system (LTQ XL/LTQ Orbitrap Discovery, UK) connected to a Thermo Instruments HPLC system (Accela PDA detector, Accela PDA autosampler and Accela Pump). The analyses were executed on a reversed-phase column (Pursuit XRs ULTRA 2.8, C18, 100 × 2 mm, Agilent Technologies, UK). Sample injection volume was 20 µL at 30 °C temperature. Mobile phase A contained 0.1% formic acid in water and mobile phase B contained 0.1% formic acid in MeOH. Gradient HPLC acquisition was performed at 1 mL/min flow rate starting with 100% solvent A with gradual increase of solvent B till reaching 100 % over 20 min, followed by additional 5 min using 100% of solvent B [29]. LC-HRESIMS of all samples was conducted using XCalibur 3.0 and allowing for M + H/M + Na adduct. The suggested compound dereplication and identification was conducted based on the generated molecular formulae and isotope patterns compared to Dictionary of Natural Products (DNP 23.1, 2015 on DVD) and Reaxys online database.

### 2.4. Microbial DNA Extraction from OMW and OMP Samples, PCR Amplification and Metagenomics Analysis of Uncultivable Bacteria and Fungi

The MA sample collected from soil mixed with dried OMW and deposited in a huge pile and the OMP sample, were used for the metagenomics analysis of uncultivable bacteria and fungi (Table 1). The detailed protocol of metagenomics extraction was described by Mefteh et al. [28]. In brief, genomic DNA was extracted via the UltraClean® Microbial DNA Isolation Kit (QIAGEN, Basel, Switzerland) according to manufacturer instructions. DNA was then used for 16S-rDNA PCR amplification using a Biometra T-one thermal cycler (Labgene, Chatel-Saint-Denis, Switzerland). Metagenomics analysis was achieved by using the PICRUSt tool which helps in predicting the marker gene functions based on the sequence analogy with the reference genome [30]. It was applied to predict the functions of the bacterial populations of OMW.

## 3. Results

### 3.1. Effects of the Disposal of OMW in the Evaporative Ponds on the Existing Fauna and Flora

Figure 1c,f belong represent only one plant species growing in the vicinity of the evaporation ponds of Agareb and Cherarda. They belong to the halophytic genus *Atriplex*.

Figure 3 clearly shows the extremely dangerous impact of the uncontrolled discharge of OMW in evaporation ponds. The skeletons of multiple insects (dragonflies and bugs) and animals (rats, hedgehogs and multiple species of birds) are dispersed all over the ponds of Agareb and Cherarda. Birds were the most threatened animals because of the camouflage created from the reflection of the sky on the evaporation pond, making it seems like water. The oily texture of the OMW hinders the movements of animals and traps them inside the pond engendering their subsequent death (Figure 3).

**Figure 3.** Pictures showing skeletons of multiple species of insects and animals dispersed all over and in the vicinity of the evaporation ponds of Agareb and Cherarda.

### *3.2. Heavy Metals Concentrations in OMW Samples*

Eleven samples were collected from different spots located within and surrounding OMW evaporation ponds localized in the regions of Cherarda and Agareb (Figure 4). Both C3 and A3 are the control spots, they represent samples collected outside the evaporation pond with no signs of contamination in the regions of Cherarda and Agareb, respectively. The above mentioned samples were tested to determine their heavy metals (HMs) composition. Results presented in Figure 4 were divided into 4 groups based on the HMs concentrations and presence or absence in the sample: (i) the most abundant metals were Calcium (Ca), Potassium (K), Aluminum (Al), Iron (Fe), Magnesium (Mg), Sodium (Na) and Strontium (Sr), (ii) the less abundant metals were Manganese (Mn), Barium (Ba), Zinc (Zn), Rubidium (Rb), Lithium (Li), Chromium (Cr), Copper (Cu), Nickel (Ni), Arsenic (As), Cobalt (Co), Led (Pb) and Tin (Sn), (iii) metals found in only one sample each were Silver (Ag) and Selenium (Se). Ag and Se exist in the sample collected from the center of dried evaporation pond (A4) and the sample collected from the center of the nearly dried evaporation pond (A5), respectively. (iv) lastly metals not detected in either OMW or OMP samples were Cadmium (Cd), Mercury (Hg) and Molybdenum (Mo) (Figure 4).

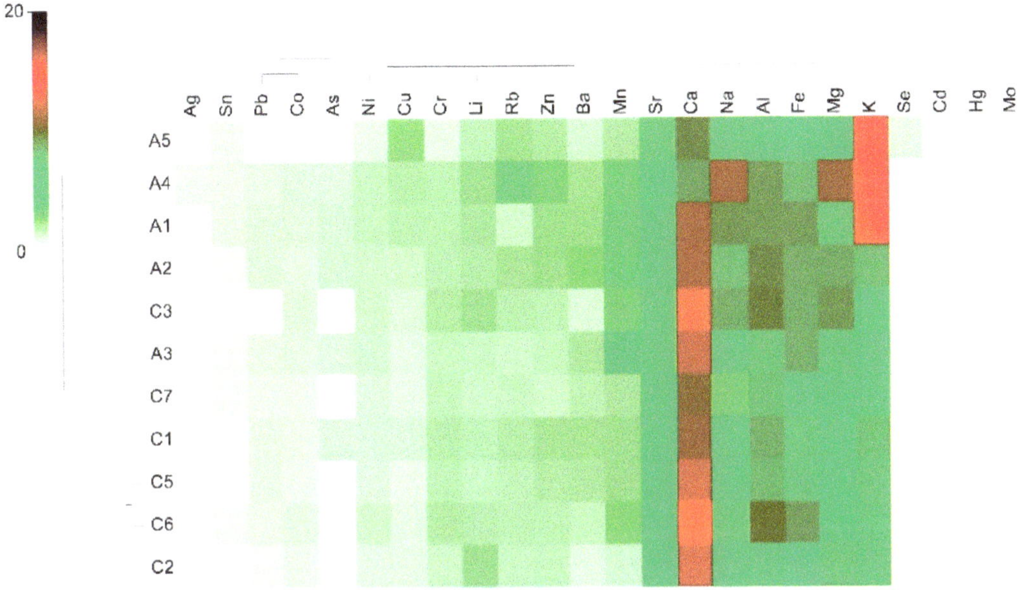


**Figure 4.** Heat map resulting from ICP-OES analysis and describing heavy metal composition of 11 samples (A1, A2, A3, A4, A5, C1, C2, C3, C5, C6 and C7) collected from different spots situated within and surrounding OMW evaporation ponds of Agareb and Cherarda. (Original data (Table S3) unit is 'ppm' but was transformed by Primer-e program with Euclidean similarity and forth root transformation during heat-map producing).

### *3.3. LC-HRMS Analysis of Secondary Metabolites in OMW and OMP*

LC-HRMS analysis of OMW and OMP samples allowed us to detect 46 (Supplementary Table S1) and 69 (Supplementary Table S2) secondary metabolites belonging to different chemical classes, mainly phenolics, terpenoids, and alkaloids. They include metabolites having several biotechnological functions and other metabolites with no reported activities.

Other than that, multiple metabolites were reported to be harmful to humans such as gallic acid, cafeic acid, scoparone, rutin, esculin, apigenin, secologanoside, oleaceran, stigmasterol, apigenin 7-*O*-D-glucoside and hydroxypinoresinol 4″-*O*-methyl ether. Additionally, two secondary metabolites named ferulic acid and maslinic acid were reported to be toxic to the environment.

### *3.4. Genus Level and Metagenomics Analysis of Indigenous Bacterial Communities in OMW and OMP Samples*

The Bar Chart (Figure 5a) presents the bacterial genera within the OMP and OMW. Results showed a remarkable difference in the bacterial genera distribution and diversification within the tested samples. The metagenomics analysis of OMP display *Acetobacter* as the most abundant genus (49.7%), followed by *Gluconobacter* (17.36%), *Gortzia* (13.77%) and *Nardonalla* (5.38%) genera. However, the OMW sample was massively dominated by *Prevotella* (82.8%) succeeded by *Spiroplasma* (6.75%), *Olsenella* (2.73%) and *Lactobacillus* (2.16%) genera (Figure 5a). The heat map in Figure 5b contains the bacterial genus level of OMW with both *Prevotella* and *Spiroplasma* being the most abundant genera and in OMP with *Acetobacter* as the most abundant genus. Furthermore, the bacterial genera within the OMW were far more abundant than those existing in OMP. Six bacterial genera named *Acetobacter*, *Gluconobacter*, *Fusobacterium*, *Catenibacterium*, *Lactobacillus* and *Prevotella* existed in common between OMW and OMP samples (Figure 5b).

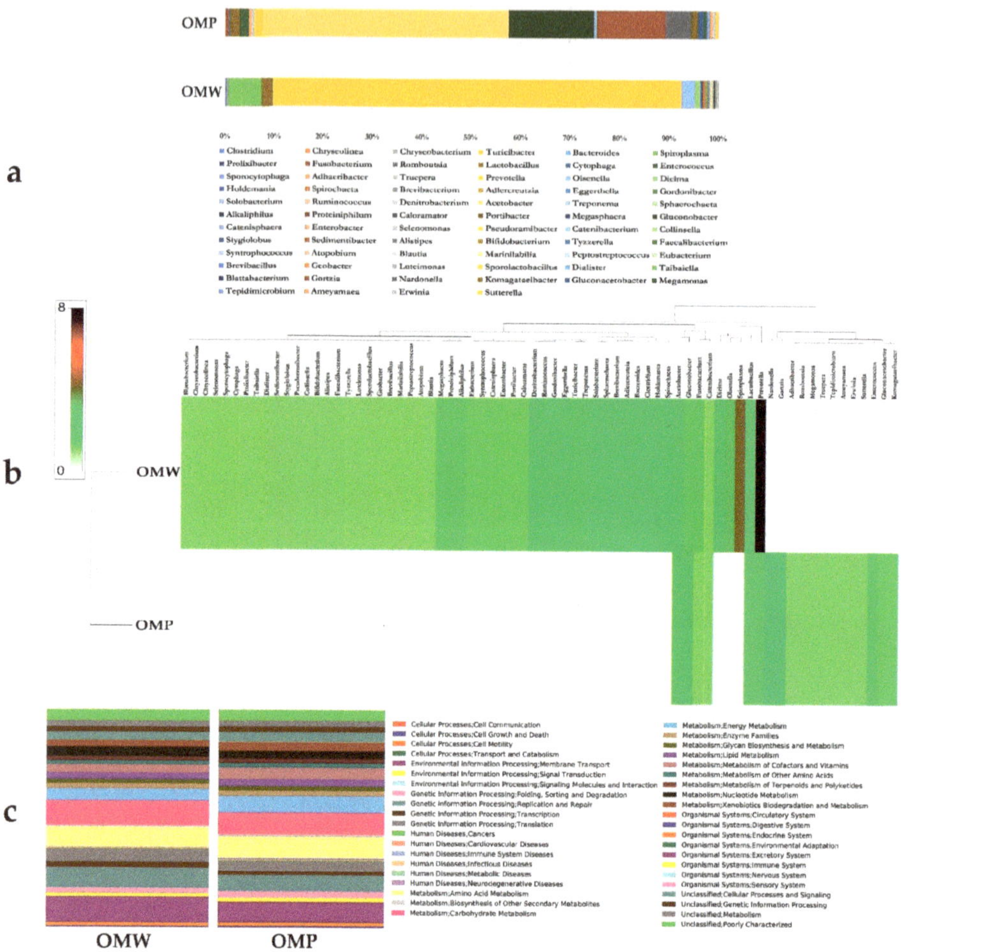


**Figure 5.** (**a**) Bar-chart (percentages) and (**b**) heat map (abundance) presenting uncultivable bacterial genera abundance within the OMP and OMW samples. Original data (Supplementary Material S3) unit is the number of bacterial genera in to OMW and OMP samples but was transformed by Primer-e program with Euclidean similarity and forth root transformation during heat-map producing. (**c**) PICRUSt classification of OMW and OMP functional category abundance.

Supplementary Material S1 contains heat-maps of the detailed bacterial classification of OMW and OMP including the phylum, class, order, family, genus and species of bacteria. Indeed, 4 *Prevotella* species (*Prevotella oris*, *Prevotella stercorea*, *Prevotella genomo* sp. and *Prevotella albensis*) and 1 *Spiroplasma* sp. prevailed the OMW. Other than that, *Acetobacter ghanensis* and *Gortzia infectiva* species prevailed the OMP (Supplementary Material S1).

Bacterial functions in OMW are very similar to those of OMP (Figure 5c). They are divided into 7 categories: cellular processes, organismal systems, metabolism, genetic information processing, environmental information processing, human diseases and few unclassified functions. The main bacterial functions were dedicated to the carbohydrate metabolism (10.7%), the amino acid metabolism (9.6%), and the energy metabolism (6.6%). Otherwise, a considerable percentage (12.5%) was attributed to the categories of environmental information processing and human diseases (Figure 5c).

### *3.5. Genus Level and Metagenomics Analysis of Indigenous Fungal Communities in OMW and OMP Samples*

The fungal genera composition of OMW and OMP was described in Figure 6a. It clearly showed the profusion of 4 fungal genera named *Nakazawaea*, *Saccharomyces*, *Lachancea* and *Candida* in both OMW and OMP and that the fungal genera abundance was much higher in the OMP sample than the OMW sample. The obtained results also demonstrated that almost 35% of fungal genera were unclassified (Figure 6a). Genome sequencing results presented in the Supplementary Material S2 showed that both OMW and OMP fungal diversity derived from the phylum Ascomycota and families of Pichiaceae and Saccharomycetaceae (Supplementary Material S2). Figure 6b showed that the OMP encompassed 36 different fungal genera and the OMW included 30 diverse fungal genera. These 2 related niches share in common 16 fungal genera, with *Nakazawaea molendini-olei* being the most abundant genus in both OMP and OMW (Figure 6b and Supplementary Material S2). The remaining OMW fungal genera existed in small amounts as clearly shown in Figure 6b. Besides the *Nakazawaea molendini-olei*, three other fungal species named *Saccharomyces cerevisiae*, *Lachancea fermentati* and *Candida diddensiae* existed generously in OMP (Figure 6b, Supplementary Material S2).

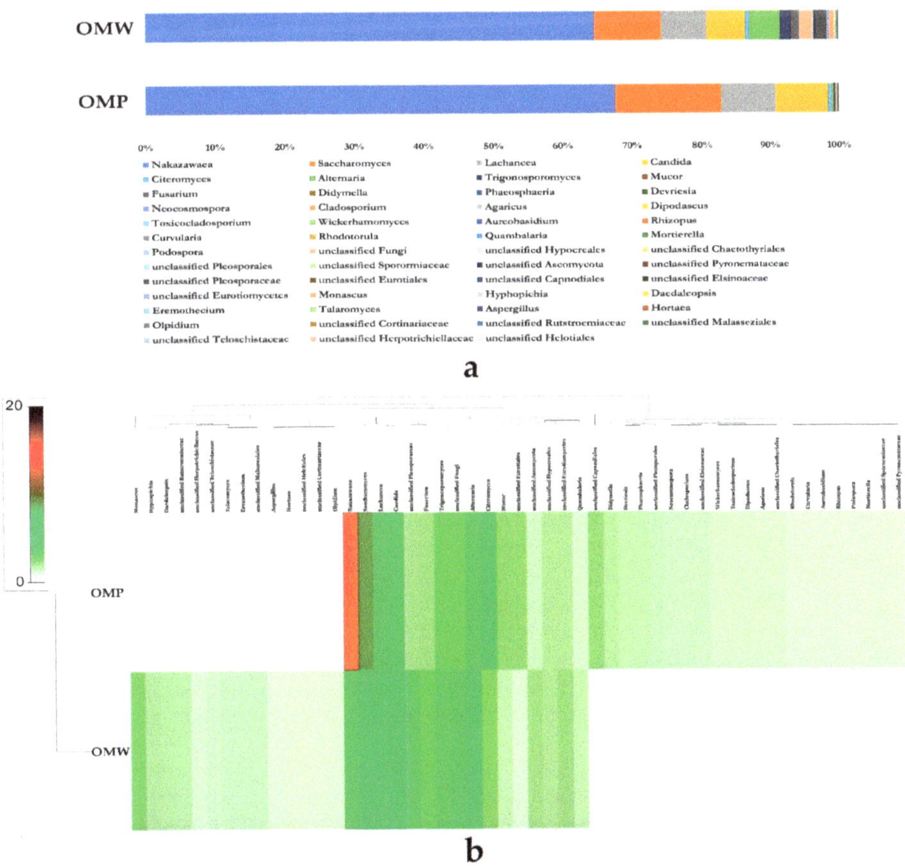


**Figure 6.** (**a**) Bar-chart (percentages) and (**b**) heat map (abundance) presenting uncultivable fungal genera within the OMP and OMW samples. Original data Supplementary Material S4 unit is the number of fungal genera in to OMW and OMP samples but was transformed by Primer-e program with Euclidean similarity and forth root transformation during heat-map producing.

## 4. Discussion

Olive oil production is a staple agricultural and economic field in the Mediterranean countries. Solid (OMP) and liquid (OMW) wastes are composed of several organic and inorganic compounds, they contain high chemical oxygen demand (COD), biochemical oxygen demand (BOD) and heavy metals [7,31].

One of the most conventional treatment technologies is the disposal of OMW in evaporative ponds due to its simple and low-cost construction process [32]. The results from Figure 2 clearly showed the drawbacks of the use of uncontrolled evaporation ponds. Multiple insects and animals skeletons were stuck in the vicinity and inside the OMW ponds. This is confirmed by Michael et al. [24] who mentioned that evaporation ponds could reduce the amount of water and not the toxicity of OMW. Moreover, Lee et al. [18] indicated that the distinct and unpleasant odors derived from the evaporative ponds, attract insects during the processes of the by-products decomposition.

The analysis of the toxicity level of OMW, allowed us to detect the presence of various heavy metals (HMs) in 11 sampling sites situated within and around the OMW evaporation ponds of Agareb and Cherarda. We divided our results into 4 groups based on the HMs concentrations and presence or absence in the sample: (i) major metals (Ca, K, Al, Fe, Mg, Na and Sr), (ii) minor metals (Mn, Ba, Zn, Rb, Li, Cr, Cu, Ni, As, Co, Pb and Sn), (iii) metals found in only one sample each were Ag and Se, (iv) and lastly metals not detected in either OMW or OMP samples were Cd, Hg and Mo. Mineral salts such as Na, Mg and K were considered beneficial when spreading OMW in the agricultural soils as previously reported by multiple researchers [16,33]. Additionally, metal cations such as Al, Fe, Mn, and Ni could be beneficial when present in modest concentrations. However, they became dangerous when they exist in high concentrations [34]. Pb, Cu, Cr, As, Hg, and Cd could inhibit certain indigenous microbes' functions or cause environmental problems even at small concentrations [35]. Taken together, it is clear that OMW uncontrolled discharge is risky to the ecosystem especially that it is generated in huge amounts for a short period [1,36,37].

LC-HRMS analysis of OMW and OMP samples revealed multiple secondary metabolites endowed with several biotechnological functions, but also, with harmful effects to humans or to the environment. For example, gallic acid was classified as a teratogen and mutagen [38], scoparone is an immunosuppressive agent [39,40], and stigmasterol is responsible for cardiac injury and it promotes mortality [41]. Moreover, two environmentally harmful secondary metabolites were detected in both OMW and OMP. The ferulic acid responsible for the initiation of the virulence region of *Agrobacterium tumefaciens*, inducing it to infect plant cells [42] and the maslinic acid having weak inhibitory activities on cytochrome P450 (CYP) isoforms, suggesting that it has low potential to cause possible toxicity and drug interactions involving CYP enzymes [43,44].

Multiple technologies have been investigated to treat the olive mill discharges [5,13,32,45]. Particularly, bioremediation was often used in toxicity alleviation or pathogens inhibition by either applying external [22] or indigenous microbiota [27]. In the present study, we conducted an in-depth work on bacteria and fungal species residing within OMW and OMP using culture-independent approaches. Such indigenous microbiota could be further exploited in the remediation of OMW and OMP toxic wastes. For instance, we have implemented the metagenomics approach to identify and elucidate the role of the detected microbiota. Other researchers have used the DNA high-throughput sequencing and omics technologies to identify microbial communities and explain their mechanisms of OMW degradation [25,26,46].

We investigated the phyla, classes, orders, families, genera and species of bacteria and fungi inhabiting OMW and OMP evaporation ponds. Our results indicated that the bacterial diversity in OMW is larger than that of OMP. Furthermore, the most abundant phyla existing in OMW were Bacteroidetes, Tenericutes, Actinobacteria and Firmicutes; and in OMP Proteobacteria phylum was dominating. Previous studies have reported that the main phyla revealed using high-throughput sequencing technologies in different organic wastes including OMW are members of Bacteroidetes, Actinobacteria, Proteobacteria and Firmicutes [4,26,47–49]. OMW's dominating families were those of Prevotellaceae, Spiroplasmataceae, Atopobiaceae, Lactobacillaceae and Erysipelotrichaceae. Whereas, OMP's most abundant families were Acetobacteriaceae and Holosporaceae. Interestingly, Kavroulakis and Ntougias [50] have found that the Prevotellaceae family existed in the OMW because of the anaerobic condition created in case of harvested olive accumulation (before olive oil extraction). Additionally, the Spiroplasmataceae family was mainly represented by the *Spiroplasma* sp., which was reported to contribute to the degradation of multiple categories of organic wastes as mentioned by Gupta and Garg [51]. The non-$H_2$ producing bacteria *Olsenella* sp. was abundantly present in the OMW, and it was identified to turn hexose into propionic acid and acetic acid in food wastes [52]. Other than that, previous works proved the major role of indigenous microbiota in the degradation of OMW hazardous components during the composting process [18,53–55]. Lastly, PICRUSt was used to identify gene function abundance based on the sequence similarity with the input marker gene. The major bacterial functions were dedicated to crucial functions such as amino acid, carbohydrate and energy metabolisms. Similar findings indicated that bacterial genetic composition was mainly devoted to the vital bacterial functions [28,56]. Surprisingly, bacterial functions of OMW and OMP involved human diseases (cardiovascular, infectious, and metabolic diseases). Similarly, Doula et al. [57] cited that intense OMW discharge near urban areas affects the water surface leading to human health problems.

Concerning the fungal microbiota, we demonstrated that the fungal communities of OMW and OMP were very similar to each other. They were mainly composed of Ascomycota, Basidiomycota and unclassified fungi. These findings accorded with the fungal sequences deposited in GenBank, which indicated that Ascomycota, Basidiomycota and unclassified fungi were mainly abundant in the OMW environments [25]. The predominant species in our study were *Nakazawaea molendini-olei*, *Saccharomyces cerevisiae*, *Lachancea fermentati*, *Candida diddensiae*, *Candida adriatica* and *Alternaria* sp. Many of them were widespread in olive fruits, olive oil, or olive mill solid and liquid wastes derived from several countries including Morocco and Italy [58–60]. Researchers proved that yeasts such as *Saccharomyces* and *Candida*, and fungi such as *Lachancea*, *Alternaria* and *Pichia* were able to reduce multiple recalcitrant compounds, thus contributing to the bioremediation of the olive mill discharges [61–63].

## 5. Conclusions

The OMW effluent discharge is continuously rising with the rise of the global demand for olive oil. Our study provided valuable information about the OMW and OMP toxicity levels with the existence of several heavy metals in the evaporation ponds of Agareb and Cherarda, which constitute an extreme threat to our ecosystem. Despite these findings, the metagenomic identification of the indigenous microbial community, allowed us to deduce their prospective role in alleviating the toxicity of olive mill by-products. In perspective, researchers and olive oil industries are considering working on specified and lucrative treatment methods to manage these wastes.

**Supplementary Materials:** The following are available online at https://www.mdpi.com/article/10.3390/app11115131/s1, Table S1: LCMS analysis results of OMW. Table S2: LCMS analysis results of OMP., Table S3: ICP-OES analysis and describing heavy metal composition. Supplementary Material S1: Different taxonomical levels of bacteria from OMW and OMP. Supplementary Material S2: Different taxonomical levels of fungi from OMW and OMP. Supplementary Material S3: Bacterial genera numbers in OMW and OMP samples. Supplementary Material S4. Fungal genera numbers in OMW and OMP samples.

**Author Contributions:** Conceptualization, H.B.S., A.C.B., A.K., L.L., F.N.A. and L.B.; methodology, L.B.; software, A.C.B., M.E.R.; validation, L.B.; formal analysis, A.C.B., M.E.R., L.B.; investigation, H.B.S., A.C.B., L.L., L.B.; resources, A.V. and L.B.; data curation, A.C.B.; writing—original draft preparation, H.B.S. and L.B.; writing—review and editing, H.B.S., L.B., M.E.R. and A.C.B.; visualization, L.L.; supervision, L.B.; project administration, L.B.; funding acquisition, T.O. and L.B. All authors have read and agreed to the published version of the manuscript.

**Funding:** This research received no external funding.

**Institutional Review Board Statement:** Not applicable.

**Informed Consent Statement:** Not applicable.

**Conflicts of Interest:** The authors declare no conflict of interest.

## References


1. Gomez-Munoz, B.; Hatch, D.J.; Bol, R.; Garcia-Ruiz, R. The compost of olive mill pomace: From a waste to a resource—environmental benefits of its application in olive oil groves. In *Sustainable Development: Authoritative and Leading Edge Content for Environmental Management*; Curkovic, S., Ed.; InTechOpen Ltd.: Rijeka, Croatia, 2012. [CrossRef]
2. Federici, E.; Massaccesi, L.; Pezzolla, D.; Fidati, L.; Montalbani, E.; Proietti, P.; Nasini, L.; Regni, L.; Scargetta, S.; Gigliotti, G. Short-term modifications of soil microbial community structure and soluble organic matter chemical composition following amendment with different solid olive mill waste and their derived composts. *Appl. Soil Ecol.* **2017**, *119*, 234–241. [CrossRef]
3. Piotrowska, A.; Rao, M.A.; Scotti, R.; Gianfreda, L. Changes in soil chemical and biochemical properties following amendment with crude and dephenolized olive mill waste water (OMW). *Geoderma* **2011**, *161*, 8–17. [CrossRef]
4. Tsiamis, G.; Tzagkaraki, G.; Chamalaki, A.; Xypteras, N.; Andersen, G.; Vayenas, D.; Bourtzis, K. Olive-mill wastewater bacterial communities display a cultivar specific profile. *Curr. Microbiol.* **2012**, *64*, 197–203. [CrossRef] [PubMed]
5. Doğan, K.; Sarıoğlu, A.; Coşkan, A. Contribution of green manure, *Rhizobium* and humic+ fulvic acid on recovering soil biologic activity of olive mill wastewater contaminated soil. *Sci. Pap. A. Agron.* **2016**, *59*, 63–68.
6. Meftah, O.; Guergueb, Z.; Braham, M.; Sayadi, S.; Mekki, A. Long term effects of olive mill wastewaters application on soil properties and phenolic compounds migration under arid climate. *Agric. Water Manag.* **2019**, *212*, 119–125. [CrossRef]
7. El-Bassi, L.; Azzaz, A.A.; Jellali, S.; Akrout, H.; Marks, E.A.N.; Ghimbeu, C.M.; Jeguirim, M. Application of olive mill waste-based biochars in agriculture: Impact on soil properties, enzymatic activities and tomato growth. *Sci. Total Environ.* **2021**, *755*, 142531. [CrossRef]
8. Milanović, V.; Osimani, A.; Cardinali, F.; Taccari, M.; Garofalo, C.; Clementi, F.; Ashoor, S.; Mozzon, M.; Foligni, R.; Canonico, L.; et al. Effect of inoculated Azotobacteria and Phanerochaete *Chrysosporium* on the composting of olive pomace: Microbial community dynamics and phenols evolution. *Sci. Rep.* **2019**, *9*, 16966. [CrossRef] [PubMed]
9. Ioannou-Ttofa, L.; Michael-Kordatou, I.; Fattas, S.C.; Eusebio, A.; Ribeiro, B.; Rusan, M.; Amer, A.R.B.; Zuraiqi, S.; Waismand, M.; Linder, C.; et al. Treatment efficiency and economic feasibility of biological oxidation, membrane filtration and separation processes, and advanced oxidation for the purification and valorization of olive mill wastewater. *Water Res.* **2017**, *114*, 1–13. [CrossRef]
10. Kontos, S.; Iakovides, I.; Koutsoukos, P.; Paraskeva, C. Isolation of purified high added value products from olive mill wastewater streams through the implementation of membrane technology and cooling crystallization process. *Chem. Eng. Trans.* **2016**, *47*, 337–342.
11. Chiavola, A.; Farabegoli, G.; Antonetti, F. Biological treatment of olive mill wastewater in a sequencing batch reactor. *Biochem. Eng. J.* **2014**, *85*, 71–78. [CrossRef]
12. Annab, H.; Fiol, N.; Villaescusa, I.; Essamri, A. A proposal for the sustainable treatment and valorisation of olive mill wastes. *J. Environ. Chem. Eng.* **2019**, *7*, 102803. [CrossRef]
13. Benamar, A.; Mahjoubi, F.Z.; Barka, N.; Kzaiber, F.; Boutoial, K.; Ali, G.A.M.; Oussama, A. Olive mill wastewater treatment using infiltration percolation in column followed by aerobic biological treatment. *SN Appl. Sci.* **2020**, *2*, 655. [CrossRef]
14. Yay, A.S.E.; Oral, H.V.; Onay, T.T.; Yenigün, O. A study on olive oil mill wastewater management in Turkey: A questionnaire and experimental approach. *Resour. Conserv. Recycl.* **2012**, *60*, 64–71.
15. Poerschmann, J.; Weiner, B.; Baskyr, I. Organic compounds in olive mill wastewater and in solutions resulting from hydrothermal carbonization of the wastewater. *Chemosphere* **2013**, *92*, 1472–1482. [CrossRef] [PubMed]
16. Aharonov-Nadborny, R.; Tsechansky, L.; Raviv, M.; Graber, E.R. Impact of spreading olive mill waste water on agricultural soils for leaching of metal micronutrients and cations. *Chemosphere* **2017**, *179*, 213–221. [CrossRef]
17. Benavente, V.; Fullana, A.; Berge, N.D. Life cycle analysis of hydrothermal carbonization of olive mill waste: Comparison with current management approaches. *J. Clean. Prod.* **2017**, *142*, 2637–2648. [CrossRef]
18. Lee, Z.S.; Chin, S.Y.; Lim, J.W.; Witoon, T.; Cheng, C.K. Treatment technologies of palm oil mill effluent (POME) and olive mill wastewater (OMW): A brief review. *Environ. Technol. Innov.* **2019**, *15*, 100377. [CrossRef]
19. Amor, C.; Lucas, M.S.; Garcia, J.; Dominguez, J.R.; De Heredia, J.B.; Peres, J.A. Combined treatment of olive mill wastewater by Fenton's reagent and anaerobic biological process. *J. Environ. Sci. Health A* **2015**, *50*, 161–168. [CrossRef]
20. Frascari, D.; Bacca, A.E.M.; Zama, F.; Bertin, L.; Fava, F.; Pinelli, D. Olive mill wastewater valorisation through phenolic compounds adsorption in a continuous flow column. *Chem. Eng. J.* **2016**, *283*, 293–303. [CrossRef]
21. Reis, P.M.; Martins, P.J.M.; Martins, R.C.; Gando-Ferreira, L.M.; Quinta-Ferreira, R.M. Integrating Fenton's process and ion exchange for olive mill wastewater treatment and iron recovery. *Environ. Technol.* **2018**, *39*, 308–316. [CrossRef]
22. Sáez, J.A.; Pérez-Murcia, M.D.; Vico, A.; Martínez-Gallardo, M.R.; Andreu-Rodriguez, F.J.; López, M.J.; Bustamante, M.A.; Sanchez-Hernandez, J.C.; Moreno, J.; Moral, R. Olive mill wastewater-evaporation ponds long term stored: Integrated assessment of in situ bioremediation strategies based on composting and vermicomposting. *J. Hazard. Mat.* **2021**, *402*, 123481. [CrossRef]

23. Cassano, A.; Conidi, C.; Drioli, E. Comparison of the performance of UF membranes in olive mill wastewaters treatment. *Water Res.* **2011**, *45*, 3197–3204. [CrossRef]
24. Michael, I.; Panagi, A.; Ioannou, L.A.; Frontistis, Z.; Fatta-Kassinos, D. Utilizing solar energy for the purification of olive mill wastewater using a pilot-scale photocatalytic reactor after coagulation-flocculation. *Water Res.* **2014**, *60*, 28–40. [CrossRef]
25. Ntougias, S.; Bourtzis, K.; Tsiamis, G. The Microbiology of Olive Mill Wastes. Available online: https://www.hindawi.com/journals/bmri/2013/784591/ (accessed on 29 October 2020).
26. Tortosa, G.; Castellano-Hinojosa, A.; Correa-Galeote, D.; Bedmar, E.J. Evolution of bacterial diversity during two-phase olive mill waste ("Alperujo") composting by 16S-rRNA gene pyrosequencing. *Bioresour. Technol.* **2017**, *224*, 101–111. [CrossRef]
27. Martínez-Gallardo, M.R.; López, M.J.; López-González, J.A.; Jurado, M.M.; Suárez-Estrella, F.; Pérez-Murcia, M.D.; Sáez, J.A.; Moral, R.; Moreno, J. Microbial communities of the olive mill wastewater sludge stored in evaporation ponds: The resource for sustainable bioremediation. *J. Environ. Manag.* **2021**, *279*, 111810. [CrossRef]
28. Mefteh, F.; Chenari Bouket, A.; Daoud, A.; Luptakova, L.; Alenezi, F.; Gharsallah, N.; Lassaad, B. Metagenomic insights and genomic analysis of phosphogypsum and its associated plant endophytic microbiomes reveals valuable actors for waste bioremediation. *Microorganisms* **2019**, *7*, 382. [CrossRef]
29. Daoud, A.; Mefteh, F.; Mnafgui, K.; Turki, M.; Jmal, S.; Amar, R.; Ayadi, F.; ElFeki, A.; Abid, L.; Rateb, M.; et al. Cardiopreventive effect of ethanolic extract of date palm pollen against isoproterenol induced myocardial infarction in rats through the inhibition of the angiotensin-converting enzyme. *Exp. Toxicol. Pathol.* **2017**, *69*. [CrossRef] [PubMed]
30. Douglas, G.M.; Beiko, R.G.; Langille, M.G.I. Predicting the functional potential of the microbiome from marker genes using PICRUSt. *Methods Mol. Biol.* **2018**, *1849*, 169–177. [CrossRef] [PubMed]
31. Karpouzas, D.G.; Ntougias, S.; Iskidou, E.; Rousidou, C.; Papadopoulou, K.K.; Zervakis, G.I.; Ehaliotis, C. Olive mill wastewater affects the structure of soil bacterial communities. *Appl. Soil Ecol.* **2010**, *45*, 101–111. [CrossRef]
32. Alfano, G.; Lustrato, G.; Lima, G.; Ranalli, G. *Present and Future Perspectives of Olive Residues Composting in the Mediterranean Basin (CompMed). Dynamic Soil, Dynamic Plant*; Global Science Books: Isleworth, UK, 2009; pp. 39–56.
33. Tajini, F.; Ouerghi, A.; Hosni, K. Effect of irrigation with olive-mill waste-water on physiological and biochemical parameters as well as heavy-metal accumulation in common bean (*Phaseolus vulgaris* L.). *J. New Sci.* **2019**, *66*, 6.
34. Martinez-Garcia, G.; Bachmann, R.T.; Williams, C.J.; Burgoyne, A.; Edyvean, R.G. Olive oil waste as a biosorbent for heavy metals. *Int. Biodeter. Biodegrad.* **2006**, *58*, 231–238. [CrossRef]
35. Guo, Q.; Majeed, S.; Xu, R.; Zhang, K.; Kakade, A.; Khan, A.; Hafeez, F.Y.; Mao, C.; Liu, P.; Li, X. Heavy metals interact with the microbial community and affect biogas production in anaerobic digestion: A review. *J. Environ. Manag.* **2019**, *240*, 266–272. [CrossRef]
36. Vuppala, S.; Bavasso, I.; Stoller, M.; Di Palma, L.; Vilardi, G. Olive mill wastewater integrated purification through pre-treatments using coagulants and biological methods: Experimental, modelling and scale-up. *J. Clean. Prod.* **2019**, *236*, 117622. [CrossRef]
37. El Hanandeh, A. Energy recovery alternatives for the sustainable management of olive oil industry waste in Australia: Life cycle assessment. *J. Clean. Prod.* **2015**, *91*, 78–88. [CrossRef]
38. Chanwitheesuk, A.; Teerawutgulrag, A.; Kilburn, J.D.; Rakariyatham, N. Antimicrobial gallic acid from *Caesalpinia mimosoides* Lamk. *Food Chem.* **2007**, *100*, 1044–1048. [CrossRef]
39. Huang, H.-C.; Lee, C.-R.; Weng, Y.-I.; Lee, M.-C.; Lee, Y.-T. Vasodilator effect of Scoparone (6,7-Dimethoxycoumarin) from a Chinese herb. *Eur. J. Pharmacol.* **1992**, *218*, 123–128. [CrossRef]
40. Huei-Chen, H.; Shu-Hsun, C.; Chao, P.-D.L. Vasorelaxants from Chinese herbs, Emodin and Scoparone, possess immunosuppressive properties. *Eur. J. Pharmacol.* **1991**, *198*, 211–213. [CrossRef]
41. Tao, C.; Shkumatov, A.A.; Alexander, S.T.; Ason, B.L.; Zhou, M. Stigmasterol accumulation causes cardiac injury and promotes mortality. *Commun. Biol.* **2019**, *2*, 1–10. [CrossRef]
42. Kalogeraki, V.S.; Zhu, J.; Eberhard, A.; Madsen, E.L.; Winans, S.C. The phenolic *vir* gene inducer ferulic acid is O-Demethylated by the *virH2* protein of an *Agrobacterium tumefaciens* Ti Plasmid. *Mol. Microbiol.* **1999**, *34*, 512–522. [CrossRef]
43. Lozano-Mena, G.; Sánchez-González, M.; Juan, M.E.; Planas, J.M. Maslinic Acid, a natural phytoalexin-type triterpene from olives—a promising nutraceutical? *Molecules* **2014**, *19*, 11538–11559. [CrossRef] [PubMed]
44. Mokhtari, K.; Rufino-Palomares, E.E.; Pérez-Jiménez, A.; Reyes-Zurita, F.J.; Figuera, C.; García-Salguero, L.; Medina, P.P.; Peragón, J.; Lupiáñez, J.A. Maslinic Acid, a triterpene from olive, affects the antioxidant and mitochondrial status of B16F10 melanoma cells grown under stressful conditions. *Evid-Based Compl. Alt. Med.* **2015**, *2015*. [CrossRef] [PubMed]
45. Hodaifa, G.; Ochando-Pulido, J.M.; Rodriguez-Vives, S.; Martinez-Ferez, A. Optimization of continuous reactor at pilot scale for olive-oil mill wastewater treatment by fenton-like process. *Chem. Eng. J.* **2013**, *220*, 117–124. [CrossRef]
46. Hultman, J.; Kurola, J.; Rainisalo, A.; Kontro, M.; Romantschuk, M. Utility of molecular tools in monitoring large scale composting. In *Microbes at Work: From Wastes to Resources*; Insam, H., Franke-Whittle, I., Goberna, M., Eds.; Springer: Berlin/Heidelberg, Germany, 2010; pp. 135–151.
47. Storey, S.; Chualain, D.N.; Doyle, O.; Clipson, N.; Doyle, E. Comparison of bacterial succession in green waste composts amended with inorganic fertilizer and wastewater treatment plant sludge. *Bioresour. Technol.* **2015**, *179*, 71–77. [CrossRef]
48. Morillo, J.A.; Aguilera, M.; Antízar-Ladislao, B.; Fuentes, S.; Ramos-Cormenzana, A.; Russell, N.J.; Monteoliva-Sánchez, M. Molecular microbial and chemical investigation of the bioremediation of two-phase olive mill waste using laboratory-scale bioreactors. *Appl. Microbiol. Biotechnol.* **2008**, *79*, 309–317. [CrossRef] [PubMed]

49. Neher, D.A.; Weicht, T.R.; Bates, S.T.; Leff, J.W.; Fierer, N. Changes in bacterial and fungal communities across compost recipes, preparation methods, and composting times. *PLoS ONE* **2013**, *8*, e79512. [CrossRef] [PubMed]
50. Kavroulakis, N.; Ntougias, S. Bacterial and β-proteobacterial diversity in *Olea europaea* var. *mastoidis* and *O. europaea* var. *koroneiki* generated olive mill wastewaters: Influence of cultivation and harvesting practice on bacterial community structure. *World J. Microbiol. Biotechnol.* **2011**, *27*, 57–66. [CrossRef]
51. Gupta, R.; Garg, V.K. 5—Vermitechnology for organic waste recycling. In *Current Developments in Biotechnology and Bioengineering*; Wong, J.W.-C., Tyagi, R.D., Pandey, A., Eds.; Elsevier: Amsterdam, The Netherlands, 2017; pp. 83–112.
52. Jung, J.-H.; Sim, Y.-B.; Baik, J.-H.; Park, J.-H.; Kim, S.-H. High-rate mesophilic hydrogen production from food waste using hybrid immobilized microbiome. *Bioresour. Technol.* **2021**, *320*, 124279. [CrossRef] [PubMed]
53. Federici, E.; Pepi, M.; Esposito, A.; Scargetta, S.; Fidati, L.; Gasperini, S.; Cenci, G.; Altieri, R. Two-phase olive mill waste composting: Community dynamics and functional role of the resident microbiota. *Bioresour. Technol.* **2011**, *102*, 10965–10972. [CrossRef]
54. Chowdhury, A.K.M.M.B.; Akratos, C.S.; Vayenas, D.V.; Pavlou, S. Olive mill waste composting: A review. *Int. Biodeterior. Biodegrad.* **2013**, *85*, 108–119. [CrossRef]
55. Agnolucci, M.; Cristani, C.; Battini, F.; Palla, M.; Cardelli, R.; Saviozzi, A.; Nuti, M. Microbially-enhanced composting of olive mill solid waste (wet husk): Bacterial and fungal community dynamics at industrial pilot and farm level. *Bioresour. Technol.* **2013**, *134*, 10–16. [CrossRef]
56. Slama, H.; Cherif-Silini, H.; Chenari Bouket, A.; Qader, M.; Silini, A.; Yahiaoui, B.; Alenezi, F.; Luptakova, L.; Triki, M.; Vallat, A.; et al. Screening for *Fusarium* antagonistic bacteria from contrasting niches designated the endophyte *Bacillus halotolerans* as plant warden against *Fusarium*. *Front. Microbiol.* **2019**, *9*. [CrossRef] [PubMed]
57. Doula, M.K.; Moreno-Ortego, J.L.; Tinivella, F.; Inglezakis, V.J.; Sarris, A.; Komnitsas, K. Olive mill waste: Recent advances for the sustainable development of olive oil industry. In *Olive Mill Waste: Recent Advances for Sustainable Management*; Galanakis, C.M., Ed.; Academic Press: Cambridge, MA, USA, 2017; pp. 29–56.
58. Sassi, A.B.; Ouazzani, N.; Walker, G.M.; Ibnsouda, S.; El Mzibri, M.; Boussaid, A. Detoxification of olive mill wastewaters by Moroccan yeast isolates. *Biodegradation* **2008**, *19*, 337–346. [CrossRef] [PubMed]
59. Sinigaglia, M.; Di Benedetto, N.; Bevilacqua, A.; Corbo, M.R.; Capece, A.; Romano, P. Yeasts isolated from olive mill wastewaters from Southern Italy: Technological characterization and potential use for phenol removal. *Appl. Microbiol. Biotechnol.* **2010**, *87*, 2345–2354. [CrossRef] [PubMed]
60. Ciafardini, G.; Zullo, B.A. Use of selected yeast starter cultures in industrial-scale processing of brined taggiasca black table olives. *Food Microbiol.* **2019**, *84*, 103250. [CrossRef]
61. Bleve, G.; Lezzi, C.; Chiriatti, M.A.; D'Ostuni, I.; Tristezza, M.; Di Venere, D.; Sergio, L.; Mita, G.; Grieco, F. Selection of non-conventional yeasts and their use in immobilized form for the bioremediation of olive oil mill wastewaters. *Bioresour. Technol.* **2011**, *102*, 982–989. [CrossRef]
62. Jarboui, R.; Baati, H.; Fetoui, F.; Gargouri, A.; Gharsallah, N.; Ammar, E. Yeast performance in wastewater treatment: Case study of *Rhodotorula mucilaginosa*. *Environ. Technol.* **2012**, *33*, 951–960. [CrossRef]
63. Bevilacqua, A.; Cibelli, F.; Raimondo, M.L.; Carlucci, A.; Lops, F.; Sinigaglia, M.; Corbo, M.R. Fungal bioremediation of olive mill wastewater: Using a multi-step approach to model inhibition or stimulation. *J. Sci. Food Agric.* **2017**, *97*, 461–468. [CrossRef]

applied sciences

MDPI

Article

# Analysis of Quality of Backyard Compost and Its Potential Utilization as a Circular Bio-Waste Source

**Apolka Ujj [1], Kinga Percsi [1,*], Andras Beres [2], Laszlo Aleksza [2], Fernanda Ramos Diaz [3,*], Csaba Gyuricza [4] and Csaba Fogarassy [1,*]**

[1] Institute of Sustainable Development and Farming, Hungarian University of Agriculture and Life Sciences (MATE), Pater Karoly Street-1, 2100 Gödöllő, Hungary; ujj.apolka@uni-mate.hu
[2] Institute of Environmental Sciences, Hungarian University of Agriculture and Life Sciences (MATE), Pater Karoly Street-1, 2100 Gödöllő, Hungary; beres.andras@uni-mate.hu (A.B.); aleksza.laszlo@uni-mate.hu (L.A.)
[3] Doctoral School of Economics and Regional Sciences, Hungarian University of Agriculture and Life Sciences (MATE), Pater Karoly Street-1, 2100 Gödöllő, Hungary
[4] Institute of Agronomy, Hungarian University of Agriculture and Life Sciences (MATE), Pater Karoly Street-1, 2100 Gödöllő, Hungary; gyuricza.csaba@uni-mate.hu
* Correspondence: percsi.kinga@uni-mate.hu (K.P.); Maria.Fernanda.Ramos.Diaz@hallgato.uni-szie.hu (F.R.D.); fogarassy.csaba@uni-mate.hu (C.F.)

**Abstract:** The use and quality analysis of household compost have become very important issues in recent years due to the increasing interest in local food production and safe, self-produced food. The phenomenon was further exacerbated by the COVID-19 pandemic quarantine period, which gave new impetus to the growth of small home gardens. However, the knowledge associated with making high-quality compost is often lacking in home gardeners. Therefore, the objective of this research was to find answers to the following questions: can the quality of backyard compost be considered safe in terms of toxicity and nutrient content? Can weed seed dispersion affect the usability of backyard compost? In general, can the circulation of organic matter be increased with the spread of home composting? In this study, 16 different house composts were analysed for stability, weed seed contamination, toxic elements, and nutrient content using analysis of variance. The results of the research showed that the quality properties of the composts (including their weed seed dispersion effect) were greatly influenced by the different techniques and raw materials used. The toxicity levels, as well as the content of macro and microelements, were within the parameters of safe-quality compost. The specific macronutrient (Ca, Mg) and micronutrient (Fe, Mn) contents of the tested composts have a similar and, in some cases, more favorable nutrient supply capacity in crop production than the frequently-used cow manure-based composts. With a plan of basic education on composting, there is potential to encourage farmyard composting.

**Keywords:** compost quality; backyard compost; toxins; nutrient; oxygen consumption; AT4 value; weed control; hygienisation; compost stability; waste circulation

**Citation:** Ujj, A.; Percsi, K.; Beres, A.; Aleksza, L.; Diaz, F.R.; Gyuricza, C.; Fogarassy, C. Analysis of Quality of Backyard Compost and Its Potential Utilization as a Circular Bio-Waste Source. *Appl. Sci.* **2021**, *11*, 4392. https://doi.org/10.3390/app11104392

Academic Editor: Carlos Rico de la Hera

Received: 13 April 2021
Accepted: 6 May 2021
Published: 12 May 2021



## 1. Introduction

The growing popularity of the concept of circular economy gives favorable background support to environmentally friendly waste-management solutions. It is different from any of the policies of the European Union; therefore, it is worth treating as a special phenomenon. Its main goal is to manage waste through the circular design of material use, product use, and system applications [1]. However, positive tendencies can be observed in the sorting of biowaste due to the principles of circular economy and the assumptions of the zero waste program [2]. The majority of this biowaste is not collected selectively or utilized, but collected together with the municipal solid waste and placed in waste disposal sites or waste incinerators [3]. Thus, the nutrient contents of these biowastes, which could be used in agriculture, are lost and, in most cases, are replaced by fertilizers which can have negative long-term effects on the soil and the environment [4]. Furthermore, these solutions unnecessarily decrease the available waste disposal capacity and increase the environmental risks of waste incineration and waste disposal [5].

Composting is a safe method of waste management [6] and has the potential to improve soil structure, texture, and aeration, as well as improve the water retention capacity of the soil [7]. In addition, the Environmental Protection Agency (EPA) of the United States declared that composting reduces methane emissions in landfills, lowering their carbon footprint [8]. Therefore, composting is an environmentally friendly alternative applied to recycle organic waste and obtain products used as amendments in agriculture [9–14]. Organic fertilizers are originated from agricultural by-products (manures, liquid manures, straw, plant residues, etc.), residues and waste from the food industry (e.g., sewage, waste from slaughterhouses, etc.), and biologically degradable municipal waste [4,15–17]. By selective collection and composting of these biowastes, separated from the municipal solid wastes, it is possible to obtain valuable, secondary raw materials that can be utilized locally, in gardens, and are produced in an economically feasible way.

### 1.1. Backyard Composting Methods

Backyard composting cannot be considered as a uniform method since the ingredients of the compost, the applied technical solutions (composting drums or bins, composting tanks, nets, or prisms), and the treatment of the materials (grinding, turning, aeration) differ considerably [17,18]. It is important to emphasize that backyard composting is considered as a closed system as the used raw materials are created, processed, and utilized on-site by the compost maker themselves [19]. Hence, the available information regarding the quality of backyard composts is limited, occasional, and based on a small number of samples. However, there are numerous available studies regarding industrial composting [20–23]. In Hungary, there has not been any such comprehensive examination carried out related to backyard composts. The novelty of the current research is that we did not find any similar research in the literature, because the method of backyard composting differs from household to household and the technological descriptions cannot be followed in practice; conducting scientific research in the field is, therefore, a very difficult task. Consequently, there is limited information about the quality and physical, chemical, and biological parameters of backyard compost.

### 1.2. Determination of Compost Quality and ITS Measurement

According to Frickte [24], the quality of compost is mainly determined by the kind and composition of input materials, the degree of purity and quality of soil, and emissions in the specific area of collection. He also highlighted that compost quality is substantially influenced by the type of technology applied during the composting process. However, his findings related to the importance of treatments during the composting were not substantiated by Jakubus's [25] examination. The determination and evaluation of compost quality can be carried out by examining physical (e.g., color, odor), chemical (e.g., C/N ratio, nitrification), and microbiological parameters (e.g., self-heating test, respiration), and parameters that influence its effect on the grown crops (e.g., germination test) [26]. Considering the abovementioned, in the case of the quality of backyard compost, these parameters are advisable to examine and are determinant during backyard compost use. The parameters are as follows: maturity and stability, which has a direct effect on plant development; macro and micronutrient contents, which contribute to the nutrient replenishment; toxic element contents, which determine the safe use of composts; and the weed seed content, which influences the long-term weeding effect of backyard composts and as such is a very practical quality parameter. Maturity can be characterized by physical, chemical, and biological stability [27]. Stability refers to the microbial activity in compost, and is evaluated by respirometric measurements or by analysis of the changes in chemical properties [28]. Furthermore, stability can be evaluated by using biological methods such as the germination index (GI) [29–31]. It should be noted that the principle of microbiological maturity tests depends on the assumptions that the maturity of composts is strongly related to stability, and that the microbial activity of composts determines their maturity [32,33]. The different maturity levels of composts depend on the alteration of the different components, the extent and quality of humification and mineralization processes, and the degree of decomposition of the original organic materials [34].The examination of the respiration of composts is an adequate method that can be applied as a test for microbiological compost maturity [26]. The respiration examination can measure either the carbon dioxide evolution or the consumed oxygen during the biological decomposition of the compost samples. Several studies considered oxygen consumption as a reliable variable for evaluating the development of composting due to its correlation with the metabolic potential of compost [6,35,36]. Additionally, oxygen is directly responsible for the oxidation of organic matter [8]. This can be measured through respirometry, a recognized method for measuring the oxygen consumption of microorganisms [37]. In the case of this technique, two methods can be differentiated, i.e., the static and the dynamic methods. The differentiation depends on the whether the measurements are carried out with or without a continuous oxygen supply [35,38,39]. The so-called $AT_4$ value is an internationally accepted method to determine the maturity, stability, and usability of composts. It measures the total amount of oxygen that is consumed by the microorganisms during the biological decompositions of the biologically degradable organic matter contained in the samples within four days of examination (examined parameter: respiration activity, oxygen demand, unit: mg $O_2/g$ dry matter) [39–41].

### 1.3. Macro and Micro Nutrient Content of Compost

Jakubus [25] examined the chemical composition of different compost samples and found that the differences in the chemical composition of composts primarily resulted from the raw materials used in composting, and to a lesser extent from the processes and technology used. One of the important results of the abovementioned examination was that compost with low nutrient and organic matter content could be characterized by favorable stability and maturity parameters, as organic matter content decreases during composting. In addition, the author found that the composts based on yard trimmings and household wastes prepared in home composters, while highly-recommended for small-scale vegetable cultivation, had less desirable chemical composition compared with composts prepared with a share of sewage sludge, whilst the mixture of municipal sewage sludge and biowastes is recommended to build soil production capacity.

### 1.4. Toxicity Levels and Chemical Impact of Compost

Another important quality aspect of the compost is the chemical and environmental impact caused by the application. The pollution caused by potentially toxic elements (PTEs) (e.g., Sb, As, Cd, Cr, Cu, Pb, Hg, Ni, Se, Ag, Sn, Zn) have particularly relevant ecological and human health issues. Potentially toxic elements (PTEs) are a major concern from food safety, environmental, and ecological points of view [42]. There is a worldwide increase in heavy metal or potentially toxic element (PTE) contamination in agricultural soils caused mainly by human and industrial action, which leads to food contamination through crops [43]. Toxic elements, especially the heavy metals of the periodic table, are normal elements found in the environment, and trace amounts of them are always found in foodstuffs [44]; however, foods from contaminated areas may contain higher amounts. Toxic elements primarily enter foodstuffs through contact with the environment. Cadmium is one of the most mobile elements among all toxic heavy metals. Because of its high mobility, cadmium is readily taken up by plants from the soil and transferred to the aerial parts of the plant, where it can accumulate to a high level. Because of its high mobility in soil, the bioaccumulation of cadmium in plant-based food is usually high compared with the other trace elements [45,46]. Organic amendments can contain undesirable components, such as potentially toxic elements, that affect both humans and the environment. PTEs (and several chemical compounds) can also affect soil microbial communities and, in turn, soil functionality and fertility [47]. The soil, by itself, is not able to biodegrade the heavy metals, which have the tendency to persist and accumulate. It is therefore necessary for the assessment of the parameters of the quality of compost to consider both the aspects connected to agronomic efficiency and those of environmental compatibility [48].

### 1.5. Weed-Seed Dispersal Effect and Biological Quality

During quality analysis of compost, the weed seed dispersal effect is an important biological qualitative parameter to consider [49]. In the case of application in the field of a compost containing a significant amount of viable weed seeds and plant propagules, the emerging weeds can cause significant damage to the crop in terms of quality and yield. These disadvantageous effects can be reduced or even completely eliminated by properly performed composting. Studies addressing these issues showed that inappropriate composting practices, specifically related to short composting time and to a lesser extent to temperature, result in weed seeds not losing their viability by the end of the composting process [50]. The survival of weed seeds and viable plant propagules depends on different processes taking place during composting and the characteristics of resistance of each weed variety. The most important parameters to consider are the following: length of time at the appropriate temperature within the active phase of composting, and moisture content. Based on the experimental results, the different weed species lose their viability at different temperatures and periods of time [51]. The fate of weed seeds is influenced considerably by the moisture content as well: higher moisture content causes certain weed seeds to lose their viability at a higher rate [52,53]. According to the parameters mentioned above, certain weeds, such as *Solanum nigrum* L. [51], *Sorghum halepense* (L.) *Pers.*, and *Convolvulus arvensis* L. are considered very resistant [54]. Strict regulations have been established in the compost-related national regulations and standards to ensure the appropriate sterility of the produced composts and to eliminate the viability of the weed seeds. The temperature that needs to be achieved in the active phase of the composting process and the related time period are also set strictly in these regulations. In addition, these regulations are not unified in Europe; they can differ in the expected highest temperature and the time interval as well. In Austria, for example, 55–65 °C has to be maintained for 14 days; in Germany or Denmark, at least 55 °C has to be maintained for 14 days. In France, at least 60 °C has to be maintained for 4 days [49]. However, the determination of the number of weed plants (on a weight or volume basis) that emerge from an appropriately prepared and properly produced compost sample is a common feature of the currently available standard methods for the examination of the weed seed dispersal effect of composts.

On the whole, it is important to examine the quality of the composts that are utilized in farming or household gardening to ensure quality, even if they are only used in small quantities.

This research analyzed the potential that backyard composting offers as an alternative to remove some of the pressure that biowaste has on traditional waste-management systems and its potential usability at a household level.

In order to determine the quality and potential utilization of backyard compost, this research intends to answer the following questions:

Is the quality of backyard compost, in terms of toxicity and nutrient content, considered safe to use at a household and farm level? Do the different treatments and ingredients (raw materials) applied in the composting process have a significant impact on the quality of the final product? Are the techniques and treatments used in individual backyard composting adequate enough for agricultural purposes, or should the raw materials be collected by an external organization and delivered to a common composting site for stricter control and unified quality? Could weed seed dispersion affect the usability of backyard compost? Do the values and species identified in the weed seed dispersion effect evaluation fall within the acceptable parameters at a household and farm level? Is backyard composting a feasible circular alternative to conventional waste management and environmental issues?

According to the objectives abovementioned, the research hypothesis is that, based on the complex qualitative analysis of backyard composting, it is highly recommended to use composting in backyard gardening because the quality level reaches the level defined by the related regulations and national norms.

## 2. Materials and Methods

### 2.1. The Examined Composts Materials

The samples of this research had two distinctive origins: backyard compost at a household level and at an ecological farm level. In higher proportion, we examined composts that were mainly produced in gardens of family houses by utilizing biowaste products as main ingredients. The ecological farms, from which three of the examined compost samples originated, are predominantly characterized by the production of vegetables, herbs, and fruits. The samples produced therein contained walnut leaves, bean stems, plant waste, and horse manure as main ingredients. The raw materials, the technology of composting, and the age of the produced composts varied considerably from one sample to another. It is necessary to highlight that all the factors mentioned above were decided by the compost makers themselves and were not subject to our interference. The composts included in the examinations were considered mature and ready for application by the compost producers in all cases. The most important characteristics of the samples used for this research are summarized in Table 1.

**Table 1.** The properties of the examined composts (BYC—backyard compost; ECOL—ecological farm).

| Code | Type | Method of Composting | Raw Materials | Preparation | Treatment | Age |
|---|---|---|---|---|---|---|
| A | BYC | Pile | Garden green waste | Grinding | Monthly turning | 2 years |
| B | BYC | Pile | Garden green waste | Grinding | Monthly turning | 1 year |
| C | BYC | Closed composting bin | Garden green waste, kitchen waste | Grinding | Monthly turning, moistening | 2 years |
| D | BYC | Open composting | Garden green waste, kitchen waste | Grinding | Monthly turning, moistening | 1.5 years |
| E | BYC | Open composting | Garden green waste, wood ash, poultry manure, soil | - | - | 3 years |
| F | BYC | Open composting | Garden green waste, wood ash, soil | - | - | 3 years |
| G | BYC | Open composting | Garden green waste, kitchen waste | Grinding | Addition of earthworms | 3 years |
| H | BYC | Pile | Garden green waste, kitchen waste, rabbit manure | Grinding | Monthly turning | 2 years |
| I | BYC | Open composting | Garden green waste, kitchen waste, wood ash | - | Periodic turning, addition of earthworms | 2 years |
| J | BYC | Open composting | Garden green waste | Grinding | Addition of earthworms | 2 years |
| K | BYC | Pile | Garden green waste, grape leaves, dishwashing water | Grinding | - | 4 years |
| L | BYC | Pile | Garden green waste, sycamore leaves, poultry manure | Grinding | - | 4 years |
| M | BYC | Open composting | Garden green waste, fruit waste | - | Addition of earthworms | 4 years |
| N | ECOL | Open composting | Walnut leaves, bean stem (50–50%) | | Periodic turning, moistening | 8 months |
| O | ECOL | Open composting | Walnut leaves | | Periodic turning, moistening | 8 months |
| P | ECOL | Open composting | Horse manure and other plant materials (80-20%) | | Periodic turning, moistening | 1.5 years |

### 2.2. Methods

Compost samples were collected from home-mixed composts that were considered mature, stable, and ready for use by their producer. The representative samples, which consisted of the mixture of several subsamples, were taken from different parts of the compost piles and weighed 10–30 kg. The samples were stored in a refrigerator or in a deep freezer until the laboratory examinations started. During preparation, the samples were chopped into 10 mm in size with a shredder and were sent through a 10 mm sieve. The samples were prepared by sample quartering from the whole samples for the examinations. The experiments were carried out in the Waste Management Laboratory, Institute of Environmental Sciences at MATE, formerly Szent István University.

#### 2.2.1. Stability Measurement

In order to measure the stability, the $AT_4$ respiration activity of the examined compost samples was determined based on the method written in ÖNORM S 2027-4:2012 standard [55]. The moisture content of the examined samples was set by hand-squeeze test; thus, the moisture content was between 45% and 55%. The moisture content of the samples was determined according to the MSZE 21420-18:2005 standard [56]. The biological activity (stability) of the compost samples was determined by measuring the respiration activity and by determining the $AT_4$ values of the samples in three replicates according to the applied standard. The $AT_4$ respiration activity values were determined by OxiTop Control measuring system. We placed a 100 $cm^3$ prepared, moistened sample in the vessels. During the determination of oxygen consumption, according to the description, the vessels were placed in a thermostat cabinet, and 20 °C was maintained during the experiment. During the measurements, the samples placed in aerobic conditions in the vessels started to biologically decompose depending on their biological activity (stability). During this decomposition process, the oxygen in the vessel was partly consumed and its quantity decreased. The carbon dioxide that was produced during the aerobic biological decomposition of the samples was absorbed in the air of the vessel by an absorber, which resulted in a decrease in air pressure in the vessel that was then in proportion with the oxygen consumption and the respiration activity. A measuring device (controller) was part of the measuring system, by which the basic parameters of the samples and the measurement itself could be set (e.g., volume of the sample, measuring time, air pressure, etc.) in the beginning of the measurements. The measured data could be stored and read in the measuring heads or downloaded to the computer. The $AT_4$ respiration activity values were determined on the dry-matter basis of the samples (mg $O_2/g$ dry matter) using the primer measuring results downloaded from the measuring heads, applying the method written in the standard (average value and standard deviation of the three repetitions). During the examinations, the so-called lag-phase was not noticed in any of the examined samples [57].

#### 2.2.2. Weed Seed Contamination Effects Determination

The weed seed dispersion effect of the samples was determined according to the method written in the MSZ-08-0012/4-79 standard (physical, chemical, and biological testing of peat and peat products, examination of weed expansion and germ killing effect) [58]. Based on that standard, we prepared a mixture of compost (with its original moisture content) and soil (Arenosol with 1.6–1.9% humus content, pH 6.5–7.5, $CaCO_3$ content is maximum 1%) in a 1:3 ratio by weight. From this mixture, a similar number of samples (3-3 samples) were measured into plastic containers (15 cm in diameter). The height of the samples was approximately 20 mm in the containers. Based on the ratio of the weighted mixtures, the weight of the compost and the soil was determined in the plastic containers. In three containers, the same amount of soil (control samples) was measured in three containers in order to examine the weed seed dispersal effect of the soil itself. The mixtures were moistened and kept in a well-lit place at room temperature for 14 days. The evaporated water was replaced daily by measuring the weight loss of the samples. After 14 days, the number of emerged weeds was counted in all the different containers. Then, the number of weeds emerged in the soil samples was subtracted (control samples) from the number of weeds in the compost mixture. The result was determined to be 100 g of compost material. According to the standard, the number of couch grass (*Agropyron repens* (L.) P. B.) was determined separately, on the basis of 100 g of compost.

### 2.2.3. Toxic Elements and Nutrient Contents Analysis

The sampling, storing, preparation, grinding, and sample quartering of the composts were carried out by the previously described method. For the examination of the potentially toxic element (Cd, Cr, Cu, Hg, Ni, Pb, Zn) and macro and micronutrient content, 500 g of subsample was air dried in the laboratory. All unsuitable substances (glass, metal, plastic, gravel, etc.) were removed from the samples, and all the compost samples were then chopped into less than 0.25 mm in size by a shredder and homogenized. The samples for the examinations (0.25–0.5 g) were further prepared by sample quartering from the whole, homogenized, <0.25 mm sized samples for the examinations. The digestion of the samples was carried out by a CEM MARS 5 (CEM Microwave Corporation, Matthews, NC, USA) microwave digester by applying a mixture of 5 mL 65 $m/m\%$ $HNO_3$ and 2 mL 30 $m/m\%$ $H_2O_2$. The examination of the samples was carried out by HORIBA JOBIN YVON ACTIVA-M ICP-OES (Horiba Jobin Yvon, Bernsheim, Germany). The moisture content of the samples was determined according to the MSZE 21420-18:2005 standard in a drying oven (Characterization of wastes. Part 18: Determination of humidity and dry matter content).

### 2.2.4. Statistical Factor Analysis

The effects of the different factors, such as methods (pile, closed bin, open composting), treatments (turning, moistening, adding earthworms), type of the samples (backyard, ecological), composting periods, and raw materials were subjected to evaluation with the assistance of one-way ANOVA. Each of the factors was tested independently using the F-test at the significance level $\alpha = 0.95$. The dependent variables were the AT4 values and the effects of the similar factors were tested also on the weed seed content, potentially toxic elements, and macro and micro nutrient contents of the different samples. The null hypothesis that the average values of the examined parameter would be equal for each of the 16 composts was tested against the alternative hypothesis stating that not all averages are equal. Scheffe's analysis was also performed, wherein it was possible to distinguish homogeneous groups among the 16 composts. The data were analysed using the SPSS 25 software.

## 3. Results

The age of the examined composts differed considerably; they were between 8 months and 4 years old. The reason why such aged composts were chosen is that backyard composting is a slower process than controlled large-scale composting. Fresh compost can be created in 3 months, but mature composting often takes 12 months (depending on the composting treatments such as turning, wetting, etc.) Compost makers judge the maturity of composts primarily on the basis of physical properties whilst biological properties are not typically taken into account [59]. Accordingly, compost samples were collected for analysis according to two aspects. The first is that the compost sample age exceeded 9 weeks, regardless of any treatments, as the experiment's results suggested that this is the minimum time required for the compost to become stable and to provide a potential N source [60]. The other aspect is that the compost producer considered the compost samples mature and ready for application based on their own knowledge of composting.

Thus, the different composting periods can be the explanation for the producers having different experiences in practice; they had different knowledge regarding composting and the method thereof (the type of composting device where the process occurred, preparation and the treatment of the raw materials during composting) [17,59]. The composts that originated from ecological farms had shorter composting periods (0.8–1.5 years), which can be attributed to the expertise and proper knowledge of long-term farmers.

### 3.1. Compost Maturity

The results regarding the maturity of the examined composts, the determined $AT_4$ respiration activity values, are shown in Figure 1, on which the determined respiration activity ($AT_4$) is recorded.

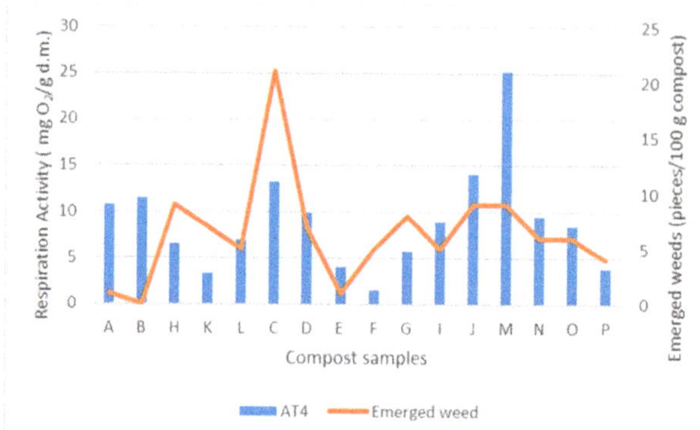


**Figure 1.** The determined respiration activity (AT4) and the number of weeds that emerged within 14 days.

Based on the determined respiration activity values (AT4), it can be stated that the biological stability of the examined backyard composts that can be determined based on the AT4 values differed considerably (1.6–25.2 mg $O_2/g$ dry matter). In one case (compost "M"), the determined respiration activity (AT4) was extremely high (25.2 mg $O_2/g$ d.m.) There were smaller differences and lower typical ranges experienced in respiration activity in the case of the composts which originated from the ecological farms (3.9–9.5 mg $O_2/g$ d.m.) This difference in the determined biological stability values between the backyard and the ecological composts can be attributed to the fact that the producers in the ecological farms are highly aware of the environmental effects of composting and have more scientific background in the topic. There is no unified regulation in Europe regarding the stability and maturity of the treated biowaste products. In Austria and Germany, in order to determine stability, the extent of respiration activity (AT4) is taken into consideration.

For accomplishing the proper stability, the respiration activity value (AT4) has to decrease below 5 mg $O_2/g$ d.m. The suggested limit value in the EU for the same parameter is 10 mg $O_2/g$ d.m. [23]. If we consider the Austrian and the German qualification aspects among all the examined composts, three backyard composts (E, F, and K composts) and only one ecological compost (P) met the requirements for the compost maturity. However, based on the suggested, more permissive unified EU limit value, most of the examined backyard composts (8 out of the examined 13 composts) and all of the examined ecological composts show appropriate maturity and stability values. Some backyard composts (A, B, C, J, and M), however, still did not meet the requirements even in the case of the less strict limit and showed inappropriate biological stability. In the case of these composts, the beneficial effects of composts on the soils and the soil-plant system notwithstanding, unfavorable effects can also occur due to their inappropriate biological stability. The results obtained regarding the low stability of backyard composts might reinforce the concern about its quality [17,23,59,60]. It has to be emphasized that several methods are used worldwide to determine the maturity, biological stability, and the respiration activity of composts. These are the following: the Solvita test, the self-heating test, static respirometry index (SRI), dynamic respiration index (DRI), and the specific oxygen uptake rate (SOUR). The comparability of the results obtained by these different methods has been examined by several authors; as a result, correlations are available to help in the comparison of the available results [23,38,61].

### 3.2. Weed Seed Dispersion

The experiment carried out on the weed seed dispersion effect of the compost samples is shown in Figure 1, wherein the number of weeds that emerged within 14 days from the compost samples is recorded (calculated to 100 g of compost). During the 14 days of examination, the following plants emerged from the samples that we could identify at a species level: *Solanum lycopersicum* L., *Amaranthus retroflexus* L., *Stellaria media* (L.) *Vill.*, *Trifolium* sp., *Capsella bursa-pastoris* (L.) *Medik.* Regarding the permission of marketing of composts, the related examinations, and their evaluations, the Hungarian regulation (Licensing, storage, distribution, and utilization of crop increasing materials 36/2006. (V. 18.) FVM Regulation) states that the compost that is to be sold as a product cannot contain quarantine weed seeds or plant parts. These weeds are the following: *Ambrosia artemisiifolia* L., *Solanum dulcamara* L., *Asclepias syriaca* L., and *Cuscuta* sp.

Based on the results of our examinations, it can be concluded that there was not any weed species considered as quarantine weed in the case of the backyard nor of the ecological composts. However, based on the results, it can be observed that among all the examined composts, only the compost B did not contain viable weed seeds, based on the required test. Regarding the other examined composts, it can be stated that in some cases (e.g., backyard compost C), if they are applied to the field, they could cause serious weed problems due to the emerging weeds on the compost treated fields. In the European countries, there is not any unified regulation related to the weed seed dispersal effects of composts. For example, in Italy and France, there is not any regulation for this; in Belgium and Great Britain, however, the high-quality compost must be free of viable weed seeds. In the case of Germany and the Netherlands, the allowed limit of viable weed seeds is two, while in Austria this value reaches three viable weed seeds per one liter of compost sample in its original state [49]. During examinations in general, the compost samples are mixed with some other material, or medium, then the mixture is watered and stored in regulated climatic conditions with available light. After the examination period, the number of emerged weeds is determined. This is in contrast to the Austrian standard method wherein the samples need to be mixed with high-quality peat kept moist, and, after 15 days of examination, the number of emerged weeds determined per litre of compost [62]. However, the standard method that was applied regarding the setting of the mixture and the length of examination time can be considered unique. Based on the obtained results, and considering the smaller amount of compost weight (100 g) as a basis, it can be assumed that the quality of all the examined compost samples (except for compost B) does not reach the required limit of the above-mentioned countries. This means that the quality of these composts was objectionable in terms of weed seed dispersal effect. This also draws attention to the fact that, during inappropriate backyard composting, the weed seeds do not lose their germination ability at the end of the process [50].

### 3.3. Potentially Toxic Element Content

Based on the results of the examined parameters, the potentially toxic element content of the samples is shown on Table 2.

Based on the presented results, it can be stated that all the examined backyard and ecological composts met the Hungarian requirements for the potentially toxic elements. However, the copper concentration (98.16 mg $\times$ kg$^{-1}$ d.m.) of backyard compost K was close to the limit value (100 mg $\times$ kg$^{-1}$ d.m.) The reason for the relatively high copper content of backyard compost K can be partially explained by the applied raw materials and the applied dishwashing water, which was only used in this case, for moistening compost. Fernández-Delgado Juárez et al. (2015) also noticed high copper content in their research in the case of green waste composts [63]. It can be further stated that the backyard composts only partially met the requirements of the EU regarding the potentially toxic elements and the regulations for the backyard composts (e.g., Austria, hobby gardening, EU Ecolabel). Backyard compost K did not meet the requirements of some regulations (Germany, biowaste ordinance, Class I, limit value 70, measured value 98.16 mg $\times$ kg$^{-1}$

d.m.) As explained earlier, the high copper content of compost K, similarly to the high cadmium content, can be explained by the applied dishwashing water. In the case of backyard compost F, the higher Cd content can be explained by the wood ash since it may have contained high amounts of Cd (2.3 mg $\times$ kg$^{-1}$ d.m.) [63]. Based on the results, it can be emphasized that the potentially toxic element content of the examined composts, which originated from and was utilized in ecological farms, met the specific requirements (Austria, organic farming, EU Regulation on organic agriculture) for composts that can be utilized in ecological farms. All of the examined ecological composts resulted in lower potentially toxic element content and narrower concentration range compared with the backyard composts.

**Table 2.** The determined potentially toxic element content of the examined composts.

| Sample | Cd | Cr | Cu | Hg | Ni | Pb | Zn |
|---|---|---|---|---|---|---|---|
| | mg $\times$ kg$^{-1}$ d.m. | | | | | | |
| A [a] | 0.23 ± 0.03 | 10.68 ± 0.00 | 12.25 ± 0.11 | nd | 5.12 ± 0.06 | 3.49 ± 0.12 | 49.8 ± 0.3 |
| B [a] | 0.64 ± 0.03 | 17.42 ± 0.03 | 29.05 ± 0.28 | nd | 7.95 ± 0.29 | 15.21 ± 0.70 | 94.8 ± 0.1 |
| C [a] | 0.69 ± 0.05 | 16.51 ± 0.12 | 32.00 ± 0.12 | nd | 9.35 ± 0.24 | 11.80 ± 0.22 | 118.7 ± 0.1 |
| D [a] | 0.35 ± 0.01 | 12.28 ± 0.07 | 19.16 ± 0.60 | nd | 5.87 ± 0.27 | 5.30 ± 0.57 | 66.5 ± 0.3 |
| E [a] | 0.85 ± 0.01 | 22.87 ± 0.29 | 23.35 ± 0.27 | nd | 10.06 ± 0.18 | 18.68 ± 0.49 | 134.8 ± 0.9 |
| F [a] | 0.93 ± 0.05 | 30.02 ± 0.31 | 21.07 ± 0.39 | nd | 10.84 ± 0.1 | 14.17 ± 0.62 | 109.2 ± 0.2 |
| G [a] | 0.87 ± 0.01 | 20.59 ± 0.19 | 11.81 ± 0.29 | nd | 8.66 ± 0.21 | 15.55 ± 0.33 | 52.7 ± 0.0 |
| H [a] | 0.77 ± 0.03 | 11.14 ± 0.02 | 9.82 ± 0.08 | nd | 6.35 ± 0.63 | 6.56 ± 0.40 | 85.7 ± 0.7 |
| I [a] | 0.66 ± 0.04 | 17.87 ± 0.02 | 44.03 ± 0.52 | nd | 11.76 ± 0.15 | 9.21 ± 0.41 | 53.8 ± 0.3 |
| J [a] | 0.65 ± 0.01 | 15.11 ± 0.15 | 25.08 ± 0.47 | nd | 9.74 ± 0.01 | 8.53 ± 0.97 | 55.6 ± 0.3 |
| K [a] | 0.91 ± 0.06 | 29.19 ± 0.46 | 98.16 ± 2.87 | nd | 7.91 ± 0.20 | 13.70 ± 1.06 | 78.9 ± 0.4 |
| L [a] | 0.56 ± 0.03 | 10.45 ± 0.01 | 22.12 ± 0.16 | nd | 5.64 ± 0.44 | 12.12 ± 0.08 | 135.6 ± 0.1 |
| M [a] | 0.70 ± 0.11 | 17.04 ± 0.11 | 21.74 ± 0.49 | nd | 10.21 ± 0.14 | 10.14 ± 0.12 | 71.2 ± 0.3 |
| N [b] | nd | 11.1 ± 0.06 | 10.07 ± 0.39 | nd | 5.13 ± 0.15 | 5.6 ± 0.02 | 38.8 ± 0.5 |
| O [b] | nd | 13.46 ± 0.24 | 12.85 ± 0.44 | nd | 5.28 ± 0.14 | 6.02 ± 0.2 | 46.6 ± 0.7 |
| P [b] | nd | 13.42 ± 0.25 | 10.13 ± 0.53 | nd | 5.69 ± 0.13 | 5.39 ± 0.63 | 32.9 ± 0.7 |

[a] backyard compost; [b] composts from ecological farms; d.m.—dry matter; nd—not detectable.

### 3.4. Macro and Micronutrient Content

Based on the results, the macronutrient and micronutrient content of the backyard and the ecological composts typically did not differ from each other, and were in the same range. It can be stated that all the examined backyard and ecological composts met the requirements valid in Hungary for the calcium content; however, only some of the backyard composts met the requirements for the magnesium content (composts B, C, D, H, and M). Therefore, the examined composts mixed into the topsoil in a suitable quantity would provide an appropriate amount of calcium and sufficient amount of magnesium for the produced crops. Based on the macro and micronutrient content of the examined composts, it can be stated that their calcium (1.32–5 and 1.8–3.9% d.m.) and magnesium content (0.3–0.7 and 0.3–0.4% d.m.) reached and exceeded the typical range for cow manures (calcium: 2.03–2.16; magnesium: 0.42–0.46% d.m.) that is often applied in hobby gardens and household farming, which is a common application area for backyard composts. Their iron (0.6–1.1 and 0.7% d.m.) and manganese (239–667 and 296–340 mg $\times$ kg$^{-1}$ d.m.) content, however, exceeded the typical range for cow manures (iron: 0.22–0.51% d.m.; manganese: 117–161 mg $\times$ kg$^{-1}$ d.m.) [64]. Based on these findings, it can be concluded that, according to the determined macronutrient (Ca, Mg) and micronutrient (Fe, Mn) content of the examined composts, they have similar and, in some cases, more favorable nutrient-supplying capacity in crop production than the frequently-used cow manures.

### 3.5. Results of Statistical Analysis

The possible effect of the most common treatments on the potential toxic elements and macro and micronutrient content was examined. The grouping of the samples according to the treatments was as follows:

1 = Monthly turning, MT (Samples A, B, H)
2 = Monthly turning, moistening, MTM (Samples C, D)
3 = No treatment, NT (Samples E, F, K, L)
4 = Addition of earthworms (not red worms), AE (Samples G, J, M)
5 = Periodic turning, addition of earthworms, PAE (Sample I)
6 = Periodic turning, moistening, PTM (Samples N, O, P)

The variance of homogeneity and significant differences in category means are present simultaneously only for Zn and Mg: zinc as a potential toxic element, and magnesium as a macronutrient. A post hoc test could not be performed because treatment 5 (periodic turning and addition of earthworms) had only one element. As there was no post hoc test, the visual representation of the means provides a basis for judging differences between groups.

It can be seen from Figure 2 that the largest difference is between treatments 3 and 6 in terms of Zn content.

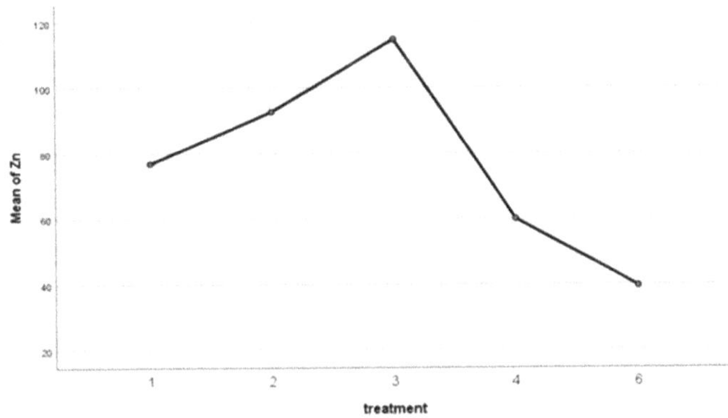


**Figure 2.** Average comparison. Note: unit of measure is mg × kg$^{-1}$ d.m.

In terms of the effect on Mg content, there was the largest difference between treatments 2 and 3 that can be seen in Figure 3.

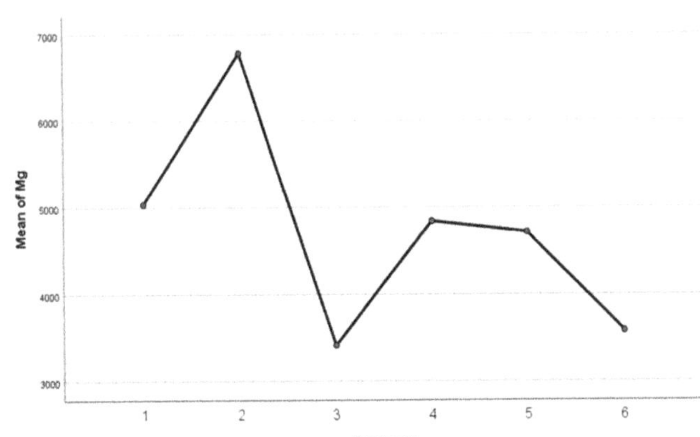


**Figure 3.** Average comparison of Mg. Note: unit of measure is mg × kg$^{-1}$ d.m.

In the cases wherein no treatment was applied, the Mg content was at the lowest level, while the zinc content, which is a potentially toxic element, was the highest. The latter was true even for lead, cadmium, and chromium. In the case of the untreated compost, however, the Ca content was at the lowest level. The effects of the raw materials on toxic element content and macro and microelement content were also examined. The ingredients of the raw materials of the examined composts were very diverse. In order to compare the samples, five categories were created, taking into account the chemical properties of the ingredients and their main components' effect on the microbial activity. Components that have a differentiating effect on the quality parameters of the composts, such as the effect on the toxic element and macro- and micronutrient content, were selected. According to this, the following categories were created:

1 = GGW-garden green waste only (Samples A, B, J)
2 = M-Use of manure (Samples H, L, P)
3 = S-Use of soil and wood ash (Samples E, F)
4 = WL-Walnut leaves (Samples N, O)
5 = GGW and others (Samples C, D, G, I, K, M)

Category averages differed significantly for Cd, Cr, Ni, and Fe. The post hoc (Scheffe) test was feasible in this case and showed a significant difference between raw material categories 3 and 4 (Figure 4). Thus, cadmium levels were highest in terrestrial raw materials, while they were lowest in walnut leaves. The latter includes the fact that the walnut leaf was of organic origin.

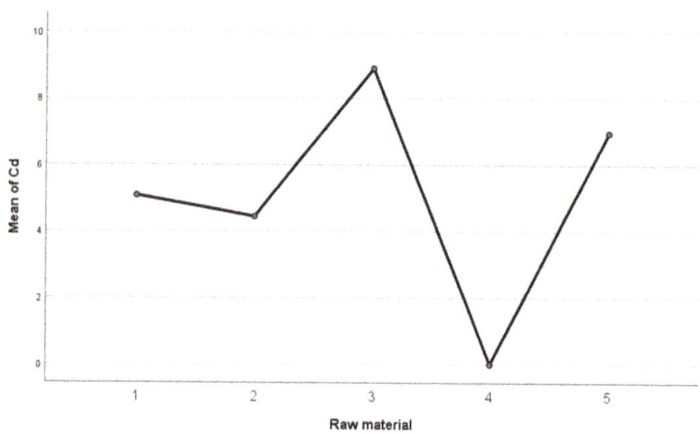


**Figure 4.** Average comparison for Cd.

The raw material had an effect on the level of the Cd (Figure 4). Category averages differed significantly for cadmium and iron when we analyzed the impact of the two main raw material categories, ecological and conventional. A post hoc test could not be performed here either, as the number of categories (organic and conventional raw material) was less than three. Overall, based on the impact assessments of the previous two raw materials, it can be stated that the use of organic raw materials (since walnut leaves were also organic) had a positive effect on the overall quality of the compost, including the cadmium level [65–67]. In addition to the abovementioned, the effects of composting time, method, treatment, and raw materials were also examined on other quality characteristics of compost, i.e., AT4 and weed content, but no significant correlation was found, probably due to the low number of items in the sample [68]. Nevertheless, the study is unique because there has never been such a detailed, multi-characteristic compost study in Hungary.

### 3.6. Practical Implication of the Study

The results of this complex study highlight the uncertainties regarding the quality of backyard compost. The unfavorable macro and micronutrient content and the possible weeding effect do not pose a risk if compost remains within the farm cycle; its use is highly recommended to support circular systems. It is only necessary to comply with the regulations if, in addition to its backyard-level use, the compost is sold. Furthermore, composting requires expertise and attention if nutrient replenishment is provided over a larger area (small scale farm level) with backyard compost. In this case, it is worth taking into account the recommendations supported by the literature (separate collection of weeds, regular turning and watering). Due to the essence of backyard composting, each compost producer uses their own raw materials, ensuring the closed backyard nutrient cycle and the regulation of external outputs. However, this implies the difficulties of preparing a comparative analysis and taking uniform samples. Apart from the raw materials of backyard composting, the applied composting processes, the frequency of the processes, and the quality of their execution are not regulated; these factors are greatly influenced by the expertise of the compost maker. Further analyses and additional samples from organic farms are also needed to draw deeper, statistically verifiable conclusions.

## 4. Discussion

### 4.1. Maturity and Stability

Based on the results of the examination, it can be stated that the composts produced in the backyard and organic farms are characterized by different quality in terms of maturity and stability values. It should be added that the collected samples were considered mature and ready to use by the compost makers themselves, regardless of the age of the compost (e.g., samples N and O were 8 months old whilst samples L and M were 4 years old). However, in most cases at a household level and small farms, the analysis of the maturity and readiness of the final product is performed by empirical tests based on the compost maker's senses and experience, or lack thereof. Physical characteristics such as texture, smell, or color are the major indicators, rather than laboratory tests, rendering it a challenge to have consistent and accurate results at all times [69]. Consequently, if the compost does not achieve optimal levels of maturity and stability, it could slow down the plant growth process and therefore be considered of inferior quality compared with store-bought compost [70]. For this reason, education on basic practices and standards of compost making to enthusiast backyard gardeners and small-scale bio farmers is highly suggested to achieve desirable performance and usability of compost.

### 4.2. Toxicity and Macro and Micro Nutrients

The quality properties of the analyzed composts were greatly influenced by the different techniques and raw materials used. The levels of toxicity and the content of macro and micro nutrients identified in all the samples fell within the Hungarian parameters for quality and safe compost. Moreover, the determined macronutrient (Ca, Mg) and micronutrient (Fe, Mn) content of the examined composts have similar and, in some cases, more favorable nutrient-supplying capacity in crop production than the frequently used cow manure-based composts. Despite the positive results, it is important to highlight the higher levels of zinc that fell close to the limit of toxicity and the low presence of magnesium, the latter being an important component in a high-quality compost. This tendency was found in samples with no treatments such as turning and watering, which would have sped up the composting process. The statistical analysis of our examinations, in part, supports Jakubus' results [25], according to which the initial raw materials of composts primarily determined their macro and microelement content and thus the quality of the finished compost. This is evidenced by the fact that in the case of untreated compost, wherein the initial raw materials included the slowly degradable sycamore and grape leaves, not enough magnesium was revealed even after 3–4 years without intervention. This can be explained in part by the low magnesium content of the initial raw materials, but also by the lack of composting treatments that slowed down the mineralization. Similarly, the higher zinc content in the non-treated composts can be explained by the initial raw materials (wood and ash, which can be even higher in the case of preservative-treated wood or surface-treated wood waste), which occurred mainly in compost samples with no treatments. Another important consideration is regarding the higher levels of cadmium content, mainly in samples that contained wood ash (samples E and F), dishwashing water (sample K), and/or kitchen waste as raw materials. This examination confirms what other authors have already described [65]: that the release of cadmium into soil and groundwater is largely attributed to anthropogenic activities, including the use of synthetic phosphorus fertilizers (which may be contaminated with cadmium depending on the location of extraction). Other potential sources of the cadmium pollution identified in the literature are drinking water and recycled wastewater [66,67]. Given the above, samples from organic farming resulted in zero cadmium content.

### 4.3. Seed Content

Regarding the weed-seed content and germination test, it can be stated that the weed-seeds species emerging from the compost samples analysed are not considered as quarantine weeds by Hungarian regulations in the case of the backyard or the ecological compost samples. It is highly advised, for backyard compost especially, to avoid the use of weeds as raw materials in the compost pile at a flowering or seed stage because the recognition of weeds and the identification of their lifeform (in order to apply e.g., pre-drying in the case of weeds with propagating roots) present a great challenge for inexperienced farmers/gardeners, as evidenced by sample B wherein no germinated weed seed was detected. In addition, knowledge regarding the raw compost materials and their behaviour during the composting process is essential since no weed was included in the raw materials. In both backyard farming and organic farming, the sorting of seeded weeds is a time-consuming activity during manual weeding. In the case of annual weeds, their prolonged seed maturation period must also be considered, as supported by other studies [70]. These findings are confirmed by the weed-test results performed in this study. However, with appropriate composting practices, excessive weed emergence can be avoided. Furthermore, knowledge of composting, especially regarding the proper preparation of weeds as ingredients, the recognition of the importance of the thermodynamics of the pile, and the role of turning the compost, is essential. It should be noted that weed germination was not prominent in the case of samples from organic farming. Interestingly, germinated weed was also observed in the case of compost which consisted of 100% walnut leaves; this can be explained by the method of collection of walnut leaves from the soil and by the random propagation of seeds due to wind.

## 5. Conclusions

Kitchen and backyard waste represent a high cost to municipal waste collection systems. The quality analysis of backyard compost and compost from ecological farms proves that using simple methods nets positive outcomes at a macro and micro level of nutrients and limits the toxicity of backyard composting. Although, in addition to the safe usage of compost, it would be worthwhile to examine the effect of compost in terms of crop yield and nutrient and soil organic matter balance at the small-farm level with a special regard to the whole nutrient cycle of organic farms wherein compost can be a valuable resource to organic fertilizers. At a household and community level, municipalities have the potential to encourage composting with a plan of basic education on composting. It is particularly important to note that for the sampled compost households, the composting process is not waste treatment, but rather the return of organic matter, providing nutrient cycling, and is therefore part of value creation. Households have also taken extra care throughout their consumption systems to ensure that contaminated or risky foods do not enter the kitchen or into the compost and retail gardens. Backyard and biofarm composts have acceptable quality and safety to be utilized by their own producers without the need of external or mechanized processes. In general, the circulation of organic matter can be increased with the spread of home composting. Measures dealing with the treatment of biowaste from mixed municipal waste can create major contributions to its dissemination in practice through concrete actions such as the introduction of a new collection system, further reduction of the amount of biodegradable waste sent to landfills, and the promotion of community based composting. It is no coincidence that the Food Systems Summit 2021 addresses food waste problems as a priority and influences the National Waste Management Plans for a Circular Economy, which clearly supports the promotion of home and community composting and translates policy measures into actions. However, local governments are also encouraged to create programs and foster circular investment to diversify the systems intended to reduce the ecological and economic impacts of conventional waste collection.

**Author Contributions:** All authors: conceived the study, involved in its design, and participated in field data collection. A.B., L.A., K.P., A.U.: prepared the data, performed the statistical analysis. A.U., K.P., C.F., F.R.D. drafted the manuscript together. C.F., C.G., F.R.D., A.U. read and commented on the first draft. All authors have read and agreed to the published version of the manuscript.

**Funding:** This research was funded by Hungarian National Research, Development and Innovation Office—NKFIH (Program ID: OTKA 131925) and KTIA_AIK 121 2013-0015 programs.

**Institutional Review Board Statement:** Not applicable.

**Informed Consent Statement:** Not applicable.

**Data Availability Statement:** Not applicable.

**Conflicts of Interest:** The authors declare no conflict of interest.

**Limitations:** The study environment was specifically in a small urban setting, where residential areas are surrounded by greenery and typically supplement their fruit and vegetable needs with food production for their own consumption. In the study areas, pollution indicators related to environmental elements are not relevant.

## References

1. Fogarassy, C.; Finger, D. Theoretical and Practical Approaches of Circular Economy for Business Models and Technological Solutions. *Resources* **2020**, *9*, 76. [CrossRef]
2. European Comission COM. *Communication from the Commission to the European Parliament, the Council, the European Economic and Social Committee and the Committee of the Region. Toward A Circular Economy: Zero Waste Programme for Europe*; European Comission: Brussels, Belgium, 2014.
3. Blumenthal, K. Generation and Treatment of Municipal Waste. Available online: https://ec.europa.eu/eurostat/documents/3433488/5579064/KS-SF-11-031-EN.PDF/00c0b3fe-db08-4076-b39a-e92015ce99e0 (accessed on 12 May 2021).
4. Alexa, L.; Dér, S. *Szakszerű Komposztálás Elmélet És Gyakorlat*; Profikomp Kft.: Budapest, Hungary, 2001; ISBN 978-963-00-5809-4.
5. Lleó, T.; Albacete, E.; Barrena, R.; Font, X.; Artola, A.; Sánchez, A. Home and Vermicomposting as Sustainable Options for Biowaste Management. *J. Clean. Prod.* **2013**, *47*, 70–76. [CrossRef]
6. Toledo, M.; Siles, J.A.; Gutiérrez, M.C.; Martín, M.A. Monitoring of the Composting Process of Different Agroindustrial Waste: Influence of the Operational Variables on the Odorous Impact. *Waste Manag.* **2018**, *76*, 266–274. [CrossRef]
7. Palaniveloo, K.; Amran, M.A.; Norhashim, N.A.; Mohamad-Fauzi, N.; Peng-Hui, F.; Hui-Wen, L.; Kai-Lin, Y.; Jiale, L.; Chian-Yee, M.G.; Jing-Yi, L.; et al. Food Waste Composting and Microbial Community Structure Profiling. *Processes* **2020**, *8*, 723. [CrossRef]
8. Bermudez, J.F.; Saldarriaga, J.F.; Osma, J.F. Portable and Low-Cost Respirometric Microsystem for the Static and Dynamic Respirometry Monitoring of Compost. *Sensors* **2019**, *19*, 4132. [CrossRef] [PubMed]
9. Gu, W.; Zhang, F.; Xu, P.; Tang, S.; Xie, K.; Huang, X.; Huang, Q. Effects of Sulphur and Thiobacillus Thioparus on Cow Manure Aerobic Composting. *Bioresour. Technol.* **2011**, *102*, 6529–6535. [CrossRef]
10. Deepesh, V.; Verma, V.K.; Suma, K.; Ajay, S.; Gnanavelu, A.; Madhusudanan, M. Evaluation of an Organic Soil Amendment Generated from Municipal Solid Waste Seeded with Activated Sewage Sludge. *J. Mater. Cycles Waste Manag.* **2016**, *18*, 273–286. [CrossRef]
11. Cai, Q.-Y.; Mo, C.-H.; Wu, Q.-T.; Zeng, Q.-Y.; Katsoyiannis, A. Concentration and Speciation of Heavy Metals in Six Different Sewage Sludge-Composts. *J. Hazard. Mater.* **2007**, *147*, 1063–1072. [CrossRef]
12. Bai, J.; Shen, H.; Dong, S. Study on Eco-Utilization and Treatments of Highway Greening Waste. *Procedia Environ. Sci.* **2010**, *2*, 25–31. [CrossRef]
13. Yu, H.; Xie, B.; Khan, R.; Shen, G. The Changes in Carbon, Nitrogen Components and Humic Substances during Organic-Inorganic Aerobic Co-Composting. *Bioresour. Technol.* **2019**, *271*, 228–235. [CrossRef]
14. Cooperband, L.R. Composting: Art and Science of Organic Waste Conversion to a Valuable Soil Resource. *Lab. Med.* **2000**, *31*, 283–290. [CrossRef]

15. Dunst, G. *Kompostierung*; Leopold Stocker Verlag: Graz-Stuttgart, Austria, 1991.
16. Gottschall, R. *Kompostierung: Optimale Aufbereitung und Verwendung Organischer Materialien im Ökologischen Landbau*; Alternative Konzepte; 4. Aufl; Müller: Karlsruhe, Germany, 1990; ISBN 978-3-7880-9798-1.
17. Roulac, J.W. *Backyard Composting*; Green Earth Books: Totnes, UK, 1996; ISBN 978-1-900322-04-1.
18. Andersen, J.K.; Boldrin, A.; Christensen, T.H.; Scheutz, C. Mass Balances and Life Cycle Inventory of Home Composting of Organic Waste. *Waste Manag.* **2011**, *31*, 1934–1942. [CrossRef]
19. Jasim, S.; Smith, S.R. The Practicability of Home Composting for the Management of Biodegradable Domestic Solid Waste. Ph.D. Thesis, Imperial College London, London, UK, 2003.
20. Colón, J.; Martínez-Blanco, J.; Gabarrell, X.; Artola, A.; Sánchez, A.; Rieradevall, J.; Font, X. Environmental Assessment of Home Composting. *Resour. Conserv. Recycl.* **2010**, *54*, 893–904. [CrossRef]
21. Martínez-Blanco, J.; Colón, J.; Gabarrell, X.; Font, X.; Sánchez, A.; Artola, A.; Rieradevall, J. The Use of Life Cycle Assessment for the Comparison of Biowaste Composting at Home and Full Scale. *Waste Manag.* **2010**, *30*, 983–994. [CrossRef]
22. Papadopoulos, A.E.; Stylianou, M.A.; Michalopoulos, C.P.; Moustakas, K.G.; Hapeshis, K.M.; Vogiatzidaki, E.E.I.; Loizidou, M.D. Performance of a New Household Composter during In-Home Testing. *Waste Manag.* **2009**, *29*, 204–213. [CrossRef]
23. Gómez, R.B.; Lima, F.V.; Ferrer, A.S. The Use of Respiration Indices in the Composting Process: A Review. *Waste Manag. Res.* **2006**, *24*, 37–47. [CrossRef] [PubMed]
24. Fricke, K.; Vogtmann, H. Compost Quality: Physical Characteristics, Nutrient Content, Heavy Metals and Organic Chemicals. *Toxicol. Environ. Chem.* **1994**, *43*, 95–114. [CrossRef]
25. Jakubus, M.A. Comparative Study of Composts Prepared from Various Organic Wastes Based on Biological and Chemical Parameters. *Agronomy* **2020**, *10*, 869. [CrossRef]
26. Hunyadi, G. Hulladékból Előállított Komposztok Degradációs Folyamatainak Nyomon Követése. Analyze of the Degradation Process of Waste Compost. Ph.D. Thesis, University of Debrecen, Debrecen, Hungary, 2012.
27. Mathur, S.P.; Owen, G.; Dinel, H.; Schnitzer, M. Determination of Compost Biomaturity. I. Literature Review. *Biol. Agric. Hortic.* **1993**, *10*, 65–85. [CrossRef]
28. López, M.; Huerta-Pujol, O.; Martínez-Farré, F.X.; Soliva, M. Approaching Compost Stability from Klason Lignin Modified Method: Chemical Stability Degree for OM and N Quality Assessment. *Resour. Conserv. Recycl.* **2010**, *55*, 171–181. [CrossRef]
29. Mushtaq, M.; Iqbal, M.K.; Khalid, A.; Khan, R.A. Humification of Poultry Waste and Rice Husk Using Additives and Its Application. *Int. J. Recycl. Org. Waste Agric.* **2019**, *8*, 15–22. [CrossRef]
30. Zhang, L.; Sun, X. Evaluation of Maifanite and Silage as Amendments for Green Waste Composting. *Waste Manag.* **2018**, *77*, 435–446. [CrossRef]
31. Kebrom, T.H.; Woldesenbet, S.; Bayabil, H.K.; Garcia, M.; Gao, M.; Ampim, P.; Awal, R.; Fares, A. Evaluation of Phytotoxicity of Three Organic Amendments to Collard Greens Using the Seed Germination Bioassay. *Environ. Sci. Pollut. Res.* **2019**, *26*, 5454–5462. [CrossRef] [PubMed]
32. Morel, T.L.; Colin, F.; Germon, J.C. Methods for the Evaluation of the Maturity of Municipal Refuse Compost. In *Composting of Agriculture and Other Wastes*; Elsevier Applied Science Publishers: London, UK, 1985; pp. 56–72.
33. Jímenez, E.I.; Garcia, V.P. Evaluation of City Refuse Compost Maturity: A Review. *Biol. Wastes* **1989**, *27*, 115–142. [CrossRef]
34. Bari, Q.H.; Koenig, A. Effect of Air Recirculation and Reuse on Composting of Organic Solid Waste. *Resour. Conserv. Recycl.* **2001**, *33*, 93–111. [CrossRef]
35. Cerda, A.; Artola, A.; Font, X.; Barrena, R.; Gea, T.; Sánchez, A. Composting of Food Wastes: Status and Challenges. *Bioresour. Technol.* **2018**, *248*, 57–67. [CrossRef]
36. Scoton, E.J.; Battistelle, R.A.G.; Bezerra, B.S.; Akutsu, J. A Sewage Sludge Co-Composting Process Using Respirometric Monitoring Method in Hermetic Rotary Reactor. *J. Clean. Prod.* **2016**, *121*, 169–175. [CrossRef]
37. du Plessis, C.A.; Barnard, P.; Naldrett, K.; de Kock, S.H. Development of Respirometry Methods to Assess the Microbial Activity of Thermophilic Bioleaching Archaea. *J. Microbiol. Methods* **2001**, *47*, 189–198. [CrossRef]
38. Adani, F.; Gigliotti, G.; Valentini, F.; Laraia, R. Respiration Index Determination: A Comparative Study of Different Methods. *Compost Sci. Util.* **2003**, *11*, 144–151. [CrossRef]
39. Sánchez Arias, V.; Fernández, F.J.; Rodríguez, L.; Villaseñor, J. Respiration Indices and Stability Measurements of Compost through Electrolytic Respirometry. *J. Environ. Manage.* **2012**, *95*, S134–S138. [CrossRef]
40. Binner, E.; Zach, A. Laboratory tests describing the biological reactivity of pretreated residual wastes. In *Bidlingmaier*; deBertoldi, M., Diaz, L., Papadimitriou, E., Eds.; Organic Recovery and Biological Treatment; Rhombos-Verlag-ORBIT 99.: Weimar, Germany, 1999; pp. 255–261.
41. Lasaridi, K.E.; Stentiford, E.I. A Simple Respirometric Technique for Assessing Compost Stabilit y. *Water Res.* **1998**, *32*, 3717–3723. [CrossRef]
42. Abdi, L.; Molaee Aghaee, E.; Nazmara, S.; reza Alipour, M.; Fakhri, Y.; Mousavi Khaneghah, A. Potentially Toxic Elements (PTEs) in Corn (*Zea mays*) and Soybean (*Glycine max*) Samples Collected from Tehran, Iran: A Health Risk Assessment Study. *Int. J. Environ. Anal. Chem.* **2020**, 1–12. [CrossRef]
43. Retamal-Salgado, J.; Hirzel, J.; Walter, I.; Matus, I. Bioabsorption and Bioaccumulation of Cadmium in the Straw and Grain of Maize (*Zea mays* L.) in Growing Soils Contaminated with Cadmium in Different Environment. *Int. J. Environ. Res. Public. Health* **2017**, *14*, 1399. [CrossRef] [PubMed]

44. Smedley, P.L.; Kinniburgh, D.G. A Review of the Source, Behaviour and Distribution of Arsenic in Natural Waters. *Appl. Geochem.* **2002**, *17*, 517–568. [CrossRef]
45. Rahman, M.A.; Hasegawa, H. High Levels of Inorganic Arsenic in Rice in Areas Where Arsenic-Contaminated Water Is Used for Irrigation and Cooking. *Sci. Total Environ.* **2011**, *409*, 4645–4655. [CrossRef]
46. Hajeb, P.; Sloth, J.J.; Shakibazadeh, S.; Mahyudin, N.A.; Afsah-Hejri, L. Toxic Elements in Food: Occurrence, Binding, and Reduction Approaches: Toxic Elements in Food. *Compr. Rev. Food Sci. Food Saf.* **2014**, *13*, 457–472. [CrossRef]
47. Picariello, E.; Baldantoni, D.; Izzo, F.; Langella, A.; De Nicola, F. Soil Organic Matter Stability and Microbial Community in Relation to Different Plant Cover: A Focus on Forests Characterizing Mediterranean Area. *Appl. Soil Ecol.* **2021**, *162*, 103897. [CrossRef]
48. Petruzzelli, G. Heavy Metals in Compost and their Effect on Soil Quality. In *The Science of Composting*; de Bertoldi, M., Sequi, P., Lemmes, B., Papi, T., Eds.; Springer: Dordrecht, The Netherlands, 1996; pp. 213–223. ISBN 978-94-010-7201-4.
49. Saveyn, H.; Eder, P. *End-of-Waste Criteria for Biodegradable Waste Subjected to Biological Treatment (Compost & Digestate): Technical Proposals*; Publications Office: Luxembourg, 2014.
50. Idelmann, M. *Hygienisierung Von Kompost: Möglichkeiten Zum Nachweis Einer Erfolgreichen Abtötung Von Pathogenen Und Unkrautsamen*; Dissertation, Schriftenreihe des Fachgebietes Abfalltechnik/Universität Kassel Dissertationen; Univ. Press: Kassel, Germany, 2006; ISBN 978-3-89958-203-1.
51. Dahlquist, R.M.; Prather, T.S.; Stapleton, J.J. Time and Temperature Requirements for Weed Seed Thermal Death. *Weed Sci.* **2007**, *55*, 619–625. [CrossRef]
52. Egley, G.H. High-Temperature Effects on Germination and Survival of Weed Seeds in Soil. *Weed Sci.* **1990**, *38*, 429–435. [CrossRef]
53. Eghball, B.; Lesoing, G.W. Viability of Weed Seeds Following Manure Windrow Composting. *Compost Sci. Util.* **2000**, *8*, 46–53. [CrossRef]
54. Wiese, A.F.; Sweeten, J.M.; Bean, B.W.; Salisbury, C.D.; Chenault, E.W. High Temperature Composting of Cattle Feedlot Manure Kills Weed Seed. *Appl. Eng. Agric.* **1998**, *14*, 377–380. [CrossRef]
55. *ÖNORM S 2027-4: Evalution of Waste from Mechanical-Biological Treatment–Part 4: Stability Parameters–Respiration Activity (AT4)*; Association of Germany: Berlin, Germany, 2012.
56. *MSZE 21420-18:2005: Hulladékok Jellemzése. A Nedvesség-És a Szárazanyag-Tartalom Meghatározása*; Characterization of Wastes. Determination of Humidity and Dry Matter Content; Hungarian Standards Institution: Budapest, Hungary, 2005.
57. Bozym, M. Analytical Issues in the Assessment of Waste Stabilisation Degree after Biological Treatment. *CHEMIK* **2012**, *66*, 1211–1218.
58. *MSZ-08-0012-4:1979: Tőzegek És Tőzegkészítmények Fizikai, Kémiai És Biológiai Vizsgálata. Gyomosító És Csírázásgátló Hatás Vizsgálata*; Physical, Chemical and Biological Testing of Peat and Peat Products. Examination of Weed Expansion and Germ Killing Effect; Hungarian Standards Institution: Budapest, Hungary, 1979.
59. Kristanto, G.A.; Rahmah, S.A. Assessment of Compost Maturity Using The Static Respirometry Index. *Reaktor* **2019**, *18*, 194–201. [CrossRef]
60. Confesor, R.B.; Hamlet, J.M.; Shannon, R.D.; Graves, R.E. Potential Pollutants from Farm, Food and Yard Waste Composts at Differing Ages: Leaching Potential of Nutrients Under Column Experiments. Part II. *Compost Sci. Util.* **2009**, *17*, 6–17. [CrossRef]
61. Barrena, R.; Font, X.; Gabarrell, X.; Sánchez, A. Home Composting versus Industrial Composting: Influence of Composting System on Compost Quality with Focus on Compost Stability. *Waste Manag.* **2014**, *34*, 1109–1116. [CrossRef]
62. Jauch, M. *Kompostieren-so Geht's: Müll Vermeiden, Kompost Sinnvoll Nutzen*; Dem Kosmos-Rat vertrauen; Franckh-Kosmos: Stuttgart, Germany, 1996; ISBN 978-3-440-07096-3.
63. Scaglia, B.; Tambone, F.; Genevini, P.L.; Adani, F. Respiration Index Determination: Dynamic And Static Approaches. *Compost Sci. Util.* **2000**, *8*, 90–98. [CrossRef]
64. *ÖNORM S 2023: Untersuchungsmethoden und Güteüberwachung von Komposten*; Österreichisches Normungsinstitut: Vienna, Austria, 1993.
65. Fernández-Delgado Juárez, M.; Gómez-Brandón, M.; Insam, H. Merging Two Waste Streams, Wood Ash and Biowaste, Results in Improved Composting Process and End Products. *Sci. Total Environ.* **2015**, *511*, 91–100. [CrossRef]
66. Capar, S.G.; Tanner, J.T.; Friedman, M.H.; Boyer, K.W. Multielement Analysis of Animal Feed, Animal Wastes and Sewage Sludge. *Environ. Sci. Technol.* **1978**, *12*, 785–790. [CrossRef]
67. Kubier, A.; Pichler, T. Cadmium in Groundwater—A Synopsis Based on a Large Hydrogeochemical Data Set. *Sci. Total Environ.* **2019**, *689*, 831–842. [CrossRef]
68. Knappe, F.; Mohler, S.; Ostermayer, A.; Lazar, S.; Kaufmann, C. *Comparative Evaluation of Substance Inputs Into Soils Via Different Input Paths*; Federal Environmental Agency (Umweltbundesamt): Dessau-Roßlau, Germany, 2008.
69. López-Núñez, R.; Ajmal-Poley, F.; Burgos-Doménech, P. Prediction of As, Cd, Cr, Hg, Ni, and Se Concentrations in Organic Amendments Using Portable X-ray Fluorescence and Multivariate Modeling. *Appl. Sci.* **2020**, *10*, 5726. [CrossRef]
70. Weaver, S.E.; McWilliams, E.L. The Biology of Canadian Weeds.: **44.** *Amaranthus Retrojkxus* L., *A. Powellii S. Wats.* and *A. Hybridus* L. *Can. J. Plant Sci.* **1980**, *60*, 1215–1234.

applied sciences

MDPI

Article

# Influence of Digester Temperature on Methane Yield of Organic Fraction of Municipal Solid Waste (OFMSW)

Gregor Sailer [1,*], Martin Silberhorn [1], Johanna Eichermüller [1], Jens Poetsch [1], Stefan Pelz [1], Hans Oechsner [2] and Joachim Müller [3]

1 Department of Bioenergy, University of Applied Forest Sciences Rottenburg, Schadenweilerhof, 72108 Rottenburg, Germany; marsilb@web.de (M.S.); eichermueller@hs-rottenburg.de (J.E.); poetsch@hs-rottenburg.de (J.P.); pelz@hs-rottenburg.de (S.P.)
2 State Institute of Agricultural Engineering and Bioenergy, University of Hohenheim, Garbenstrasse 9, 70599 Stuttgart, Germany; hans.oechsner@uni-hohenheim.de
3 Tropics and Subtropics Group, Institute of Agricultural Engineering, University of Hohenheim, Garbenstrasse 9, 70599 Stuttgart, Germany; joachim.mueller@uni-hohenheim.de
* Correspondence: sailer@hs-rottenburg.de or sbc@hs-rottenburg.de

**Abstract:** This study evaluates the anaerobic digestion (AD) of the organic fraction of municipal solid waste (OFMSW) and digested sewage sludge (DSS) at lowered temperatures. AD batch tests for $CH_4$ yield determination were carried out with DSS as inoculum between 23 and 40 °C. All results were related to organic dry matter and calculated for standard conditions (1013 hPa, 0 °C). The AD experiments at 40 °C and at 35 °C delivered specific $CH_4$ yields of 325 ± 6 mL/g and 268 ± 27 mL/g for OFMSW alone. At lower temperatures, specific $CH_4$ yields of 364 ± 25 mL/g (25 °C) and 172 ± 21 mL/g (23 °C) were reached. AD at 25 °C could be beneficial regarding energy input (heating costs) and energy output ($CH_4$ yield). Plant operators could increase AD efficiencies by avoiding heating costs. The co-digestion of OFMSW together with DSS could lead to further synergies such as better exploitation of the energy potentials of DSS, but the digestate utilization could become problematic due to hygienic requirements. Efficiency potentials through lowered operating temperatures are limited. In further research, lowered process temperatures could be applied in the AD of energy crops due to large numbers of existing plants.

**Keywords:** OFMSW; digested sewage sludge; waste characterization; anaerobic digestion; operating temperature; mesophilic; psychrophilic

**Citation:** Sailer, G.; Silberhorn, M.; Eichermüller, J.; Poetsch, J.; Pelz, S.; Oechsner, H.; Müller, J. Influence of Digester Temperature on Methane Yield of Organic Fraction of Municipal Solid Waste (OFMSW). *Appl. Sci.* **2021**, *11*, 2907. https://doi.org/10.3390/app11072907

Academic Editor: Carlos Rico de la Hera

Received: 8 March 2021
Accepted: 22 March 2021
Published: 24 March 2021



## 1. Introduction

The global municipal solid waste (MSW) generation forecast for 2025 is about 2200 million tons. Consequent to that, the organic fraction of municipal solid waste (OFMSW) is estimated to be 46% [1]. In 2015, 241 million tons of MSW with a share of 40–60% of organics were generated in the EU [2]. If OFMSW is not collected and treated separately and instead disposed of in landfills with other MSW components; degradation processes such as anaerobic digestion (AD) occur underground, producing several environmental damages such as greenhouse gas (GHG) emissions or hazardous substances causing health hazards. Therefore, reduction targets for landfilling of MSW and OFMSW in the EU have been defined and further developed since the late 1990s. In addition, the separate collection and treatment of OFMSW via AD or composting have been promoted. EU-wide, landfilling is declining but still relevant, while the depositing of untreated MSW and OFMSW is already forbidden in Germany [3–5].

Composting of OFMSW leads to GHG emissions in the form of $CO_2$ released directly into the atmosphere and can cause unpleasant odors. Due to the generation of renewable energy, AD of OFMSW is indicated with a better GHG balance, compared to composting, and odor problems are easier to control [1,2,6–11]. $CH_4$ generated in the AD of OFMSW can be used in various applications (e.g., combined heat and power or fuel) and represents a storable energy carrier [1]. AD and composting can be combined to exploit the material and energetic utilization potentials of OFMSW fully. The utilization of OFMSW via AD is in the sense of the circular economy according to the EU agenda, which envisages keeping products and materials in the market for as long as possible. However, the improvement of AD as a major component of future energy systems is the focus of politics and research facilities [1,12,13]. AD helps to achieve several of the UN sustainable development goals [14] by producing clean and affordable energy, contributing to sustainable municipalities, and supporting climate protection [15].

OFMSW is not the only waste material with untapped potentials for the generation of renewable energy. Worldwide, the $CH_4$ potentials for residues from crops (3080–3920 TWh), manure (2600–3800 TWh), food waste (880–1100 TWh), or sewage sludge (210–300 TWh) can be exploited on a larger scale via AD [2,15].

### 1.1. OFMSW as a Resource

OFMSW is an inhomogeneous mixture with characteristics depending on its origin. The composition varies in its proportions of biodegradable and non-biodegradable organics, and mineral components [1]. The different composition depends on the season, lifestyle, waste management regulations, and regional economic frameworks [16–18]. The organic fraction of the MSW is larger in lower-income countries than in higher-income countries [19]. The fluctuations in the composition of OFMSW are also reflected in the material properties. While dry matter (DM) contents of OFMSW collected from 22 different countries fluctuate between 15.0–50.2% fresh mass (FM) with a mean value of 27.2 ± 7.6% FM, contents for organic dry matter (oDM) vary between 7.4–36.1% FM (mean 22.9 ± 6.3% FM), which is equal to 43.0–95.0% DM (mean 84.6 ± 9.9% DM). The average concentrations for C (46.6 ± 4.4% DM), H (6.6 ± 0.62% DM), N (2.9 ± 0.6% DM), and S (0.3 ± 0.26% DM) are also specified, and the potential $CH_4$ yields at standard conditions fluctuate between 61 and 580 $L/kg_{oDM}$ with an average of 415 ± 138 $L/kg_{oDM}$ [1].

Due to the variety of ingredients and changing characteristics, OFMSW can serve as a resource for several treatments and utilization pathways, e.g., the share of biodegradable components in OFMSW is relevant for $CH_4$ generated in AD [7]. Furthermore, OFMSW represents a resource for the creation of value-added products such as organic fertilizers, biopesticides, and bioplastics [2]. In general, OFMSW can be recovered with technologies using biochemical, thermochemical, or physicochemical principles, either as single technology or in combination.

### 1.2. Operating Temperatures in AD

AD processes can operate at different temperature levels but definitions of temperature ranges and process optima vary in the literature (Table 1). As can be seen in Table 1, a clear classification is difficult and the gaps between the temperature levels imply smooth transitions. However, according to Fritsche and Laplace [20] and Munk et al. [21], process temperatures in AD can be subdivided into four different ranges, namely, psychrophilic (12–15 °C), psychrotolerant (20–30 °C), mesophilic (30–40 °C) and thermophilic (55–75 °C). In general, microorganisms with a metabolism based on AD are known to exist in a temperature range of −5–121 °C. Microorganisms between 70 and 85 °C are called extremely thermophilic, while the temperature range from 85–121 °C is described as hyper-thermophilic [22]. Although the higher activity of the methanogenic microorganisms at thermophilic temperature leads to a higher degradation rate and higher biogas yields, the process stability is more sensitive, and maintaining process temperatures requires more energy, which reduces the total energy yield [23–27]. An additional advantage of thermophilic process temperatures is a lower level of sludge generation and the possibility to fulfill hygienic requirements (e.g., regulated by the German Biowaste Ordinance) for substrates such as OFMSW that need hygienic treatment [22,28]. Therefore, usually, no separate sanitation systems are required at thermophilic AD plants.

**Table 1.** Temperature ranges and optima in technical anaerobic digestion processes according to different studies.

| Reference | Year | Psychrophilic (°C) | Mesophilic (°C) | Thermophilic (°C) |
|---|---|---|---|---|
| Zábranská et al. [29] | 2000 | - | 30–40 | 50–70 |
| Ahring et al. [30] | 2001 | - | - | 55–70 |
| Kashyap et al. [31] | 2003 | <20 | 32–38 | 50–55 |
| Connaughton et al. [32] | 2006 | <20 | 25–45 | 45–65 |
| Reichard [10] | 2006 | | <50 | >50 |
| LfU [33] | 2007 | <20 | 30–42 | 48–55 |
| Effenberger et al. [34] | 2008 | <25 | 32–42 | 50–57 |
| Vindis et al. [27] | 2009 | 12–16 | 35–37 | 55–60 |
| Amon et al. [35] | 2013 | <25 | 37–42 | 50–60 |
| Donoso-Bravo et al. [36] | 2013 | 15–25 | 35–37 | 50–55 |
| Jain et al. [26], Fernández-Rodríguez et al. [37] | 2015, 2013 | - | 35 | 55 |
| Szyłak-Szydłowski et al. [38] | 2016 | - | 35–37 | - |
| Kaltschmitt et al. [22] | 2016 | <25 | 35–42 | 50–55 |
| Liu et al. [39] | 2016 | - | 25–37 | 55–65 |
| Chala et al. [40] | 2019 | <20 | 20–45 | 45–60 |
| Jain et al. [15] | 2019 | - | 35–40 | 55–60 |
| Kumar and Samadder [41] | 2020 | ~20 | ~35 | ~55 |
| Rocamora et al. [9] | 2020 | - | 35–40 | 50–57 |
| Jaimes-Estévez et al. [42] | 2020 | <20 | 20–45 | - |
| Lanko et al. [43] | 2020 | - | 35–40 | 55–70 |
| Pasalari et al. [44] | 2021 | 9–25 | 25–35 | 35–70 [1] |

[1] subdivided into thermophilic (35–55 °C) and extreme-thermophilic (55–70 °C).

Cavinato et al. [45] discovered that biogas yields via thermophilic AD (55 °C) of OFMSW together with sewage sludge can be 45–50% higher compared to mesophilic AD (37 °C). Derbal et al. [46] already investigated the influence of mesophilic (35 °C) and thermophilic (55 °C) AD of OFMSW on the biogas quantity and quality with a hydraulic retention time (HRT) of 25 d. In this case, the mesophilic AD was less efficient than the thermophilic AD. Fernández-Rodríguez et al. [37] confirmed these results.

Mesophilic AD is indicated with higher process stability than thermophilic AD through greater diversity of microorganisms, and with lower resistance to foaming and lower organic loading rates. At mesophilic temperature, the amount of $CO_2$ in the biogas is reduced as a higher percentage remains dissolved in the liquid phase, but the conversion rate of cellulose and hemicellulose is lower than in thermophilic AD. Lower process temperatures lead to lower energy demands of AD plants and offer the possibility to use low-temperature waste heat for process heating [22,23,27,47].

Rajagopal et al. [48] proved that psychrophilic AD is indicated with high process stability, which was also confirmed by other studies [49,50]. At the psychrophilic level, the additional energy input to maintain AD temperature is reduced or nullified depending on the ambient temperatures [32]. The comparatively slow $CH_4$ generation in AD at lower temperatures can be compensated by larger digester volumes and an increased HRT, reaching similar final $CH_4$ yields (tested with coffee husks, pulp, and mucilage) to those of mesophilic AD [40,51]. Connaughton et al. [32] also concluded that the total $CH_4$ yields of mesophilic AD (37 °C) of brewery wastewater are similar to the yields in psychrophilic AD (15 °C). With lower temperatures, the $CH_4$ concentrations in the biogas increase. Liu et al. [39] proved that the psychrophilic AD (15 °C) of sewage sludge delivered approximately half of the $CH_4$ yield, compared to the AD at 30 °C at the same time step. The finding of Kashyap et al. [31], i.e., AD processes at psychrophilic temperature level require an HRT roughly twice as long to achieve the same $CH_4$ yield, underlines the findings of Liu et al. [39] and Chala et al. [40].

Anaerobic microorganisms can adapt to different temperatures, but the specific methanogenic activity, and thus the $CH_4$ production, decreases proportionally to process temperature [25,31,52]. Other studies [30,53] complement this correlation. It was found that an increased process temperature (up to 55 °C) leads to an increased $CH_4$ yield in the AD of cattle manure. However, further increased AD temperatures (65 °C) lead to a reduction in $CH_4$ yields. This can be explained by findings of other studies [53,54] in which it was found that the diversity of microorganisms decreases with increasing AD temperatures. Therefore, $CH_4$ yield increases through heightened process temperatures are limited (e.g., the $CH_4$ yield of waterweed was halved at 65 °C, compared to 55 °C). The diversity of methanogenic organisms also decreases at low AD temperatures, e.g., when reducing the process temperature from 37 to 15 °C [55].

### 1.3. AD of OFMSW

OFMSW can be digested in dry AD processes at DM contents >15–20% FM or wet AD processes at DM contents <15–20% FM. Luning et al. [56] found that the same $CH_4$ yields can be achieved in practice with wet and dry AD processes, which was also confirmed by Rajagopal et al. [57] and Kern and Raussen [58]. OFMSW, in its natural state, is indicated with a DM content >20% and complex composition, including substances such as lignin that hardly contribute to biogas yields. Therefore, it is often digested in dry processes such as plug-flow or batch digesters. If OFMSW is used in wet processes (e.g., via stirred tank reactors), it has to be diluted to a pumpable condition through its mixture with other substrates or water [1,22].

On a global level, biowaste (e.g., food waste as a component of OFMSW) is mostly digested at mesophilic temperatures in wet AD plants [15]. Due to the simpler construction of an unheated psychrophilic biogas plant, this method is currently an important form of supplying $CH_4$ in developing countries and emerging economies. Micro digesters (household size) can be found millions of times in China (42 million, but also 7000 AD plants with an electrical capacity of approximately 0.1–1 MW), India (5 million), and Africa, along with other Asian countries (0.7 million). The generated biogas is mostly used for cooking stoves [15,19,59]. In other regions of the world, micro digesters are barely used, and large(r) scale biogas plants are common., e.g., 2200 biogas plants operate in the USA with an installed total capacity of 977 MW, leading to an average plant size of approximately 0.45 MW. Including landfill and sewage AD plants, approximately 17,800 AD plants with an installed electrical capacity of 10.5 GW operate in the EU out of which approximately 700 are utilizing OFMSW and industrial biowaste. An increase in the number of biogas plants for OFMSW and industrial biowaste is expected as a separate collection of OFMSW is envisaged EU-wide [15,19].

By 2018, approximately 9500 biogas plants with an installed electrical capacity of 5 GW were operated in Germany, mainly based on energy crops [60,61]. Most agricultural biogas plants are equipped with stirred tank reactors at mesophilic temperatures with defined feeding intervals. The generated biogas is mostly used for the combined generation of electricity and heat. In the currently operating approximately 215 biogas plants, biogas is upgraded to biomethane and fed into the gas grid. Psychrophilic biogas plants are uncommon in Germany [22,58,62]. In 2015, 1392 biogas plants in Germany were registered to use biowaste as feedstock, but only 337 actually used it as the main substrate or in co-digestion. Then, again, 75 biogas plants are specialized on OFMSW (share of OFMSW compared to the total feedstock >90%) operating as dry AD system at a thermophilic and mesophilic temperature in equal share [58,60].

According to the literature [63–65], the heat demand of AD systems depends on substrate characteristics, operating temperature, and geographic region, and on AD parameters such as digester type or plant size. The benchmarking of German biogas plants [65] has proven that heat demands for digester heating can be between 0 and 100% of the generated heat. Especially small biogas plants with an electrical capacity below 150 kW and plants utilizing substrates with high water contents show higher heat demands. Typically, the digester heating demand is 30 ± 20% of the generated heat. A lower digester heating demand would allow heat export, and hence the profitability of biogas plants could be increased.

### 1.4. Aim of the Study

Currently, typical digester temperatures of OFMSW biogas plants are at mesophilic or thermophilic levels, causing a high heating demand [58]. The aim of this study was therefore to investigate the influence of lower-process temperatures on the quantity and quality of biogas generated from OFMSW. The main objective was to determine whether it is possible to increase the efficiency of OFMSW biogas plants by avoiding heating demands without losing energy yields. Another objective was to determine whether the combined treatment of OFMSW and DSS, which was used as inoculum, could create synergies in a wet digestion system due to large numbers of existing wastewater plants and infrastructures.

## 2. Materials and Methods

### 2.1. Substrate Characteristics

All analytic procedures were performed at the University of Applied Sciences, Rottenburg, following VDI 4630 [66] and corresponding standards.

#### 2.1.1. Sampling of OFMSW and Digested Sewage Sludge (DSS)

The feedstock for AD experiments was untreated OFMSW collected at a full-scale biowaste AD plant in southern Germany processing exclusively OFMSW. Sampling was conducted in accordance with the German Biowaste Ordinance [67]. The samples were collected with a spade from a heap stored no longer than 24 h in a closed hall. The heap was subdivided into 12 equal parts, leading to 12 sampling points out of which approximately 0.5 L were collected while discarding the upper waste layer. The resulting single samples were merged into one bulk sample (10.5 kg FM). To avoid uncontrolled degradation processes, the sample was transported and further processed within 2 h after collection.

The inoculum used for AD experiments was DSS collected at the local municipal sewage treatment plant (Rottenburg-Kiebingen, Germany), where wastewater is treated via aerobic and anaerobic steps. The terms inoculum and DSS are used synonymously in this study. DSS is regarded as suitable inoculum for AD experiments because it contains a large number of different microorganisms [66]. Each sample collection followed the same procedure [66]. The sampling was conducted in November (series 1) and in February (series 2) from the outlet valve of the digestate container where the DSS is stored at 37 °C. The digestate container also serves as a secondary digester with an identical temperature level compared to the main AD unit. However, the digestate container is operated with a low HRT. It mainly serves as a buffer prior to the solid–liquid separation of the DSS. To date, the additional biogas yields are marginal but still used in the combined heat and power unit at the treatment plant. After opening the outlet valve, the first approximately 10 L were discarded. The samples were filled into two 30-L buckets, closed airtight for transportation, and handled within 1 h to avoid excess cooling and contact with ambient air.

#### 2.1.2. DM, oDM, and Processing

DM contents were determined according to the German Biowaste Ordinance [68] through drying at 105 °C in a drying oven (UNP 700, Memmert, Schwabach, Germany) for at least 24 h. Before further processing, impurities (stones, metal, glass, plastics) in the OFMSW bulk sample were removed, and the remaining DM was milled in a cutting mill (Pulverisette 19, Fritsch, Idar-Oberstein, Germany) to particle sizes of approximately 1 mm. To achieve homogeneous particle sizes, the dry inoculum was manually crushed in a ceramic mortar; prior sorting was not required. Afterward, oDM was determined for each sample in minimum triplicate [69] through incineration of approximately 1 g DM in a ceramic crucible by a muffle furnace (AAF 1100, Carbolite, Neuhausen, Germany). This procedure [69] slightly differed from the method mentioned in the German Biowaste Ordinance [70] and was chosen due to the availability of a preprogrammed automatic furnace.

### 2.1.3. Elemental Analysis and Stoichiometric Biogas Potentials

The determination of C, H, and N was carried out with an elemental analyzer (vario MACRO cube, Elementar, Langenselbold, Germany) in a minimum of four replicates per sample [71]. Each sample (approximately 40 mg DM) was pressed into a zinc foil-coated tablet. Stoichiometric biogas and $CH_4$ yields were calculated according to Equation (1) [72,73] and related to standard conditions. O contents were calculated based on the mean values for C, H, N, and ash. S was neglected in the calculations. Therefore, the stoichiometric biogas yield equaled the sum of all products except $H_2S$ (which slightly reduced the biogas yields).

$$C_nH_aO_bN_cS_d + \left(n - \frac{a}{4} - \frac{b}{2} + \frac{3}{4}c + \frac{d}{2}\right)H_2O \rightarrow \left(\frac{n}{2} - \frac{a}{8} + \frac{b}{4} + \frac{3}{8}c - \frac{d}{4}\right)CO_2 + \left(\frac{n}{2} + \frac{a}{8} - \frac{b}{4} - \frac{3}{8}c - \frac{d}{4}\right)CH_4 + cNH_3 + dH_2S. \quad (1)$$

### 2.2. AD Experiments and Process Monitoring

Biochemical methane potential tests were carried out according to VDI 4630 [66]. The volumetric biogas production was measured using glass manometers (1-L gas storage) considering the temperature within the digester and ambient conditions. The biogas composition was analyzed with portable biogas monitors (BIOGAS 5000 and GAS 5000, Geotech, Coventry, UK) from biogas collected and stored in bags (PLASTIGAS, Linde, Pullach, Germany). The specific biogas and $CH_4$ production were related to oDM and calculated for standard conditions (1013 hPa, 0 °C, dry gas). The configuration of each digester, the installed gas measurement system, and the storage bag as used in this study is depicted in Sailer et al. [74].

The experiments were conducted in two batch test series at 25/40 °C and at 23/35 °C using 2-L insulated glass vessels with preprogrammed heating, which were automatically stirred for 60 s/h. In each test series, 12 digesters were used (thereof four blank variants). The operating temperature of 35 and 40 °C was selected to represent typical German mesophilic biogas processes. The lowered operating temperature of 25 °C in the first test series was chosen in order to achieve an adequate reduction of the heat demand and in order to remain close to the mesophilic temperature level (Table 1). The temperature of 23 °C was the lowest possible temperature in the laboratory. It was therefore chosen in the second test series. The detailed experimental setup is provided in Table 2.

**Table 2.** Experimental setup of the anaerobic digestion experiments of both test series at 25/40 °C and 23/35 °C. Digested sewage sludge (DSS) was used as inoculum (Inoc.), while dry matter (DM) of the organic fraction of municipal solid waste (OFMSW) served as feedstock.

| Variant | Series 1 T25 | | Series 1 T40 | | Series 2 T23 | | Series 2 T35 | |
|---|---|---|---|---|---|---|---|---|
| | Inoc. (blank) | Inoc. and OFMSW | Inoc. (blank) | Inoc. and OFMSW | Inoc. (blank) | Inoc. and OFMSW | Inoc. (blanc) | Inoc. and OFMSW |
| Inoculum | 0.8 L DSS 1.2 L water | 2 L DSS | 0.8 L DSS 1.2 L water | 2 L DSS | 1 L DSS 1 L water | 1 L DSS 1 L water | 1 L DSS 1 L water | 1 L DSS 1 L water |
| Feedstock | - | 15 g DM OFMSW | - | 15 g DM OFMSW | - | 10 g DM OFMSW | - | 10 g DM OFMSW |
| Retention time (d) | 56 | 56 | 56 | 56 | 77 | 77 | 35 | 35 |
| Day with gas analysis | 56 | 5; 56 | 56 | 5; 56 | 8; 24 | 8; 24 | 8; 24 | 8; 24 |
| Replicates | 2 | 4 | 2 | 4 | 2 | 4 | 2 | 4 |

In the first test series, the biogas composition was measured on day 5 (only variants with OFMSW as feedstock) and day 56 (all digesters). In the second test series, the biogas measurements of day 24 were used to extrapolate the $CH_4$ yield results. Therefore, the measured biogas yields (mL/$g_{oDM}$) from day 25 to days 35 and 77 were transferred to $CH_4$ yields (mL/$g_{oDM}$) based on the $CH_4$ concentration (%) measured on day 24. This procedure had to be conducted in accordance with SARS-CoV-2 laboratory regulations (limited laboratory access).

The inoculum was added without further degassing (already treated anaerobically at the treatment plant). However, blind variants were carried out in duplicate, determining the residual biogas potential of the DSS. According to VDI 4630 [66], DSS contains a broad variety of microorganisms. Therefore, all digesters started with the same initial conditions at 37 °C, reaching the designated operating temperature several hours later (no long-time acclimation period for microorganisms was executed). In variants with added tap water, the water was preheated to 37 °C. The experiments were stopped as soon as the biogas formation rate per digester remained close to 1 mL/h. Only the digesters at 23 °C (unheated) were operated with an increased retention time (approximately doubled compared to the digesters at 35 °C). For both test series, the same OFMSW sample was used. The specific biogas yield of the mixture of OFMSW and DSS in each digester (Table 2) was calculated as

$$SBG_{OFMSW\ and\ DSS} = \frac{BG_{OFMSW\ and\ DSS}}{m_{oDM,OFMSW} + m_{oDM,DSS}}, \quad (2)$$

where $SBG_{OFMSW\ and\ DSS}$ (mL/$g_{oDM}$) is the specific biogas yield from the mixture of OFMSW and DSS, $BG_{OFMSW\ and\ DSS}$ (mL) is the total gas yield from the mixture of OFMSW and DSS (used as inoculum), and $m_{oDM,OFMSW}$ and $m_{oDM,DSS}$ ($g_{oDM}$) are the organic mass of OFMSW and DSS, respectively.

The specific biogas yield of OFMSW alone was calculated as

$$SBG_{OFMSW} = \frac{BG_{OFMSW\ and\ DSS} - BG_{DSS}}{m_{oDM,OFMSW}}, \quad (3)$$

where $SBG_{OFMSW}$ (mL/$g_{oDM}$) is the specific biogas yield from OFMSW alone, and $BG_{DSS}$ (mL) is the biogas yield from DSS, i.e., from the corresponding blanks.

The specific $CH_4$ yield of the mixture of OFMSW and DSS was calculated as

$$SMY_{OFMSW\ and\ DSS} = \sigma_{OFMSW\ and\ DSS} \cdot SBG_{OFMSW\ and\ DSS}, \quad (4)$$

where $SMY_{OFMSW\ and\ DSS}$ (mL/$g_{oDM}$) is the specific $CH_4$ yield from the mixture of OFMSW and DSS, and $\sigma_{OFMSW\ and\ DSS}$ (-) is the measured volume concentration of $CH_4$ in the biogas from the mixture.

The specific $CH_4$ yield of the blanks was calculated as

$$SMY_{DSS} = \sigma_{DSS} \cdot SBG_{DSS}, \quad (5)$$

where $SMY_{DSS}$ (mL/$g_{oDM}$) is the specific $CH_4$ yield from DSS (blanks), $SBG_{DSS}$ (mL/$g_{oDM}$) is the specific biogas of the DSS, and $\sigma_{DSS}$ (-) is the measured volume concentration of $CH_4$ in the biogas from the blanks.

Finally, the specific $CH_4$ yield of OFMSW alone was calculated as

$$SMY_{OFMSW} = \frac{\sigma_{OFMSW\ and\ DSS} \cdot BG_{OFMSW\ and\ DSS} - \sigma_{DSS} \cdot BG_{DSS}}{m_{oDM,OFMSW}}, \quad (6)$$

where $SMY_{OFMSW}$ (mL/$g_{oDM}$) is the specific $CH_4$ yield from OFMSW alone.

In experiments with two subsequent gas analyses (Table 2), measured $CH_4$ concentrations have been applied for the corresponding time interval. For all experiments, $SMY_{DSS}$, $SMY_{OFMSW}$, and $SMY_{OFMSW\ and\ DSS}$ results were plotted based on mean values (n = 2 for DSS variants, n = 4 for variants containing OFMSW and DSS) on day 35 in order to evaluate the goodness of fit. The coefficient of determination ($R^2$) was determined with the help of a linear regression line.

# 3. Results and Discussion

## *3.1. Characteristics of Raw OFMSW and DSS*

In Table 3, basic substrate characteristics are compiled, serving as a basis for all AD experiments and for stoichiometric biogas and $CH_4$ yield calculations. The FM of the OFMSW sample was dried in separate fractions causing high standard deviations (SD), which was not relevant for this study. DM, oDM, and C, H and N contents of OFMSW are in line with the ranges presented in Campuzano and González-Martínez [1]. However, the DM and oDM contents, in particular, depend on various factors such as season or settlement structure. This is also represented by Campuzano and González-Martínez [1] with oDM contents for OFMSW of 84.24 ± 10.09% DM.

**Table 3.** Dry matter content (DM), organic dry matter content (oDM), chemical elements of digested sewage sludge (DSS) used as inoculum, and organic fraction of municipal solid waste (OFMSW) used as substrate; mean values ± standard deviation.

| Material | DM (% FM) | oDM (% DM) | C (% DM) | H (% DM) | N (% DM) | O (% DM) |
|---|---|---|---|---|---|---|
| DSS, series 1 | 3.33 ±1.84 | 58.55 ±0.40 | n.a. [1] | n.a. [1] | n.a. [1] | n.a. [1] |
| DSS, series 2 | 4.07 ±0.01 | 60.21 ±0.17 | 30.38 ±0.11 | 4.50 ±0.06 | 4.11 ±0.04 | 21.22 |
| OFMSW, series 1 and 2 | 33.28 ±5.43 | 77.88 ±1.37 | 39.49 ±2.55 | 5.29 ±0.35 | 2.13 ±0.32 | 30.97 |

[1] not analyzed.

The basic characteristics (FM and oDM) of the DSS sample used in series 1 and series 2 slightly differed from each other. This could be explained by different storage durations of the DSS at the treatment plant or seasonal variations. However, both DM and oDM contents were higher compared to Bertau et al. [75], in which a DM content of 2.40 ± 0.50% FM and an oDM content of 49.00 ± 2.00% DM were reported. It cannot be expected that the characteristics of DSS fluctuate as strong as the characteristics OFMSW, but wastewater treatment plants can operate with different treatment approaches influencing DSS characteristics. The contents for C, H, and N of DSS (series 2) are almost identical, compared to mean values in Maier and Scheffknecht [76].

Based on Equation (1), the calculated stoichiometric biogas potential of the OFMSW was 990 $L/kg_{oDM}$ (771 $L/kg_{DM}$, 257 $L/kg_{FM}$), with a stoichiometric $CH_4$ potential of 506 $L/kg_{oDM}$ (394 $L/kg_{DM}$, 131 $L/kg_{FM}$). The DSS of series 2 reached a stoichiometric biogas potential of 1051 $L/kg_{oDM}$ (633 $L/kg_{DM}$, 26 $L/kg_{FM}$), leading to a stoichiometric $CH_4$ potential of 514 $L/kg_{oDM}$ (400 $L/kg_{DM}$, 16 $L/kg_{FM}$). In the literature, the calculated stoichiometric yields of DSS were between 919 and 1533 $L_{biogas}/kg_{oDM}$ and between 462 and 772 $L_{CH4}/kg_{oDM}$, while OFMSW reached 1087 $L_{biogas}/kg_{oDM}$ and 596 $L_{CH4}/kg_{oDM}$. These values are in line with the DSS and OFMSW samples used in this study [1,75,76]. However, stoichiometric biogas and $CH_4$ yields assume complete digestion, which is not achieved in practice.

## *3.2. Influence of Process Temperature on AD of OFMSW*

For both test series, the influence of process temperatures on AD of OFMSW together with DSS is depicted in Figures 1–3. Figure 1 presents the results for $SBG_{OFMSW \text{ and } DSS}$ and $SMY_{OFMSW \text{ and } DSS}$ in the different digester configurations. The ratio of oDM from OFMSW to oDM from DSS according to the specifications of VDI 4630 [66] was 0.30 in test series 1 and 0.32 in test series 2. From a digester perspective, the influence of the OFMSW on biogas and $CH_4$ yields per digester ($SBG_{OFMSW \text{ and } DSS}$ and $SMY_{OFMSW \text{ and } DSS}$) was relatively low due to a larger share of inoculum (DSS), compared to OFMSW. Figure 2 presents the results for specific biogas yields (SBG) and methane yields (SMY) for OFMSW and DSS alone at the different process temperatures. In both test series, higher temperatures were beneficial for $SBG_{DSS}$ and $SMY_{DSS}$. Based on the results for $SBG_{DSS}$ and $SMY_{DSS}$ it can be assumed that the residual $CH_4$ potential of DSS in wastewater treatment plants could be better exploited through mesophilic temperatures in the storage tank serving as a second digester or by increasing the HRT for sewage sludge in the main AD unit.

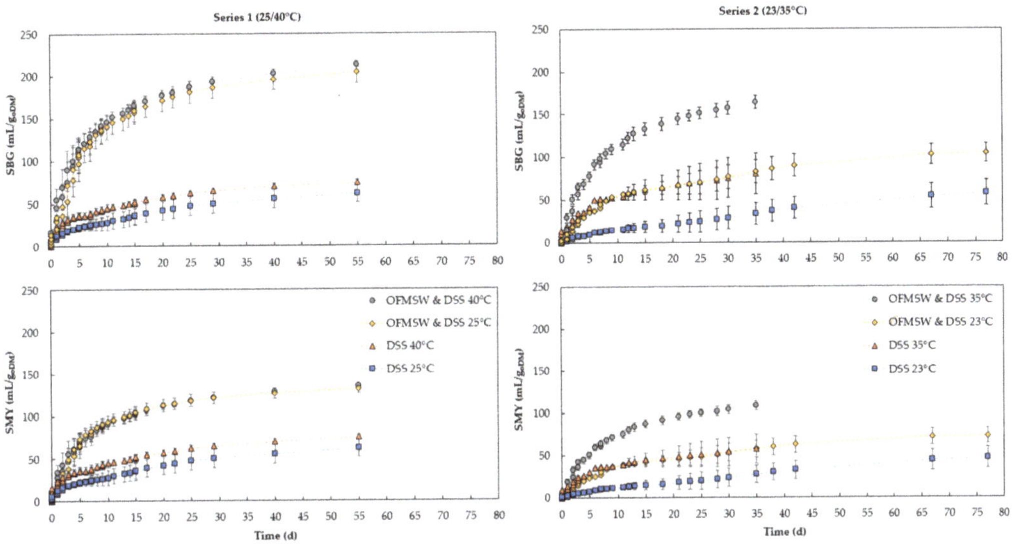


**Figure 1.** Specific biogas yield (SBG) and specific methane yield (SMY) for the mixture of the organic fraction of municipal solid waste (OFMSW) and digested sewage sludge (DSS); mean values ± standard deviation (n = 2 for DSS, n = 4 for OFMSW and DSS) based on organic dry matter (oDM), gas volume at 1013 hPa, 0 °C.

In test series 1, the variants using OFMSW as feedstock achieved similar $SBG_{OFMSW\ and\ DSS}$ and $SMY_{OFMSW\ and\ DSS}$ per digester (Figure 1). By subtracting the yields of the blank variants (DSS) for the calculation of $SBG_{OFMSW}$ and $SMY_{OFMSW}$ (Figure 2) at 25 °C and 40 °C, the lower $SBG_{DSS}$ and $SMY_{DSS}$ at 25 °C led to slightly higher yields for OFMSW at 25 °C. In general, the two test series showed strong differences regarding the final SBG and SMY. While the $SBG_{OFMSW}$ at 25 °C (after 56 d) was 623 mL/$g_{oDM}$, the $SBG_{OFMSW}$ at 23 °C (after 77 d) dropped by 57.7%. The results for the $SBG_{OFMSW}$ and $SMY_{OFMSW}$ digested at 25 and 23 °C showed that an extended HRT in test series 2 (21 d) was not sufficient to deliver similar results. However, the results (25 °C beneficial, 23 °C disadvantageous) should be questioned because all yields were lower in test series 2, even for variants that were operated at 35 °C. Nevertheless, it can be deduced that the microorganisms do not perform efficiently without heating or strongly increased HRT, which was confirmed by King et al. [51]. As mentioned, the $SMY_{OFMSW}$ at 35 °C also decreased by approximately 17.5%, compared to OFMSW digested at 40 °C, probably due to the lower activity of microorganisms [55]. Although both test series were executed with the same procedure, the differences between the OFMSW variants at 23, 25, 35, and 40 °C could also be influenced by variations in the quality of the DSS and the potential inhomogeneity of OFMSW samples.

All final SBG and SMY for DSS and OFMSW alone and corresponding $CH_4$ concentrations are presented in Table 4. The $CH_4$ concentrations of OFMSW did not vary as strongly as the $CH_4$ concentrations of the DSS at different temperatures. Between the $SBG_{DSS}$ and $SMY_{DSS}$ at 25 and 40 °C and between the DSS at 23 and 35 °C, a difference of almost 10%-points with higher $CH_4$ concentrations at lower temperatures was measured. Similar increases in $CH_4$ concentrations through reduced process temperatures were discovered in the literature [32]. It can be concluded that the inoculum is the major and liquid component in the digester, and therefore, the biogas composition can be influenced toward higher $CH_4$ concentrations through lowered temperatures.

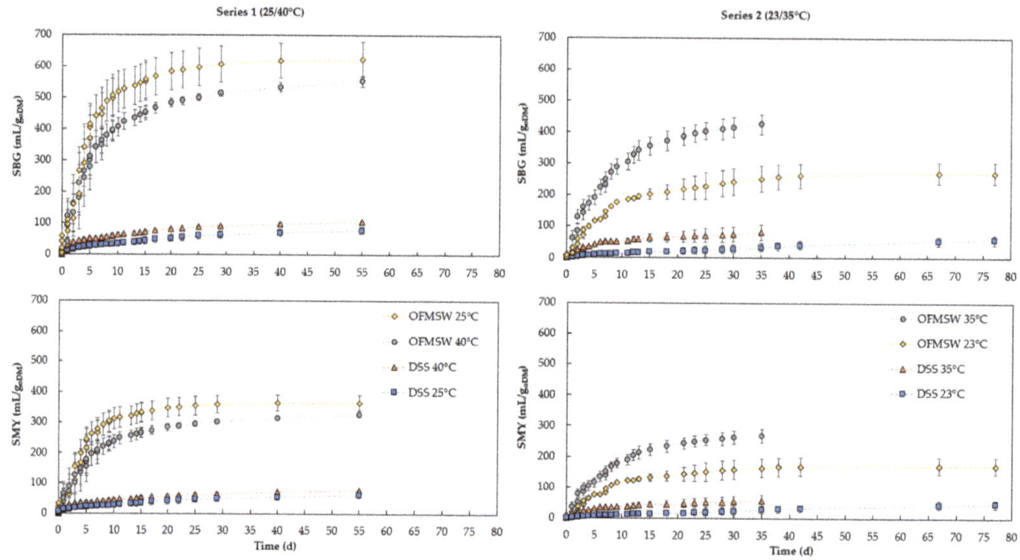


**Figure 2.** Specific biogas yield (SBG) and methane yield (SMY) for the organic fraction of municipal solid waste (OFMSW) alone and the digested sewage sludge (DSS) alone; mean values ± standard deviation (n = 2 for DSS, n = 4 for OFMSW) based on organic dry matter (oDM), gas volume at 1013 hPa, 0 °C.

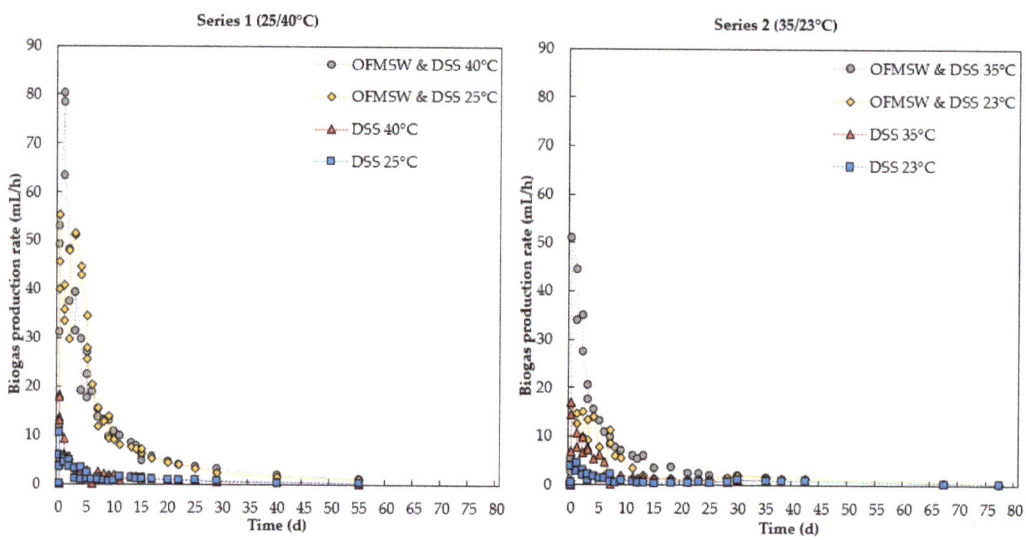


**Figure 3.** Biogas production rates for the mixtures of the organic fraction of municipal solid waste (OFMSW) and digested sewage sludge (DSS), and for DSS alone; mean values (n = 2 for DSS, n = 4 for OFMSW and DSS), gas volume at 1013 hPa, 0 °C.

The $SBG_{DSS}$ and $SMY_{DSS}$ in the blank variants varied between 58 and 105 $mL_{biogas}/g_{oDM}$ and between 47 and 75 $mL_{CH4}/g_{oDM}$, reaching the highest yields at 40 °C, followed by 35 and 25 °C at a similar level. Although reaching the highest $CH_4$ concentrations, the DSS variant at 23 °C clearly delivered the worst results for SBG and SMY. The final experimental yields equaled approximately 6–10% (SBG) or 9–15% (SMY), compared to the theoretical stoichiometric yields for DSS in series 2 (Table 3). This can be explained by the circumstance that the DSS already was AD treated, and the remaining oDM suggests a low digestibility. However, both series showed that there is still energy potential within the DSS that could be exploited better. In Germany, the average SBG production from sewage sludge amounts to 520 mL/$g_{oDM}$ [22]. The determined residual SBG potentials equaled approximately 10–20% of this average, showing that efficiency enhancements are possible. In addition, the results for $SMY_{DSS}$ at 23 °C coincide with those of Liu et al. [39], in which the transition to the psychrophilic temperature range is accompanied by a doubled HRT to reach approximately similar $CH_4$ yields.

**Table 4.** Specific biogas yields (SBG), methane yields (SMY), and $CH_4$ concentrations (σ) of the digested sewage sludge (DSS) and the organic fraction of municipal solid waste (OFMSW) at different temperatures and retention times; mean values ± standard deviation.

| Variant | SBG (mL/$g_{oDM}$) | SMY (mL/$g_{oDM}$) | σ (%) |
|---|---|---|---|
| DSS T40 (56 d) | 105.45 ± 2.16 | 74.75 ± 2.92 | 70.89 ± 4.22 |
| DSS T35 (35 d) | 80.75 ± 23.92 | 57.88 ± 17.09 | 71.69 ± 0.07 |
| DSS T25 (56 d) | 78.03 ± 10.10 | 62.06 ± 10.38 | 79.53 ± 3.05 |
| DSS T23 (77 d) | 57.67 ± 14.86 | 46.71 ± 12.50 | 80.99 ± 0.86 |
| OFMSW T40 (56 d) | 552.80 ± 17.22 | 325.17 ± 6.13 | 58.82 ± 0.84 |
| OFMSW T35 (35 d) | 424.49 ± 31.59 | 268.35 ± 26.74 | 63.22 ± 0.67 |
| OFMSW T25 (56 d) | 623.91 ± 56.22 | 364.19 ± 24.71 | 58.37 ± 1.41 |
| OFMSW T23 (77 d) | 270.19 ± 33.47 | 171.89 ± 21.19 | 63.62 ± 2.05 |

The mean values for $SBG_{OFMSW}$ and $SMY_{OFMSW}$ (Table 4) were measured between 270 and 624 $mL_{biogas}/g_{oDM}$ and was between 172 and 364 $mL_{CH4}/g_{oDM}$, corresponding to approximately 27–63% (biogas) and 34–73% ($CH_4$) of the stoichiometric yields. The variant at 25 °C delivered the best results but was indicated with the largest SD. As mentioned, this result can also be attributed to the calculation method presented in Equation (2) to Equation (6). When neglecting the two best performing digesters at 25 °C ($SBG_{OFMSW}$ of 700 and 650 mL/$g_{oDM}$), the average $SBG_{OFMSW}$ drops to 570 mL/$g_{oDM}$ equaling the $SBG_{OFMSW}$ at 40 °C. In total, an operating temperature of 25 °C could be beneficial regarding the potential ratio between saved energy costs (heating) and $CH_4$ yields at a probably similar level.

A study [51], in which an adaptation from mesophilic to psychrophilic AD temperatures took several months, could explain the results for the OFMSW variant at 23 °C. For the evaluation of $SBG_{OFMSW}$ and $SMY_{OFMSW}$ at 35 °C (Table 4), the reduced HRT (21 d) also has to be considered. Depending on the reactor design, typical OFMSW yields (SBG) in Germany vary between 80 and 120 mL/$g_{FM}$ [58]. Normalized on oDM by using the DM and oDM contents of OFMSW (Table 3), SBG yields vary between 309 and 463 mL/$g_{oDM}$. The determined gas potentials are similar considering the influence of laboratory conditions.

In addition, Figure 3 presents the biogas production rates for the digester configurations for both test series. The variants at higher temperatures showed higher production rates and thus higher methanogenic activity at the beginning of the experiment. This is also found in other studies [25,31,52]. In total, the peak values for the biogas production rates of the variants at 23 and 25 °C were lower, but from approximately day 5 to day 10, the biogas production rates were higher or at a similar level, compared to the variants at 35 and 40 °C. This could be attributed to a short adaption phase of the microorganisms as the diversity of microorganisms decreases with lowered temperatures [53,54]. For all variants, the biogas production occurred primarily in the first two weeks allowing conclusions for HRT in practice. The anaerobic treatment of DSS could be increased by one week in order to reduce the residual biogas potential of the DSS. For OFMSW, an HRT of 30 to 40 days seems appropriate in wet digestion with low shares of OFMSW in the mixture. In this study, an inoculum/substrate-ratio of approximately 0.3 in terms of oDM was applied. The HRT depends on various factors such as the organic loading rate or the digestion principle. Therefore, a recommendation of HRT of 30 to 40 days for OFMSW cannot be generalized.

The comparison of process temperature-dependent $SMY_{DSS}$ and $SMY_{OFMSW}$ and $SMY_{DSS\ and\ OFMSW}$ on day 35 of the experiments is presented in Figure 4 in order to determine correlations between process temperatures and $CH_4$ yields for OFMSW and DSS. In the literature, different authors [25,31,39,40,52] stated a linear correlation between $CH_4$ yield and process temperature in AD for other substrates. As can be seen in Figure 4, the $SMY_{DSS}$ seemed to be linear dependent on the process temperature. The $SMY_{OFMSW}$ and the $SMY_{OFMSW\ and\ DSS}$ delivered different results. While the exclusion of the results for AD of OFMSW at 25 °C led to a relatively well-fitting linear trend, the joint reflection did not deliver a linear trend. However, the co-digestion of OFMSW with its various ingredients [1], together with DSS as inoculum delivering a variety of microorganisms [66], is not yet tested extensively. When assuming a linear correlation, lower $CH_4$ yields through lower process temperatures could have been overcompensated by other process parameters in the AD of OFMSW and DSS at 25 °C. Nevertheless, it cannot be excluded that the process temperature at approximately 25 °C provides favorable overall conditions. In additional research, this should be validated by detailed process monitoring and through repetition of the experiment.

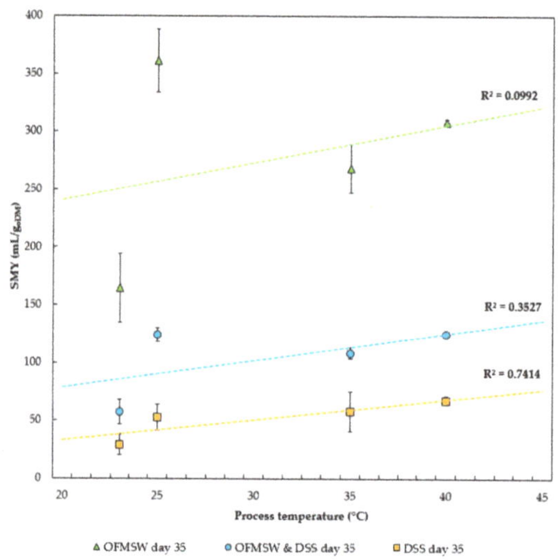


**Figure 4.** Specific methane yield (SMY) on day 35 of the experiment based on organic dry matter (oDM) for the organic fraction of municipal solid waste (OFMSW) and digested sewage sludge (DSS) alone and in the mixture. Gas volume at 1013 hPa, 0 °C; mean values ± standard deviation including trend lines with n = 2 for DSS, n = 4 for OFMSW, n = 4 for OFMSW and DSS.

### 3.3. *Efficiency Potentials through Lowered Process Temperatures*

By reducing operating temperatures, the efforts and costs for digester heating and technical equipment can be reduced or (depending on the ambient temperature) even completely eliminated. In general, the heat demand of biogas plants arises due to heat losses of the digester and because of the necessity to heat the incoming substrate (Appendix A). The main proportion of the heat demand is related to substrate heating. Therefore, it is advisable to preheat the amount of OFMSW before starting the AD process. In practice, this can be conducted by self-heating processes with a short phase of aerobic treatment, which is stopped as soon as the OFMSW reaches the designated process temperature for AD. However, this process leads to lower energy potentials of the OFMSW. Saved heating costs could possibly compensate for lower gas yields, especially if, as can be seen, for example, with DSS, the $CH_4$ concentration increases due to the lower operating temperatures. The reduction of treatment costs for biogas purification is a possible secondary effect of lower operating temperatures. However, larger digester volumes with increased HRT, and separate sanitation systems (biowaste) seem to become necessary. The balance between energy savings due to lower digester temperatures, changes in biogas yields, and infrastructural costs has to be investigated in further research based on detailed simulation to determine optimum conditions due to various interdependencies.

In this study, an operating temperature of 25 °C for the AD of OFMSW together with DSS delivered similar results, compared to an AD temperature of 40 °C, and could be beneficial. Although the experiments for AD of OFMSW were conducted in quadruplicate with relatively low SD, the results have to be verified in further experiments at 25 °C, especially due to the nonlinear correlation (Figure 4) and the findings of other authors [25,31,52] that suggest a linear correlation between SMY and process temperatures. Moreover, the exclusion of the 25 °C variants delivered well-fitting linear trends.

The estimation of an OFMSW-based AD plant with an annual throughput of 10,000 tons of OFMSW (Appendix A) can provide a basic understanding of the influence of the process temperature. When assuming the same energy yield provision at 25 °C, compared to the energy yield at 40 °C, the annual heat demand of the AD plant decreases by approximately 10%. However, the hypothesis of heightened efficiencies in the AD of OFMSW together with DSS through lowered operating temperatures cannot be answered conclusively. If an operating temperature of 25 °C proves in further research that it can deliver similar energy yields, an efficiency increase could be realistic. Otherwise, efficiency potentials through lowered operating temperatures have to be determined based on detailed models with a cost-oriented optimization approach.

Notwithstanding the above, the co-digestion of OFMSW together with DSS could also lead to further synergies. First, the $SBG_{DSS}$ and $SMY_{DSS}$ could be increased through a continued treatment in AD. Second, the DSS could serve as a suitable inoculum delivering a broad variety of microorganisms for the AD of OFMSW with its changing characteristics. Furthermore, the co-digestion of OFMSW, which has a high DM content, together with DSS, which has a low DM content, could be a beneficial option to achieve suitable DM contents for wet digestion systems such as stirred-tank reactors without adding water. The extension of existing wastewater treatment plants with AD units for co-digestion of OFMSW and DSS could be an approach worth investigating.

With or without prior AD, OFMSW typically provides compost products, mainly for application in agriculture. A major challenge in the utilization of OFMSW remains the removal of impurities, which was conducted manually in this study. Especially, effective removal of plastics is challenging, but a separated plastic fraction could be utilized in alternative treatment plants, e.g., for the production of fuels via pyrolysis [77,78].

DSS is also used in agriculture, but its utilization as fertilizer is declining or even forbidden due to problematic ingredients such as heavy metals. Incineration plants currently utilize the major portion of the DSS. Therefore, the utilization possibilities or treatment options for digestates consisting of a mixture of OFMSW and DSS is a field of further research. Creating larger amounts of potentially problematic wastes with limited application possibilities should be avoided. The potentials for decreasing operating temperatures in AD are limited due to a decreasing diversity of microorganisms, as described by McAteer et al. [55]. The reduction of process temperatures could be an approach to improve the profitability of AD plants. This could also be relevant for energy crops-based AD plants. However, in a real-life and continuous AD, a constant temperature is also necessary. The temperature fluctuations should be lower for reactors at thermophilic temperature (±1 °C) than in AD at mesophilic conditions (±3 °C). A comparison of a heated German biogas plant and an unheated American biogas plant can be found in Lansing et al. [79].

## 4. Conclusions

The co-digestion of OFMSW together with DSS could create synergy effects such as achieving required DM contents necessary for the operation of wet digestion systems. In addition to its role as inoculum, the inherent $CH_4$ potential of the DSS would be better exploited. In Germany and worldwide, the mesophilic process temperatures are most commonly used. Therefore, lowered process temperatures could positively influence biogas plants worldwide through reduced operating costs. The results were achieved with OFMSW and DSS but could also be relevant for AD of energy crops together with manure. However, lowered process temperatures could be problematic for biogas plants using substrates that need hygienic treatment such as OFMSW. A separate sanitation unit could become necessary leading to investment and operating costs. According to the German Biowaste Ordinance, only the treatment of OFMSW in AD at thermophilic temperatures fulfills the hygienic requirements, which is why biogas plants operated at mesophilic conditions are already equipped with a separate sanitation unit. From a legal perspective, lowered process temperatures are mainly relevant for biogas plants with separate sanitation units (e.g., mesophilic biogas plants using OFMSW) and for biogas plants operating at mesophilic temperatures without using critical feedstock such as energy-crop-based biogas plants.

Further experiments with thermophilic temperature levels, other AD substrates, and different AD systems should be carried out in order to validate the results of this study. In addition, detailed modeling and simulation of different biogas plants and their operation at lowered temperatures should be conducted, focusing on heat balance, energy output, and AD process parameters such as HRT, organic loading rate, or digester volumes.

**Author Contributions:** Conceptualization, G.S. and S.P.; methodology, G.S., J.P., H.O., and J.E.; validation, G.S., M.S., J.E., J.P., and J.M.; formal analysis, G.S.; investigation, G.S., M.S., and J.E.; writing—original draft preparation, G.S.; writing—review and editing, G.S. and J.M.; visualization, G.S. and J.M.; supervision, S.P., H.O., and J.M.; project administration, S.P.; funding acquisition, S.P. All authors have read and agreed to the published version of the manuscript.

**Funding:** This research was funded by the project ENsource, which was supported by the Ministry of Science, Research, and the Arts of the State of Baden-Wuerttemberg (Germany), and the European Regional Development Fund (ERDF 2014–2020). Support code: FEIH_ZAFH_562822, FEIH_ZAFH_1248932.

**Institutional Review Board Statement:** Not applicable.

**Informed Consent Statement:** Not applicable.

**Data Availability Statement:** The data presented in this study are available on request from the corresponding author.

**Acknowledgments:** The authors would like to thank Sabine Nugent for language editing.

**Conflicts of Interest:** The authors declare no conflict of interest. The funders had no role in the design of the study; in the collection, analyses, or interpretation of data; in the writing of the manuscript, or in the decision to publish the results.

## Abbreviations

| | |
|---|---|
| AD | anaerobic digestion |
| DM | dry matter |
| DSS | digested sewage sludge |
| FM | fresh mass |
| GHG | greenhouse gas(es) |
| HRT | hydraulic retention time |
| oDM | organic dry matter |
| MSW | municipal solid waste |
| OFMSW | organic fraction of municipal solid waste |
| SBG | specific biogas yield |
| SD | standard deviation |
| SMY | specific methane yield |

## Appendix A

For the estimation of efficiency potentials through lowered operating temperatures, heat demands for OFMSW-based AD plants can be estimated based on the heat demand for substrate heating and based on heat losses of the digester. The heat quantity for substrate heating was calculated as

$$Q_{substrate} = c_{water} \cdot m_{water} \cdot \Delta T, \quad \text{(A1)}$$

where $Q_{substrate}$ (J) is the thermal energy for substrate heating, $c_{water}$ (J/kg K) is the specific heat capacity of substrate, $m_{water}$ (kg) is the mass of substrate, and ΔT (K) is the temperature difference between substrate and ambient temperature. Since the specific heat capacity of the substrate was not known, the value for water was used.

The heat flow or the heat loss capacity of the digester was calculated as

$$\dot{Q}_{total} = \sum U_{component} \cdot A_{component} \cdot \Delta T, \quad \text{(A2)}$$

where $\dot{Q}_{total}$ (W) is the sum of the heat flow for all digester components, $U_{component}$ (W/m$^2$ K) is the heat transition coefficient for each digester component, and $A_{component}$ (m$^2$) is the surface area of each component.

For the OFMSW-based AD plant characteristics, an annual throughput of 10,000 tons of OFMSW treated in a digester with a total surface area of 814 m$^2$ with an overall heat transition coefficient of 0.35 W/m$^2$ K was chosen. For OFMSW, a density equal to water was assumed together with the assumption that it would take 1.16 kWh per ton of OFMSW to heighten the temperature by 1 K. In addition, an average ambient temperature of 10 °C was assumed. The combined heat and power unit was defined with a thermal efficiency of 0.45 and a biogas yield of 80 mL/$g_{FM}$ [58].

## References


1. Campuzano, R.; González-Martínez, S. Characteristics of the organic fraction of municipal solid waste and methane production: A review. *Waste Manag.* **2016**, *54*, 3–12. [CrossRef] [PubMed]
2. Vea, E.B.; Romeo, D.; Thomsen, M. Biowaste valorisation in a future circular bioeconomy. *Procedia CIRP* **2018**, *69*, 591–596. [CrossRef]
3. CD. 1999/31/EC. *Council Directive on the Landfill of Waste*. Available online: https://eur-lex.europa.eu/legal-content/EN/TXT/PDF/?uri=CELEX:31999L0031&from=DE (accessed on 26 January 2021).
4. D. 2008/98/EC. *Directive 2008/98/EC of the European Parliament and of the Council on waste and repealing certain Directives*. Available online: https://eur-lex.europa.eu/legal-content/EN/TXT/PDF/?uri=CELEX:32008L0098&from=EN (accessed on 26 January 2021).
5. D. 2018/850. *Directive (EU) 2018/850 of the European Parliament and of the Council amending Directive 1999/31/EC on the Landfill of Waste*. Available online: https://eur-lex.europa.eu/legal-content/EN/TXT/PDF/?uri=CELEX:32018L0850&from=EN (accessed on 26 January 2021).
6. Lin, L.; Xu, F.; Ge, X.; Li, Y. Improving the sustainability of organic waste management practices in the food-energy-water nexus: A comparative review of anaerobic digestion and composting. *Renew. Sustain. Energy Rev.* **2018**, *89*, 151–167. [CrossRef]
7. Hungría, J.; Gutiérrez, M.C.; Siles, J.A.; Martín, M.A. Advantages and drawbacks of OFMSW and winery waste co-composting at pilot scale. *J. Clean. Prod.* **2017**, *164*, 1050–1057. [CrossRef]
8. Mata-Alvarez, J.; Macé, S.; Llabrés, P. Anaerobic digestion of organic solid wastes. An overview of research achievements and perspectives. *Bioresour. Technol.* **2000**, *74*, 3–16. [CrossRef]
9. Rocamora, I.; Wagland, S.T.; Villa, R.; Simpson, E.W.; Fernández, O.; Bajón-Fernández, Y. Dry anaerobic digestion of organic waste: A review of operational parameters and their impact on process performance. *Bioresour. Technol.* **2020**, *299*, 122681. [CrossRef]
10. Reichard, T. Biogasanlagen in der Steiermark: Eine Bestandsaufnahme im Juli 2005; 2006. Available online: https://www.abfallwirtschaft.steiermark.at/cms/dokumente/10212870_46555/a8aebd7f/Gesamtwerk_Biogasanlage_in_der_Steiermark_Internetversion.pdf (accessed on 11 May 2020).
11. Sankoh, F.P.; Yan, X.; Tran, Q. Environmental and health impact of solid waste disposal in developing cities: A case study of Granville Brook Dumpsite, Freetown, Sierra Leone. *JEP* **2013**, *4*, 665–670. [CrossRef]
12. Malinauskaite, J.; Jouhara, H.; Czajczyńska, D.; Stanchev, P.; Katsou, E.; Rostkowski, P.; Thorne, R.J.; Colón, J.; Ponsá, S.; Al-Mansour, F.; et al. Municipal solid waste management and waste-to-energy in the context of a circular economy and energy recycling in Europe. *Energy* **2017**, *141*, 2013–2044. [CrossRef]
13. Awiszus, S.; Meissner, K.; Reyer, S.; Müller, J. *Gärrestverwertung in Einer Warmlufttrocknungsanlage mit Integrierter Stickstoffrückgewinnung*; Landtechnik: Darmstadt, Germany, 2018. [CrossRef]
14. United Nations General Assembly. The Future We Want. Available online: https://www.un.org/en/development/desa/population/migration/generalassembly/docs/globalcompact/A_RES_66_288.pdf (accessed on 11 May 2020).

15. Jain, S.; Newman, D.; Nzihou, A.; Dekker, H.; Le Feuvre, P.; Richter, H.; Gobe, F.; Morton, C.; Thompson, R. Global Potential of Biogas; 2019. Available online: https://www.worldbiogasassociation.org/wp-content/uploads/2019/09/WBA-globalreport-56ppa4_digital-Sept-2019.pdf (accessed on 4 March 2021).
16. Hussein, A.I.; Mansour, M.S.M. Solid waste issue: Sources, composition, disposal, recycling, and valorization. *Egypt. J. Pet.* **2018**, *27*, 1275–1290. [CrossRef]
17. Saini, R.; Osorio-Gonzalez, C.S.; Hegde, K.; Brar, S.K.; Magdouli, S.; Vezina, P.; Avalos-Ramirez, A. Lignocellulosic biomass-based biorefinery: An insight into commercialization and economic standout. *Curr. Sustain. Renew. Energy Rep.* **2020**, *7*, 122–136. [CrossRef]
18. Sara, M.; Rouissi, T.; Brar, S.K.; Blais, J.F. Life Cycle Analysis of Potential Substrates of Sustainable Biorefinery. In *Platform Chemical Biorefinery*; Elsevier: Amsterdam, The Netherlands, 2016; pp. 55–76. ISBN 9780128029800.
19. Wilken, D.; Bontempo, G.; Fürst, M.; Hofmann, F.; Strippel, F.; Kramer, A.; Ricci-Jürgensen, M. Biowaste to Biogas. Available online: https://www.biogas.org/edcom/webfvb.nsf/id/DE-biowaste-to-biogas_eng/$file/biowaste-to-biogas.pdf (accessed on 3 March 2021).
20. Fritsche, W.; Laplace, F. *Mikrobiologie*; Spektrum Akad. Verl.: Berlin/Heidelberg, Germany, 2002; ISBN 3827411076.
21. Munk, K.; Dersch, P.; Eikmanns, B.; Eikmanns, M.; Fischer, R.; Jahn, D.; Jahn, M.; Nethe-Jaenchen, R.; Requena, N.; Schultzem, B. *Mikrobiologie: Taschenlehrbuch Biologie*; Thieme: Stuttgart, Germany, 2008; ISBN 978-3-13-144861-3.
22. *Energie aus Biomasse: Grundlagen, Techniken und Verfahren*; Kaltschmitt, M.; Hartmann, H.; Hofbauer, H. (Eds.) Springer Vieweg: Berlin/Heidelberg, Germany, 2016; ISBN 9783662474389.
23. Kämpfer, P.; Weißenfels, W.D. *Biologische Behandlung Organischer Abfälle*, 1st ed.; Springer: Berlin/Heidelberg, Germany, 2012; ISBN 3642626238.
24. Mähnert, P. Kinetik der Biogasproduktion aus Nachwachsenden Rohstoffen und Gülle. 2007. Available online: https://edoc.hu-berlin.de/bitstream/handle/18452/16303/maehnert.pdf?sequence=1&isAllowed=y (accessed on 5 March 2021).
25. Westerholm, M.; Isaksson, S.; Sun, L.; Schnürer, A. Microbial community ability to adapt to altered temperature conditions influences operating stability in anaerobic digestion. *Energy Procedia* **2017**, *105*, 895–900. [CrossRef]
26. Jain, S.; Wolf, I.T.; Lee, J.; Tong, Y.W. A comprehensive review on operating parameters and different pretreatment methodologies for anaerobic digestion of municipal solid waste. *Renew. Sustain. Energy Rev.* **2015**, *52*, 142–154. [CrossRef]
27. Vindis, P.; Mursec, B.; Janzekovic, M.; Cus, F. The impact of mesophilic and thermophilic anaerobic digestion on biogas production. *J. Achiev. Mater. Manuf. Eng.* **2009**, *36*, 192–198.
28. BMJV. Verordnung über die Verwertung von Bioabfällen auf landwirtschaftlich, forstwirtschaftlich und gärtnerisch genutzten Böden (Bioabfallverordnung): BioAbfV. Available online: https://www.gesetze-im-internet.de/bioabfv/BioAbfV.pdf (accessed on 29 January 2021).
29. Zábranská, J.; Štěpová, J.; Wachtl, R.; Jeníček, P.; Dohányos, M. The activity of anaerobic biomass in thermophilic and mesophilic digesters at different loading rates. *Water Sci. Technol.* **2000**, *42*, 49–56. [CrossRef]
30. Ahring, B.K.; Ibrahim, A.A.; Mladenovska, Z. Effect of temperature increase from 55 to 65 °C on performance and microbial population dynamics of an anaerobic reactor treating cattle manure. *Water Res.* **2001**, *35*, 2446–2452. [CrossRef]
31. Kashyap, D.R.; Dadhich, K.S.; Sharma, S.K. Biomethanation under psychrophilic conditions: A review. *Bioresour. Technol.* **2003**, *87*, 147–153. [CrossRef]
32. Connaughton, S.; Collins, G.; O'Flaherty, V. Psychrophilic and mesophilic anaerobic digestion of brewery effluent: A comparative study. *Water Res.* **2006**, *40*, 2503–2510. [CrossRef] [PubMed]
33. LfU. Biogashandbuch Bayern. Available online: https://www.lfu.bayern.de/energie/biogashandbuch/index.htm (accessed on 11 May 2020).
34. Effenberger, M.; Kaiser, F.; Metzner, T.; Gronauer, A. *Sicherung der Prozessstabilität in Landwirtschaftlichen Biogasanlagen*; Information Bayerische Landesanstalt für Landwirtschaft: Freising-Weihenstephan, Germany, 2008.
35. Amon, T.; Behrendt, U.P.; Daniel-Gromke, J. *Leitfaden Biogas: Von der Gewinnung zur Nutzung*; FNR: Gülzow, Germany, 2013; ISBN 3-00-014333-5.
36. Donoso-Bravo, A.; Bandara, W.M.K.R.T.W.; Satoh, H.; Ruiz-Filippi, G. Explicit temperature-based model for anaerobic digestion: Application in domestic wastewater treatment in a UASB reactor. *Bioresour. Technol.* **2013**, *133*, 437–442. [CrossRef]
37. Fernández-Rodríguez, J.; Pérez, M.; Romero, L.I. Comparison of mesophilic and thermophilic dry anaerobic digestion of OFMSW: Kinetic analysis. *Chem. Eng. J.* **2013**, *232*, 59–64. [CrossRef]
38. Szyłak-Szydłowski, M.; Kulig, A.; Miaśkiewicz-Pęska, E. Seasonal changes in the concentrations of airborne bacteria emitted from a large wastewater treatment plant. *Int. Biodeterior. Biodegrad.* **2016**, *115*, 11–16. [CrossRef]
39. Liu, D.; Zhang, L.; Chen, S.; Buisman, C.; Ter, H.A. Bioelectrochemical enhancement of methane production in low temperature anaerobic digestion at 10 °C. *Water Res.* **2016**, *99*, 281–287. [CrossRef]
40. Chala, B.; Oechsner, H.; Müller, J. Introducing temperature as variable parameter into kinetic models for anaerobic fermentation of coffee husk, pulp and mucilage. *Appl. Sci.* **2019**, *9*, 412. [CrossRef]
41. Kumar, A.; Samadder, S.R. Performance evaluation of anaerobic digestion technology for energy recovery from organic fraction of municipal solid waste: A review. *Energy* **2020**, *197*, 117253. [CrossRef]

42. Jaimes-Estévez, J.; Zafra, G.; Martí-Herrero, J.; Pelaz, G.; Morán, A.; Puentes, A.; Gomez, C.; Castro, L.d.P.; Escalante Hernández, H. Psychrophilic Full scale tubular digester operating over eight years: Complete performance evaluation and microbiological population. *Energies* **2021**, *14*, 151. [CrossRef]
43. Lanko, I.; Flores, L.; Garfí, M.; Todt, V.; Posada, J.A.; Jenicek, P.; Ferrer, I. Life cycle assessment of the mesophilic, thermophilic, and temperature-phased anaerobic digestion of sewage sludge. *Water* **2020**, *12*, 3140. [CrossRef]
44. Pasalari, H.; Gholami, M.; Rezaee, A.; Esrafili, A.; Farzadkia, M. Perspectives on microbial community in anaerobic digestion with emphasis on environmental parameters: A systematic review. *Chemosphere* **2021**, *270*, 128618. [CrossRef]
45. Cavinato, C.; Bolzonella, D.; Pavan, P.; Fatone, F.; Cecchi, F. Mesophilic and thermophilic anaerobic co-digestion of waste activated sludge and source sorted biowaste in pilot- and full-scale reactors. *Renew. Energy* **2013**, *55*, 260–265. [CrossRef]
46. Derbal, K.; Bencheikh-Lehocine, M.; Meniai, A.H. Study of biodegradability of organic fraction of municipal solids waste. *Energy Procedia* **2012**, *19*, 239–248. [CrossRef]
47. Moya, D.; Aldás, C.; López, G.; Kaparaju, P. Municipal solid waste as a valuable renewable energy resource: A worldwide opportunity of energy recovery by using waste-to-energy technologies. *Energy Procedia* **2017**, *134*, 286–295. [CrossRef]
48. Rajagopal, R.; Bellavance, D.; Rahaman, M.S. Psychrophilic anaerobic digestion of semi-dry mixed municipal food waste: For North American context. *Process Saf. Environ. Protect.* **2017**, *105*, 101–108. [CrossRef]
49. Lettinga, G. Challenge of psychrophilic anaerobic wastewater treatment. *Trends Biotechnol.* **2001**, *19*, 363–370. [CrossRef]
50. Saady, N.M.C.; Massé, D.I. Starting-up low temperature dry anaerobic digestion of cow feces and wheat straw. *Renew. Energy* **2016**, *88*, 439–444. [CrossRef]
51. King, S.M.; Barrington, S.; Guiot, S.R. In-storage psychrophilic anaerobic digestion of swine manure: Acclimation of the microbial community. *Biomass Bioenergy* **2011**, *35*, 3719–3726. [CrossRef]
52. Hussain, A.; Dubey, S.K. Specific methanogenic activity test for anaerobic degradation of influents. *Appl. Water Sci.* **2017**, *7*, 535–542. [CrossRef]
53. Watanabe, K.; Koyama, M.; Ueda, J.; Ban, S.; Kurosawa, N.; Toda, T. Effect of operating temperature on anaerobic digestion of the Brazilian waterweed Egeria densa and its microbial community. *Anaerobe* **2017**, *47*, 8–17. [CrossRef]
54. Gaby, J.C.; Zamanzadeh, M.; Horn, S.J. The effect of temperature and retention time on methane production and microbial community composition in staged anaerobic digesters fed with food waste. *Biotechnol. Biofuels* **2017**, *10*, 302. [CrossRef] [PubMed]
55. McAteer, P.G.; Christine Trego, A.; Thorn, C.; Mahony, T.; Abram, F.; O'Flaherty, V. Reactor configuration influences microbial community structure during high-rate, low-temperature anaerobic treatment of dairy wastewater. *Bioresour. Technol.* **2020**, *307*, 123221. [CrossRef]
56. Luning, L.; van Zundert, E.H.M.; Brinkmann, A.J.F. Comparison of dry and wet digestion for solid waste. *Water Sci. Technol.* **2003**, *48*, 15–20. [CrossRef] [PubMed]
57. Rajagopal, R.; Ghosh, D.; Ashraf, S.; Goyette, B.; Zhao, X. Effects of low-temperature dry anaerobic digestion on methane production and pathogen reduction in dairy cow manure. *Int. J. Environ. Sci. Technol.* **2019**, *16*, 4803–4810. [CrossRef]
58. Kern, M.; Raussen, T. *Biogas-Atlas*; 1. Aufl.; Witzenhausen-Institut für Abfall, Umwelt und Energie GmbH: Witzenhausen, Germany, 2014; ISBN 392867367X.
59. Thien, T.; Cu, T.; Cuong, P.H.; Le, H.T.; van Chao, N.; Le Anh, X.; Trach, N.X.; Sommer, S.G. Manure management practices on biogas and non-biogas pig farms in developing countries–using livestock farms in Vietnam as an example. *J. Clean. Prod.* **2012**, *27*, 64–71. [CrossRef]
60. Daniel-Gromke, J. *Anlagenbestand Biogas und Biomethan-Biogaserzeugung und-nutzung in Deutschland: (FKZ 37EV 16 111 0)*; DBFZ Deutsches Biomasseforschungszentrum Gemeinnützige GmbH: Leipzig, Germany, 2017; ISBN 978-3-946629-24-5.
61. Fachverband Biogas. *Branchenzahlen 2018 und Prognose der Branchenentwicklung 2019*. 2019. Available online: https://www.biogas.org/edcom/webfvb.nsf/id/de_branchenzahlen (accessed on 11 May 2020).
62. Völler, K. Branchenbarometer Biomethan 2020. Available online: https://www.dena.de/fileadmin/dena/Publikationen/PDFs/2020/Brachenbarometer_Biomethan_2020.pdf (accessed on 26 January 2021).
63. Guss, H.; Pertagnol, J.; Hauser, E.; Wern, B.; Baur, F.; Gärtner, S.; Rettenmaier, N.; Reinhardt, G. *Biogas-Quo Vadis?* Saarbrücken: Heidelberg, Germany, 2016; Available online: https://www.ifeu.de/fileadmin/uploads/landwirtschaft/pdf/Biogas_quo_vadis_final_report_2016.pdf (accessed on 27 February 2021).
64. Scheftelowitz, M.; Rensberg, N.; Denysenko, V.; Daniel-Gromke, J.; Stinner, W.; Hillebrand, K.; Naumann, K.; Peetz, D.; Hennig, C.; Thrän, D.; et al. Stromerzeugung aus Biomasse-Vorhaben IIa Biomasse: Zwischenbericht Mai 2015; Leipzig. 2015. Available online: https://www.dbfz.de/fileadmin/eeg_monitoring/berichte/01_Monitoring_ZB_Mai_2015.pdf (accessed on 27 February 2021).
65. Barchmann, T.; Pohl, M.; Denysenko, V.; Fischer, E.; Hofmann, J.; Lenhart, M.; Postel, J.; Liebetrau, J.; Effenberger, M.; Kissel, R.; et al. Biogas-Messprogramm III, Gülzow. 2021. Available online: https://biogas.fnr.de/biogasmessprogramm-iii/ (accessed on 27 February 2021).
66. VDI 4630. *Fermentation of Organic Materials: Characterisation of the Substrate, Sampling, Collection of Material Data, Fermentation Tests*; Verein Deutscher Ingenieure e.V.: Düsseldorf, Germany, Sweden, 2016.
67. DIN EN 12579. *Bodenverbesserungsmittel und Kultursubstrate-Probenahme 65.080*; Beuth Verlag: Berlin, Germany, 2014.
68. DIN EN 13040. *Bodenverbesserungsmittel und Kultursubstrate-Probenherstellung für Chemische und Physikalische Untersuchungen, Bestimmung des Trockenrückstands, des Feuchtigkeitsgehaltes und der Laborschüttdichte 65.080*; Beuth Verlag: Berlin, Germany, 2008.
69. DIN EN 14775. *Feste Biobrennstoffe-Bestimmung des Aschegehaltes 75.160.10*; Beuth Verlag: Berlin, Germany, 2012.

70. DIN EN 13039. *Bodenverbesserungsmittel und Kultursubstrate-Bestimmung des Gehaltes an organischer Substanz und Asche*; Beuth Verlag: Berlin, Germany, 2012.
71. DIN EN ISO 16948. *Biogene Festbrennstoffe-Bestimmung des Gesamtgehaltes an Kohlenstoff, Wasserstoff und Stickstoff 75.160.10*; Beuth Verlag: Berlin, Germany, 2015.
72. Boyle, W.C. Energy recovery from sanitary landfills-a review. In *Microbial Energy Conversion, Proceedings of a Seminar Sponsored by the UN Institute for Training and Research (UNITAR) and the Ministry for Research and Technology of the Federal Republic of Germany*; Schlegel, H.G., Barnea, J., Eds.; Elsevier Science: Burlington, VT, USA, 1977; pp. 119–138. ISBN 9780080217918.
73. Buswell, A.M. Anaerobic Fermentations. Available online: https://www.ideals.illinois.edu/bitstream/handle/2142/94555/ISWSB-32.pdf?sequence=1 (accessed on 12 April 2020).
74. Sailer, G.; Eichermüller, J.; Poetsch, J.; Paczkowski, S.; Pelz, S.; Oechsner, H.; Müller, J. Optimizing anaerobic digestion of organic fraction of municipal solid waste (OFMSW) by using biomass ashes as additives. *Waste Manag.* **2020**, *109*, 136–148. [CrossRef] [PubMed]
75. Bertau, M.; Simbach, B.; Aubel, I.; Kiehle, R.; Kaiser, D.; Tröbs, R. Verbesserung der Klärschlammentwässerung durch den Abbau der Extrazellulären Polymeren Substanzen; Freiberg. 2018. Available online: https://www.dbu.de/projekt_32909/01_db_2409.html (accessed on 23 May 2020).
76. Maier, J.; Scheffknecht, G. Systematische Untersuchungen zur Rückgewinnung von Phosphor aus Klärschlamm unter besonderer Berücksichtigung von Feuerungparametern. 2007. Available online: https://fachdokumente.lubw.baden-wuerttemberg.de/servlet/is/40276/BWT24004SBer.pdf?command=downloadContent&filename=BWT24004SBer.pdf&FIS=203 (accessed on 23 May 2020).
77. Sherwood, J. The significance of biomass in a circular economy. *Bioresour. Technol.* **2020**, *300*, 122755. [CrossRef] [PubMed]
78. Constantinescu, M.; Bucura, F.; Ionete, R.-E.; Niculescu, V.-C.; Ionete, E.I.; Zaharioiu, A.; Oancea, S.; Miricioiu, M.G. Comparative study on plastic materials as a new source of energy. *Mater. Plast.* **2019**, *56*, 41–46. [CrossRef]
79. Lansing, S.; Hülsemann, B.; Choudhury, A.; Schueler, J.; Lisboa, M.S.; Oechsner, H. Food waste co-digestion in Germany and the United States: From lab to full-scale systems. *Resour. Conserv. Recycl.* **2019**, *148*, 104–113. [CrossRef]

applied sciences

MDPI

Review

# Development and Characterization of Bioadsorbents Derived from Different Agricultural Wastes for Water Reclamation: A Review

Julián Aguilar-Rosero [1,†], María E. Urbina-López [1,†], Blanca E. Rodríguez-González [1,†], Sol X. León-Villegas [1,†], Itza E. Luna-Cruz [2] and Diana L. Cárdenas-Chávez [1,*]

[1] Tecnologico de Monterrey, School of Engineering and Science, Vía Atlixcáyotl 5718, Puebla 72453, Mexico; A00832705@itesm.mx (J.A.-R.); A01422987@itesm.mx (M.E.U.-L.); berodriguez@exatec.tec.mx (B.E.R.-G.); A01274156@itesm.mx (S.X.L.-V.)

[2] Laboratory of Immunology and Virology, Biological Sciences Faculty, University Autonomus of Nuevo León, Avenida Pedro de Alba, San Nicolás de los Garza 66450, Mexico; itza.lunacrz@uanl.edu.mx

* Correspondence: diana.cardenas@tec.mx

† These authors contributed equally to this work.

**Abstract:** The presence of dangerous pollutants in different water sources has restricted the availability of this natural resource. Thus, the development of new low-cost and environmentally-friendly technologies is currently required to ensure access to clean water. Various approaches to the recovery of contaminated water have been considered, including the generation of biomaterials with adsorption capacity for dangerous compounds. Research on bioadsorbents has boomed in recent years, as they constitute one of the most sustainable options for water treatment thanks to their abundance and high cellulose content. Thanks to the vast amount of information published to date, the present review addresses the current status of different biosorbents and the principal processes and characterization methods involved, focusing on base biomaterials such as fruits and vegetables, grains and seeds, and herbage and forage. In comparison to other reviews, this work reports more than 60 adsorbents obtained from agricultural wastes. The removal efficiencies and/or maximum adsorption capacities for heavy metals, industrial contaminants, nutrients and pharmaceuticals are presented as well. In addition to the valuable information provided in the literature investigation, challenges and perspectives concerning the implementation of bioadsorbents are discussed in order to comprehensively guide selection of the most suitable biomaterials according to the target contaminant and the available biowastes.

**Keywords:** biowastes; bioadsorbents; activated carbon; raw wastes; heavy metals; industrial contaminants; nutrients; pharmaceuticals

**Citation:** Aguilar-Rosero, J.; Urbina-López, M.E.; Rodríguez-González, B.E.; León-Villegas, S.X.; Luna-Cruz, I.E.; Cárdenas-Chávez, D.L. Development and Characterization of Bioadsorbents Derived from Different Agricultural Wastes for Water Reclamation: A Review. *Appl. Sci.* **2022**, *12*, 2740. https://doi.org/10.3390/app12052740

Academic Editor: Carlos Rico de la Hera

Received: 1 February 2022
Accepted: 24 February 2022
Published: 7 March 2022



## 1. Introduction

Water pollution has increased in recent decades as a consequence of the uncontrolled disposal of agricultural and industrial residues and domestic discharge directly into water bodies and soil, which can then reach underground sources via filtration. Two of the most common contaminants are textile dyes and heavy metals; approximately 20% of industrially-employed dyes are discharged in effluents, while heavy metals are often the result of waste from hospitals and factories. Both types of pollutants are considered important hazards for human and environmental health even at trace levels. These non-biodegradable pollutants are found in surface and ground water sources and accumulated in aquatic organisms, primarily because conventional wastewater treatments cannot completely remove them from the effluent [1–5]. There are other water contaminants, such as organic matter, pharmaceuticals, pesticides, fertilizers, chemicals from personal care products, and many micro-organic pollutants; these are known as emerging contaminants. They are characterized by the majority being unregulated substances generated from anthropogenic activities, and can only be detected by chromatography methods coupled to mass spectrometry. Even though they are present in the environment at very low concentrations (nanograms or pictograms), their adverse effects are significant. The consequences of these pollutants are diverse, generating long-term effects on aquatic life and human health such as endocrine disruption, carcinogenesis, mutagenesis, and reproductive and embryonic toxicity, among others [6–8]. Reports from the United Nations Educational, Scientific and Cultural Organization state that every day over two million tons of sewage and other effluents drain into water bodies, forcing one out of nine people in the world to consume water from unsafe sources [9]. Therefore, the necessity of environmentally friendly and affordable treatments is a priority.

Different technologies have been developed to treat wastewater before it enters rivers, oceans, or other water sources, including chemical, physical, and biological methods. Their application depends on punctual permissible levels of the effluent, the cost of the process, and environmental compatibility. These techniques include sedimentation, filtration, ion exchange, flotation, electrolysis, microbial reduction, bioremediation, and adsorption [10–12]. Biosorption is described as the process that occurs when the adsorbate or pollutant is retained on the surface of the adsorbent (biomaterial), creating a molecular or atomic film owing to residual interactions such as Van Der Waals forces or covalent bonds [13–15]. When biosorption is applied on wastewater treatment, it involves the interaction of the liquid phase component with the surface of the solid by either a chemical (chemisorption) or physical (physisorption) process, generating a mono- or multilayer (Figure 1). Similar to synthetic resins, heavy metals or other charged contaminants bind to the adsorbent biomaterial through electrostatic interactions and ion exchange.

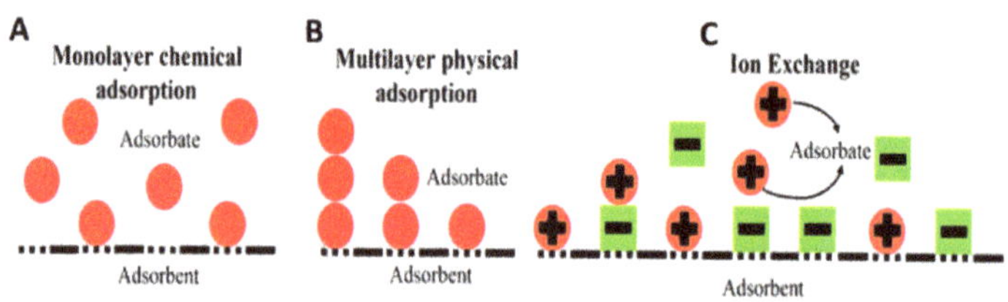


**Figure 1.** Adsorption mechanisms: (**A**) chemisorption monolayer; (**B**) physisorption multilayer; and (**C**) ion exchange adsorption.

Agricultural residues have been proposed as low-cost unconventional adsorbents due to their high content of cellulose and lignocellulose [1]. Grains, seeds, fruits, vegetables, herbage, and forage, which represent the most abundant wastes in food and agricultural industries, have great potential as adsorbent materials capable of remediating polluted effluents [3]. However, several conditions must be considered for the optimal removal of contaminants in addition to the production volume, such as the chemical and physical characteristics of the bioresidues and their operation parameters (pH, contact time, adsorbent dose, particle size, contaminant concentration, temperature, adsorption kinetics, and isothermal behavior). pH is one of the most important parameters, as it can improve or weaken removal efficiency by affecting the solubility of metal ions as well as the physical properties, structure, and availability of active sites on the surface of the adsorbent [11,15,16]. A high adsorbent dosage provides a higher number of active sites, and consequently increases the adsorption capacity until equilibrium is reached at which an excess adsorbent does not further improve removal efficiency [15]. The adsorption capacity can be further increased by reducing the particle size of the adsorbent, as there is then a higher surface area and the limitations caused by mass and diffusional transfer are decreased. This allows molecules to reach sorption sites more easily, and the capture of the adsorbate occurs in a shorter time [15,17,18]. Furthermore, the adsorption capacity increases as the contaminant concentration rises, which is calculated by isothermal models that estimate the maximum sorption capacity of the adsorbent ($q_m$) at a given temperature given the possible adsorption mechanisms [17,19]. Langmuir and Freundlich isotherms are two of the most common models employed to describe different bioadsorbents. However, the Dubinin–Radushkevic, Temkin, and other models have been used in specific cases, e.g., where the carbonaceous material capacity is defined by neither layer-by-layer adsorption nor a constant adsorption potential, or when the adsorption behavior is characterized by uniform energy distribution [20–22]. The adsorption kinetics indicate the influence of the contact time, which depends on the adsorbate concentration, temperature, and adsorbent characteristics (surface and pore size). There are several kinetic models; however, two that are generally applied in bioadsorbent studies are the pseudo-first and pseudo-second order models [23]. Both models deduce the sorption occurring on localized sites without interaction among the adsorbed ions until it reaches a maximum point on the saturated monolayer [24]. In the context of the current relevance and application of bioadsorbents for water bioremediation, although many authors have reported on several options for the removal of a wide range of contaminants, at present this information is distributed across many studies focused on a single type of contaminant or bioresidue. The present paper seeks to organize this vast amount of information into three major categories of low-cost agricultural adsorbents, namely, grains and seeds, fruits and vegetables, and herbage and forage waste, for the removal of the main groups of pollutants (heavy metals, industrial chemicals, nutrients, and pharmaceuticals) found in different water sources. Bioadsorbents derived from these three categories are described and compared here along with information on both adsorbents and the key parameters that influence the adsorption process. We sought to compile the largest collection of information on bioadsorbents reported to date in a single bibliographical source in order to allow future researchers to easily determine the most suitable alternative in each case.

## 2. Methods

The methodological approach followed to this literature review is briefly described below. The main topics were waste-based adsorbents and water remediation. These residues were categorized into three main groups: fruits and vegetables, grains and seeds, and herbage and forage. The search based on the Science Direct database was carried out from May to August 2021 using the keywords "seeds", "fruit", "vegetable", "forage", "herbage", "silage", "cereal" "water treatment", "grain", "beans contaminant", "wastewater", "by-product", "adsorption", "micronutrients" and different combinations of the same. Commands such as "AND" and "OR" were included in order to improve the investigation. Approximately 120 publications (scientific articles and reviews) from the last ten years were screened and examined. The selection parameters included relevant information about removal efficiency, adsorption and kinetic models, waste sources, and removal of specific groups of contaminants (pharmaceuticals, dyes, heavy metals, and nutrients). Book chapters and reports of international organizations were included as references for specific topics and concepts. It is important to explain that the inclusion of the herbage and forage category resulted from a meticulous search for residues derived from leaves and stems such as sugarcane bagasse (SCB), tea, date palm tree leaves and fibers, sunflower stalks and leaves, cauliflower leaf powder, potato stem and leaf powder, grape stalks, corn stalks, and straw of cereals such as rice, wheat, barley, and soybeans. The corn cob was considered a vegetable due to its peculiar way of sprouting from the stalk plant.

## 3. Obtaining Waste-Based Bioadsorbents

The use of grains and seeds for water treatment can be classified into three categories: flocculation–coagulation processes [25,26]; generation of activated carbon (AC) [27]; and production of bioadsorbent materials [28]. Hibiscus seeds have been used as a worthy example for the removal of organic matter from water. It is first necessary to dry and grind the seeds before resuspending the product in NaCl solution; it must then be defatted with hexane to obtain extracts and proteins with coagulant activity [25]. Other seeds have been tested, such as *Moringa oleifera*, common beans, and mustard seeds [29,30]. Several examples from the other two categories are described below, highlighting the type of physical or chemical pre-treatment employed in their use as bioadsorbents.

### 3.1. Activated Carbon Production

The synthesis of AC from low-cost waste materials and its subsequent use as a biosorbent has become more common. Generally, AC implementation requires pre-treatment before application to contaminated waters; the use of microwaves [31], ultrasound techniques [32], or chemical agents for biochar activation have all shown promising results. Furthermore, the use of different chlorides such as NaCl, KCl, and $CaCl_2$ has resulted in AC with better adsorption capacity [25,33]. The common way to activate biowaste-based carbon is after carbonization, although in certain cases the activation is carried out before. This can be physical (e.g., pressurization, pyrolysis, or gasification) and/or chemical (e.g., by using acid, alkali, or salt compounds) depending on the nature of the waste and its further application [34–36].

The removal of heavy metals with AC using seed and grain wastes has been widely explored, as in the case of the buttons (female flowers) that are one of the most abundant residues in coconut plantations (comprising about 55–95% of remains). This coconut waste was evaluated by Anirundhan and Sreekumari [37] for the adsorption of different heavy metals from industrial effluents. The buttons were treated with sulfuric acid and placed inside a graphite tube in a furnace to generate AC; the collected carbonized material showed high metal removal performance (100%). Spent coffee waste is another material commonly used to adsorb a wide range of contaminants present in fresh water and wastewater through carbon activation by NaOH [38]. AC from fruit waste such as papaya peel can be an effective adsorbent to remove lead from contaminated water. This biowaste was dried (105 °C) and carbonized, followed by a chemical activation process using $H_3PO_4$ as the oxidant agent [39]. Concerning AC derived from forage and herbage sources (e.g., cereal byproducts and palm tree leaves), these must first pass through a drying process, then be soaked in acid solution, and finally carbonized into a particle size between 300–425 μm. This has the aim of adjusting the kinetic model at 56 °C in order to effectuate the capacity to remove heavy metals such as Pb(II) and Cr(VI) from aqueous solutions [40,41].

The AC sources employed to remove dyes and other industrial pollutants comprise a long list of several types of vegetable wastes. Peanut shells have been transformed into AC through chemical treatment with $H_3PO_4$ and used to adsorb acid yellow 36 [42]. The same chemical modification was performed with *Acacia erioloba* seed biochar to increase its surface area and porosity and thereby improve the adsorption yield for methylene blue (MB) and iodine [43]. Other authors processed coffee waste with KOH and pyrolysis to obtain granular AC into calcium alginate beads for the treatment of dye contaminants, finding that the material could be used for up to seven cycles [44]. The efficiency of $H_3PO_4$ as a carbon activator was compared with that of $ZnCl_2$ in the context of siriguela seeds and cocoa shells for the adsorption of bovine serum albumin (BSA) and α-lactalbumin (α-Lac), which are food industry wastes; $H_3PO_4$ was more effective for α-Lac, while $ZnCl_2$ performed better in BSA removal [45]. Coir pith subjected to a carbonization treatment at 700 °C was used in a single and a multi-component system to adsorb congo red, rhodamine-B, and acid violet [46,47]. Another fruit-based biochar that has been widely studied is that resulting from pineapple waste (crowns, leaves and stems). These residues were subjected to pyrolysis and permeated with $ZnCl_2$ to turn them into AC with a prominent adsorption capacity for dyes (e.g., MB) thanks to a large surface area (914.67 $m^2/g$) [48].

A wide variety of other agricultural wastes have been tested to produce AC for the removal of nutrients and pharmaceuticals such as sodium diclofenac, diclofenac, carbamazepine, sulfamethoxazole, and ibuprofen, among others. As an example, rice husk and lemon juice residue (solid after juice extraction) were explored as adsorbents for phosphate sequestering. The dried residues were submerged in NaOH followed by $H_2SO_4$ for their respective activation; both were then dried, carbonized, and finally sieved to a 250–350 μm particle size [49]. Likewise, rubber pod husk showed favorable results for removing phosphates through chemisorption and a multilayer process after undergoing chemical treatment with $H_3PO_4$ to activate its adsorbent capacity [50]. A low percentage of cacao pod husk generated during chocolate production is recycled as a fertilizer and the rest is discarded. This massive generation facilitates its transformation into AC when treated with $H_2SO_4$ for the removal of sodium diclofenac from aqueous solutions [51]. Chemical activation with $H_3PO_4$ was further implemented by El Mouchtari et al. [52] on a carbon composite based on *Argania spinosa* tree nutshells and $TiO_2$. The latter modified the properties of the adsorbent by increasing the surface area and easily retaining diclofenac, carbamazepine, and sulfamethoxazole as well as the photolysis of the pharmaceuticals. Chakraborty et al. [53] and Cabrita et al. [54] used ACs from *Aegle marmelos* (wood apple) shells and peach stones with physical and chemical activation, respectively, to remove ibuprofen and acetaminophen from water sources. The pretreatment conditions, including the carbonization and pyrolysis temperatures employed to obtain the different activated carbons from biowastes, are summarized in Table 1.

**Table 1.** Generation and characterization of activated carbon derived from agricultural residues (grains, seeds, fruits, vegetables, herbage, and forage) to remove heavy metals, industrial contaminants, ions, and pharmaceuticals.

| Biowaste | Pretreatment | Characterization | Contaminant | Ref. |
|---|---|---|---|---|
| *Acacia erioloba* seed | Chemical activation ($H_2SO_4$), Pyrolysis (600 °C) | SEM, XRD, BET, FTIR, EDX | MB, iodine | [43] |
| *Aegle marmelos* shell | Carbonization (650 °C), Steam activation (800 °C) | SEM, BET, FTIR, PZC | IBU | [53] |
| *Argania spinosa* tree nutshells | Chemical activation ($H_3PO_4$), Carbonization (500 °C)—$TiO_2$ Impregnation | FTIR, SEM-EDS, TGA, XRD, BET | CBZ, SMX, DCF | [52] |
| Cacao pod husk | Chemical activation ($H_2SO_4$), Carbonization (600 °C) | SEM, FTIR, EDX | SD | [51] |
| Cacao shells and Siriguela seeds | Chemical activation ($ZnCl_2$, $H_3PO_4$), Carbonization (500 °C, $N_2$) | FTIR, DTA/TG, BET, BJH | α-Lac I, BSA | [45] |
| Cereal byproducts | Carbonization (600 °C) | N.R | Cr(VI) | [40] |
| Coconut buttons | Chemical activation ($H_2SO_4$), Steam carbonization (400 °C) | FTIR, XRD, SEM, TGA, PZC, BET | Pb(II), Hg(II), Cu(II) | [37] |
| Coir pith | Carbonization (700 °C) | N.R | CR, RB, AV | [46,47] |
| Palm tree leaves | Chemical activation ($H_2SO_4$), Carbonization (250–450 °C) | FTIR, SEM-EDX, BET | Pb(II) | [41] |
| Papaya Peel | Carbonization (450 °C), $H_3PO_4$ Oxidation | FESEM, SEM-EDX, FTIR, BET, XRD | Pb(II) | [39] |
| Peach stones | Chemical activation ($K_2CO_3$), Carbonization (700 °C) | $N_2$ and $CO_2$ adsorption, PZC, thermal analysis | ACP | [54] |
| Peanut Shell | Chemical activation ($H_3PO_4$), Pyrolysis (650 °C, $N_2$) | TGA, FESEM, EDS, BET, FTIR | AY-36 | [42] |
| Pineapple waste | Chemical activation ($ZnCl_2$), Pyrolysis (500 °C) | BET, BJH, FTIR, SEM | MB | [48] |
| Rice husk and Lemon juice residue | Chemical activation (NaOH, $H_2SO_4$), Carbonization (650 °C) | FTIR, BET, BJH | Phosphates | [49] |
| Rubber pod husk | Chemical activation ($H_3PO_4$), Pyrolysis (500 °C) | BET, SEM, EDX, FTIR, PZC | Phosphates | [50] |
| Spent Coffee | Chemical activation (NaOH), Pyrolysis (800 °C) | UHR-SEM, FTIR | NPX, DCF, IBU | [38] |
| Spent Coffee (Granular) | Chemical activation (KOH), Pyrolysis (700 °C), Granulation | SEM, FTIR, BET, BJH, Horvath-Kawazoe | AO7, MB | [44] |

N.R, not reported; MB, Methylene blue; IBU, Ibuprofen; CBZ, Carbamazepine; SMX, Sulfamethoxazole; DCF, Diclofenac; SD, Sodium diclofenac; CR, Congo red; RB, Rhodamine B; AV, Acid violet; ACP; Acetaminophen; AY-36, Acid yellow 36; NPX, Naproxen; AO7, Acid orange 7; EDS, Energy Dispersive Spectrometer; UHR-SEM, Ultra-High Resolution Scanning Electron Microscope.

### 3.2. Raw Wastes as Bioadsorbents

Many of the biosorbents that have been studied for the removal of pollutants were employed in raw form with or without chemical, physical, or even magnetic pretreatment (Table 2). Residues such as exhausted coffee can be used without any chemical modification to adsorb hazardous contaminants such as Hg(II) found in industrial effluents [55]. Raw tamarind seeds have been presented as a remarkable alternative for the removal of Pb(II) ions because of their high content of lignocellulose along with surface groups such as phosphorus, magnesium, vitamin C, potassium, calcium, and proteins, which can be chemically modified by $H_2SO_4$ (1:2 weight ratio) and ultrasonic waves (24 kHz) to increase their surface area [32]. Araújo et al. [56] and Çelekli et al. [57] used *Moringa oleifera* seed powder (75–500 μm particle size) without pretreatment to adsorb silver and reactive red 120 dye. Because the presence of surface functional groups is crucial to synthesize materials with higher adsorption capacity, Edathil et al. [28] developed a magnetic spent coffee nanocomposite with $Fe_3O_4$ nanoparticles to increase its affinity to Pb. Coconut is one of the most abundant agro-industrial products, and is able to remove several heavy metals by taking advantage of the fact that approximately 62–65% of the whole fruit (coir pith, coconut bunch, and husk) is considered waste [10,58]. For instance, coir pit was employed without chemical pre-treatment for the removal of Co(II), Cr(III) and Ni(II) after being air-dried, ground, and sieved using a 300–600 μm mesh [59]. Banana peels were used by Memon et al. [60,61] for Cd(II) and Cr(VI) adsorption. Dried slices of banana peel were crushed and passed through a mesh of 125 μm, re-dried in an air oven at 100 °C, and finally esterified with acidic methanol. Watermelon rinds have characteristics suitable for the removal of metal elements (Zn, Pb and Cr) in their native form and after employing calcium hydroxide and citric acid treatment (Cu), followed by a drying and crushing process a particle diameter between 150 and 300 μm can be achieved [62–66]. Certain functional groups such as carboxyl and hydroxyl are responsible for the sorption of pollutants [3].

**Table 2.** Pretreated and non-treated raw biowastes implemented as bioadsorbents to remove a wide variety of heavy metals, industrial contaminants, ions, and pharmaceuticals.

| Biowaste | Pretreatment | Characterization | Contaminant | Ref. |
|---|---|---|---|---|
| Banana peel | No treatment, Esterification (MeOH-HCl) | FTIR, BET, SEM, PZC, EDX | MO, MB, RB, CR, MV, AB-10, F, Cd(II), Cr(VI) | [60,61,67,68] |
| Barley straw | No treatment, Citric acid–NaOH treatment, Magnetic modification | FTIR, SEM | BBY, CV, MB, SO | [69] |
| Cauliflower leaf | No pretreatment | FTIR, SEM | MB | [70] |
| Coffee husk | No pretreatment | BET, FTIR, PZC, SEM-EDS | NFX | [71] |
| Coconut coir dust | No pretreatment | FTIR | MB | [72] |
| Coir pith | No pretreatment | N.R | Co(II), Cr(III), Ni(II) | [59] |
| Corn cob and stalk | No treatment, Formaldehyde, $NaOH-H_2SO_4$ treatment | N.R | Cu, Ni, Cd, Pb | [73] |
| Durian peels | HCl treatment | N.R | AG25 | [74] |
| Grapefruit peel | No pretreatment | FTIR | CV | [75] |
| Jackfruit peel | No pretreatment | N.R | MB | [76] |
| Kiwi and Tangerine peels | NaOH treatment | N.R | Cd(II), Cr(III), Zn(II) | [77] |
| Mango peel | No pretreatment | PZC, FTIR, SEM, EDX | Cd(II), Pb(II) | [78] |
| Mangosteen pericarps | No Pretreatment | SEM, FTIR, EDX, XPS, XAS | $I^-$ | [79] |
| *Moringa oleifera* seeds | No pretreatment | FTIR, SEM | RR-120, Ag(I) | [56,57] |
| Orange peel | No pretreatment | BET, SEM | CR, PO, RB, MO, MB, MV, AB-10, DR23, DR80 | [67,80,81] |
| Passion fruit rinds | No Pretreatment | SEM, FTIR, EDX, XPS, XAS | $I^-$ | [79] |
| Peanut husk | Chemical modification ($Fe_3O_4$-IA-Zr) | BET, SEM, FTIR, XPS, VSM, XRD | Phosphates | [82] |
| Pomelo peel | No pretreatment | ZP, FESEM, FTIR, BET | RB-114 | [83] |

**Table 2.** *Cont.*

| Biowaste | Pretreatment | Characterization | Contaminant | Ref. |
|---|---|---|---|---|
| Potato leaf/stem | No pretreatment | FTIR, SEM | MB, MG | [84] |
| Red onion and Red dragon fruit peels | No Pretreatment | SEM, FTIR, EDX, XPS, XAS | $I^-$ | [79] |
| Rice husk | No Pretreatment | BET, FTIR, PZC, SEM-EDS | NFX | [71] |
| Rice straw | No Pretreatment | FTIR, BET, FTIR, SEM, EDX | CFA, CBZ, Pb(II) | [16,85] |
| Soybean and wheat straw | No treatment, Formaldehyde, NaOH and $H_2SO_4$ treatment. | N.R | Cu, Ni, Cd, Pb | [73] |
| Spent ground coffee | No treatment, Magnetic modification ($Fe_3O_4$ NPs) | BET, XRD, FTIR, SEM-EDX, ZP | Hg(II), Pb(II) | [28,55] |
| Sugarcane bagasse (SCB) | No treatment, NaOH, HCl treatments | FTIR, SEM | Fe(III), Cu(II), Pb(II), Zn(II), Cd(II), Co(II), Mn(II) | [86,87] |
| SCB and beet pulp | NaOH treatment | FTIR, SEM | Mn(II) | [88] |
| Sunflower leaf/stalk | No treatment, NaOH activation (Ni removal) | PZC, SEM | Fe, Mn, Zn, Ni, Cu, Cd | [89] |
| Tamarind seeds | $H_2SO_4$ treatment., Ultrasonic modification | FTIR, SEM | Pb(II) | [32] |
| Tea waste (Black tea) | No treatment, Sulfonation ($H_2SO_4$) | BET, BSLLD, SEM-EDX, FTIR, TGA, Raman, XPS, ESR, SAXS | Cu, Pb, MB, Tet, Cr(VI) | [90,91] |
| Tea waste (mixed tea) | No pretreatment | SEM, TEM, EDS, BET, FTIR, XPS | Mn(II), Zn(II), Cr(VI) | [92,93] |
| Watermelon rinds | No treatment, $Ca(OH)_2$, Citric acid treatments | SEM-EDX, BET, MP, PZC, FTIR | Zn, Pb, Cu(II), Cr(III) | [62–66] |

N.R, not reported; MeOH, Methanol; $Fe_3O_4$ NPs, $Fe_3O_4$ nanoparticles; MO, Methyl orange; MB, Methylene blue; RB, Rhodamine B; CR, Congo red; MV, Methyl violet; AB-10, Amido black 10B; F, Fluoride; BBY, Bismarck brown Y; CV, Crystal violet; SO, Safranin O; NFX, Norfloxacin; AG25, Acid Green 25; RR-120, Reactive red 120; PO, Procion orange; DR23, Direct red 23; DR80, Direct red 80; RB-114, Reactive blue 114; MG, Malachite green; CFA, Clofibric acid; CBZ, Carbamazepine; Tet, Tetracycline; PZC, point of zero charge; EDS, Energy dispersive X-ray spectroscopy; XPS, X-ray photoelectron spectroscopy; XAS, X-ray absorption spectroscopy; XRD, X-ray diffraction; ZP, zeta potential; FESEM, Field Emission Scanning Electron Microscope; BSLLD, back-scattered laser light diffraction; TGA, Thermogravimetric analysis; ESR, Electron spin resonance spectroscopy; SAXS, Small-angle X-ray scattering; TEM, Transmission electron microscopy; MP, mercury porosimetry.

Mango is another abundant fruit grown in tropical and subtropical regions across the world; unfortunately, the peel constitutes 7–24% of the fruit's weight and is not a recyclable byproduct. For this reason, mango peel waste was evaluated as an adsorbent for Cd(II) and Pb(II) ions. It was oven-dried at 70 °C, ground and sieved until obtaining a particle size of 0.85–1.0 mm [78]. Another interesting fruit waste used to remove Cd(II), Cr(III), and Zn(II) from water was the kiwi peel, ground to two particle sizes of 1 mm and 2 mm and treated with NaOH [77]. Tangerine peel processed in the same way proved to be an excellent source of biomass for the removal of these metals from surface and groundwater due to its chemical composition, which is rich in cellulose and other polysaccharides [77,94]. Šćiban, et al. [73] evaluated the efficiency of soybean straw, wheat straw, corn stalks, and corn cobs for the adsorption of Cu, Ni, Cd, and Pb from wastewater both with formaldehyde, NaOH and $H_2SO_4$ modification and without chemical modification. In another study, Amer et al. [16] dried and ground rice straw into different particle sizes and found the 75–150 μm size to be the most optimal for Pb removal. Sugar cane baggage (SCB) and beet pulp have been tested as effective adsorbents for other heavy metals such as Fe(III), Zn(II), Co(II), and Mn(II). These biowastes were soaked in NaOH and $CH_3COOH$ to remove hydroxide traces before being dried, powdered, and sieved to an average particle size of 0.75 mm [86,88]. SCB was evaluated in raw conditions and after chemical pre-treatment with hydrochloric acid; the latter had a significant positive effect on its capacity to remove Mn(II) [87]. Other studies have evaluated black tea and mixed tea wastes as low-cost adsorbents for the removal of several heavy metals; for instance, sulphonate-treated tea waste was an effective biosorbent for the removal of Cr(VI), MB, and tetracycline [90–93]. Different amounts of Fe, Mn, Zn, Ni, Cu, and Cd were removed from aqueous solutions by employing a biosorbent from sunflower stalks and leaves, both without pretreatment and activated with NaOH to improve Ni adsorption [89].

Bioadsorbents have been implemented as one possible solution to solve the problem of colored wastewater discharges from the textile industry. For example, durian peel was collected as solid waste and used directly for the uptake of acid green 25. Durian peels were ground and screened to a particle size range of 351 to 589 μm before being treated with HCl and dried at 80 °C [74]. Another excellent adsorbent to remove dyes such as crystal violet (CV) is non-pretreated grapefruit peel, which is first separated from leaves, twigs, and other debris that interfere into the adsorption process and then dried at 70 °C, ground, and sieved to a particle size of 0.85–1.0 mm [75]. Etim et al. [72] employed coconut coir dust without chemical modification to remove MB, resulting in pH-dependent monolayer adsorption behavior. Hameed [76] studied the MB adsorption process employing a jackfruit peel biobed. This biowaste was sliced and dried at 70 °C, then ground and sieved to obtain a particle size range of 0.5–1 mm. Banana, orange, and pomelo peels represent another low-cost method, adsorbing MB and other dye pollutants such as methyl orange, rhodamine-B, congo red, methyl violet, amido black 10 B, procion orange, violet 17 (V-17), direct red 23, direct red 80, and reactive blue 114 without requiring chemical treatment. The waste peels in these studies were dried in the sunlight and a hot air oven at 60 or 120 °C, and the resulting material was crushed with a mill and sieved to obtain a particle size < 500 μm [67,80,81,83,95,96].

Low-cost adsorbents derived from herbage and forage have been studied for the removal of dyes generated from different industries. Cauliflower leaves were dried and pulverized to achieve a higher adsorption capacity for MB in synthetic aqueous solutions [70]. Gupta et al. [84] evaluated the leaves and stems from potato plants as biodegradable material for the removal of MB and malachite green; leaves boiled and dried at 60 °C showed better adsorption efficiency than stems for both dyes at same particle sizes (100–150 μm). Baldikova et al. [69] employed raw and chemically, magnetically, and non-magnetically modified versions of barley straw to remove four water-soluble dyes (bismarck brown Y, crystal violet, MB, and safranin O). The barley straw was cut and sieved into fine particles (~0.15–2 mm diameter) and a fraction was magnetically modified by microwave-synthetized magnetic iron oxides followed by treatment with citric acid, then dried at 50 °C until reaching a constant weight.

Macro- and micronutrients that alter the microbial dynamics in ecosystems have been removed by employing bioadsorbents. Fluoride, one of these pollutants, was removed from contaminated underground water using natural banana-peel powder in a fixed-bed design. This biomaterial was dried at 50 and 60 °C and then ground to obtain a particle size of 200 μm for the batch experiments [68]. Mangosteen pericarps, passion fruit rinds, red onion peels, and red dragon fruit peels were used to make an anthocyanin-based adsorbent to attract iodide ions ($I^-$). These raw materials were dried at 60 °C, ground, and sieved to obtain particles of 0.5–0.711 mm [79]. Additionally, phosphates were extracted employing peanut husk improved with $Fe_3O_4$ as an adsorption assistant, as this increases ionic attraction and provides a cheaper recovery method than centrifugation or filtration. This oxide was combined with iminodiacetic acid (IA) and zirconium (Zr) to increase the efficiency and selectivity of the peanut husk-based magnetic material [82].

As for pharmaceuticals, coffee and rice husks were tested by Paredes-Laverde et al. [71] in a variety of particle sizes (from <75 to 500 μm) to retain the antibiotic norfloxacin in distilled and municipal water. Both natural adsorbents were dried at 60 °C, ground to powder, and sieved; particles < 75 μm proved to be the best option for norfloxacin adsorption. Another cereal byproduct, rice straw, was dried at the same temperature, pulverized, and passed through a mesh sieve (<150 μm) to test its ability to remove clofibric acid and carbamazepine from aqueous solutions at different pH values and adsorbent concentrations [85].

## 4. Characterization of Bioadsorbents

There are several characterization techniques used to understand the physical and chemical properties of materials, their adsorption capacity, and their interactions with pollutants; these include scanning electron microscopy (SEM), the Brunauer–Emmett–Teller method (BET), the Barrett–Joyner–Halenda method (BJH), Fourier transform infrared spectroscopy (FTIR), and energy-dispersive X-ray (EDX), among others. Adsorbents from fruits, vegetables, grains, seeds, herbage, and forage wastes have mainly been characterized by applying FTIR (Tables 1 and 2). This technique makes it possible to identify the molecules and functional groups on the surface of the biomaterials [51] that interact with the contaminants. Some of these correspond to hydroxyl groups generated by the cellulose, hemicellulose, and lignin present in agricultural waste [42]. An SCB-based adsorbent was characterized using this analysis to demonstrate the presence of hydroxyl, C-H, C-O, and -$OCH_3$ groups as well as other oxygen-containing functional groups on its surface [86]. The structure of mango peel waste was analyzed by FTIR spectroscopy, demonstrating that carboxyl and hydroxyl groups were the main groups responsible for sorption of metals [10,78]. Yadav et al. [49] used this characterization method to evaluate the adsorbent obtained from rice husk and lemon juice residues, showing that O-H, N-O, and C-N groups bind to phosphates. All the biowastes and adsorbents reported here have been analyzed using this technique except for peach stones, sunflower leaves/stalks, and several others not characterized by any analytical method (coir pith, corn cob and stalk, soybean, wheat straw, kiwi, tangerine peels, and jackfruit peels). Peach stones were characterized by $N_2/CO_2$ adsorption, point of zero charge, and thermal analysis, whereas sunflower residues were characterized by SEM and point of zero charge. The pH at the point of zero charge indicates that above or below that value the net surface charge of the adsorbent is predominantly negative or positive, respectively. Gas adsorption and SEM studies have revealed that peach stone-derived AC presents different pore features and that sunflower leaves/stalks have an agglomerated shape with a porous, heterogeneous, and uneven structure [54,89].

Another, less conventional method is Vibrating-sample magnetometry, which involves the vibration of magnetic materials operating under Faraday's Law, using sensing coils to detect the difference in the voltage variation in proportion to the magnetic moment of the sample [97]. A magnetic peanut husk adsorbent was characterized with this technique, allowing the observation of its properties before and after phosphate adsorption and its recovery using a magnet for practical applications [82]. Other analytical techniques include Zeta potential, X-ray photoelectron spectroscopy, and X-ray absorption spectroscopy. The first technique was employed to determine the electrostatic affinity for Pb(II) and the dye reactive blue 114 on the surface of magnetic coffee waste and pomelo peel sorbents [28,83], whereas X-ray photoelectron spectroscopy, and X-ray absorption spectroscopy were used to study the iodine form, oxidation state, and structure of adsorbents derived from mangosteen pericarps, passion fruit rinds, red onion peels, and red dragon fruit peels [79].

Brunauer–Emmett–Teller and Barrett–Joyner–Halenda (BET/BJH) are techniques employed to estimate certain surface characteristics of bioadsorbents such as area, pore size, and volume [45]. BET uses the equilibrium adsorption isotherm at the adsorbate boiling point, whereas BJH uses an inverse relation between pore radius and multilayer thickness [98]. BJH was used to determine the micropore and mesopore volume of pineapple waste-based AC, obtaining values of 289 $m^2/g$ and 253 $m^2/g$ respectively. Both types of pores enlarge the surface area, resulting in better adsorption because ion interactions occur more smoothly [48]. BET permits before and after comparisons of any modifications, aiding understanding of ion integration in the adsorption process; for example, the reduction of the surface area by the element zirconium in a peanut husk adsorbent [82]. This method was chosen to characterize a banana peel adsorbent with a surface area of 13 $m^2/g$, which indicates a small pore size [60]. AC derived from coconut buttons and rice straw biosorbents were characterized by BET; the former had a surface area of 479 $m^2/g$, evidencing a wider pore size distribution and better adsorption capacity, whereas the rice straw presented a specific surface area of 1.95 $m^2/g$ and average pore size of 12.1 nm [16,37]. As can be seen in Tables 1 and 2, the majority of adsorbents and raw materials have been analyzed by BET and/or BJH, including *Acacia erioloba* seeds, *Aegle marmelos* shell, *Argania spinosa* tree nut shells, cacao shells, siriguela seeds, peanut shell, black tea waste, spent and granular coffee, and papaya, orange, and pomelo peels. These have been additionally characterized using X-ray diffraction, thermogravimetric analysis, differential thermal analysis and thermogravimetry, SEM, field emission scanning electron microscopy, energy-dispersive X-ray, back-scattered laser light diffraction, Raman spectroscopy, electron spin resonance spectroscopy, small-angle X-ray scattering, and the Horvath–Kawazoe method.

SEM is another common characterization technique, and allows identification of the morphology of the bioadsorbents before and after adsorption. It is useful for the simultaneous appreciation of the number and size of pores, as well as the verification of whether a fibrous structure is replaced by a smooth one [50]. An improvement on this method is ultra-high-resolution SEM, which is suitable for the visualization of the surface texture of spent coffee by scanning it with a high-energy beam of electrons with higher sensitivity [38,99]. SEM is usually coupled with EDX, as in an analysis performed on mangosteen pericarp adsorbent where the changes in composition and anthocyanin quantity determined the removal efficiency of radioactive iodine [79]. Amer et al. [16] characterized the surface area, pore volume, size, and chemical composition of rice straw with a particle size between 76–150 μm using EDX before and after lead adsorption; EDX showed that the presence of organic elements such as Na, Ca, and Mg decreased after heavy metal adsorption, which promoted effective ion exchange and retention of the contaminant. Many other adsorbent materials, including cacao pod husk, palm tree leaves, rubber pod husk, banana peel, barley straw, cauliflower leaves, coffee husk, potato leaves and stem, rice husk, SCB, tamarind seeds, tea mixed waste, and watermelon rinds have been analyzed with SEM-EDX or only SEM (Tables 1 and 2).

## 5. Water Treatment Employing Bioadsorbents

### 5.1. Bioadsorbents Based on Grains and Seeds

#### 5.1.1. Nutrients and Heavy Metals

Well-known contaminants in water that need to be continuously monitored include the nutrients nitrogen and phosphorus and the various heavy metals. Aryee et al. [82] studied a bioadsorbent based on peanut husk functionalized with zirconium, iminodiacetic acid, and $Fe_3O_4$ that was able to remove up to 88.5% of phosphates in six hours at a temperature of 30 °C and pH 3. The addition of zirconium was crucial due to its interaction with hydroxyl, amines, and carboxyl groups; this enhanced $PO_4^{3-}$ sequestration at the maximum sorption capacity of 13.2 mg/g. Rubber pod husk is another byproduct without commercial value that can serve as an adsorbent to remove phosphates dissolved in water (0.1–0.5 g/L), showing favorable results ($q_m$ = 39.9 mg/g) after six hours (contact time) at pH 7 and 29 °C [50]. Raw tamarind seeds are a residue of interest due to their high content of lignocellulose and surface groups; these change under chemical ($H_2SO_4$) and physical (ultrasonic waves) treatment, resulting in $q_m$ values around 18 mg/g and a 50% reduction in the reaction time to adsorb Pb(II) [32]. Ultrasound-modified residues present a large surface area with large pores and cavities. Additionally, variations in pH charge the hydroxyl groups, increasing their affinity for positive ions and achieving 99.5% removal of Pb(II) at pH 6 and 30 °C [32]. Edathil et al. [28] analyzed Pb(II) adsorption by employing a modified magnetic-coffee membrane. The assays of sorption capacity showed better results with 0.025 g of the nanocomposite at 25 °C and pH 7.3, removing 99.5% of the contaminant in 30 min. On the other hand, the non-activated exhausted coffee was able to remove 97.6% of Hg(II), adjusting the adsorbent and mercury concentrations to 0.4 and 77.9 mg/L, respectively, with a contact time of 192.4 min at 33 °C and pH 7 [55].

Another metal that contaminates water is silver. According to Araújo et al. [56], it is possible to remove up to 99% ($q_m$ = 23.13 mg/g) of silver using moringa powder thanks to its high protein content which, depending on pH, provides a high amount of charged amine and carboxyl groups. The optimal conditions for Ag retention were obtained using an adsorbent with a <500 mm particle size and 20 min of contact time at pH 6.5. Feizi and Jalali [89] demonstrated the feasible removal of Fe, Mn, Zn, Ni, Cu, and Cd from aqueous solutions by implementing raw and modified walnut shells. In general, the maximum sorption capacity for the six metals was 44.1 ± 18.5 mg/g, while a removal efficiency of 96.5% was observed only for Mn at pH 8.

Table 3 summarizes the phosphorus and heavy metals removal efficiencies previously described and the isothermal models that fitted the experimental data; the Langmuir, Freundlich, and pseudo-second order models best explained the adsorption behavior in the biomaterials.

**Table 3.** Adsorption parameters and removal efficiencies of different biosorbents derived from grains and seeds for the removal of pollutants in aqueous matrices.

| Biowaste | Contaminant | Isothermal Model Adjusted | Kinetic Model Adjusted | $q_m/q_e$ | Removal Efficiency (%) | Ref. |
|---|---|---|---|---|---|---|
| *Argania Spinosa* tree nutshells ($TiO_2$ composite AC) | DFC<br>CBZ<br>SMX | Langmuir<br>Langmuir<br>Langmuir | Pseudo-first order<br>Pseudo-first order<br>Pseudo-first order | 153.8 mg/g<br>105.3 mg/g<br>125.0 mg/g | 100.00<br>85.00<br>67.00 | [52] |
| Cacao pod husk | SD | Freundlich | Pseudo-second order | 5.53 mg/g | 93.6 | [51] |
| Cacao shell ($H_3PO_4$-activated) | α-Lac<br>BSA | Toth<br>Langmuir | Pseudo-second order<br>Pseudo-second order | 179.91 mg/g<br>41.02 mg/g | 91.67<br>21.25 | [45] |
| Cacao shell ($ZnCl_2$-activated) | α-Lac<br>BSA | Toth<br>Langmuir | Pseudo-second order<br>Pseudo-second order | 141.69 mg/g<br>147.84 mg/g | 70.33<br>86.86 | [45] |
| Coffee husk | NFX | Langmuir and Redlich-Peterson | Pseudo-second order | 33.56 mg/g | 99.66 | [71] |
| Exhausted coffee | Hg(II) | Langmuir | Pseudo-second order | 31.75 mg/g | 97.69 | [55] |
| Magnetic coffee waste ($Fe_3O_4$ NPs) | Pb(II) | Langmuir | Pseudo-second order | 48.78 mg/g | 99.56 | [28] |
| *Moringa oleifera* seeds | Ag(I)<br>RR-120 | Langmuir<br>Langmuir and Freundlich | N.R<br>Pseudo-second order | 23.13 mg/g<br>413.32 mg/g | 99.00<br>N.R | [56]<br>[57] |
| Peanut husk (Magnetic husk) | Phosphates | Freundlich | Elovich | 13.2 mg/g | 88.5 | [82] |
| Peanut shell | AY-36 | Langmuir and Freundlich | Pseudo-second order | 66.7 mg/g | 98.55 | [42] |
| Raw tamarind seeds | Pb(II) | Langmuir | Pseudo-first order | 16.0 mg/g | 99.52 | [32] |
| Rice husk | NFX | Langmuir and Redlich-Peterson | Pseudo-second order | 20.12 mg/g | 96.95 | [71] |
| Rubber pod husk | Phosphates | Freundlich | Pseudo-second order | 39.98 mg/g | N.R | [50] |
| Siriguela seeds ($ZnCl_2$-activated) | α-Lac<br>BSA | Toth<br>Langmuir | Pseudo-second order<br>Pseudo-second order | 173.05 mg/g<br>188.29 mg/g | 87.42<br>92.29 | [45] |
| Siriguela seeds ($H_3PO_4$-activated) | α-Lac<br>BSA | Toth<br>Langmuir | Pseudo-second order<br>Pseudo-second order | 193.54 mg/g<br>130.31 mg/g | 96.67<br>81.67 | [45] |
| Spent coffee ground (Calcium-alginate beads) | AO7<br>MB | Sips<br>Sips | Pore diffusion<br>Pore diffusion | 665.9 mg/g<br>986.8 mg/g | 99.90<br>100 | [44] |
| Spent coffee waste biochar (NaOH-activated) | $NPX^{LK}$<br>$DCF^{LK}$<br>$IBU^{LK}$<br>$NPX^{WW}$<br>$DCF^{WW}$<br>$IBU^{WW}$ | Langmuir<br>Langmuir<br>Langmuir<br>Langmuir<br>Langmuir<br>Langmuir | Pseudo-second order<br>Pseudo-second order<br>Pseudo-second order<br>Pseudo-second order<br>Pseudo-second order<br>Pseudo-second order | 269.01 μmol/g<br>97.17 μmol/g<br>76.10 μmol/g<br>263.34 μmol/g<br>97.12 μmol/g<br>74.07 μmol/g | 30.7–97.1 | [38] |
| Spent coffee waste biochar (Pristine-activated) | $NPX^{LK}$<br>$DCF^{LK}$<br>$IBU^{LK}$<br>$NPX^{WW}$<br>$DCF^{WW}$<br>$IBU^{WW}$ | Freundlich<br>Freundlich<br>Freundlich<br>Freundlich<br>Freundlich<br>Freundlich | Pseudo-second order<br>Pseudo-second order<br>Pseudo-second order<br>Pseudo-second order<br>Pseudo-second order<br>Pseudo-second order | 107.53 μmol/g<br>91.74 μmol/g<br>86.21 μmol/g<br>344.48 μmol/g<br>202.92 μmol/g<br>124.14 μmol/g | 7.5–10.3 | [38] |
| Sulfonated coffee waste | SMX<br>BPA | Langmuir and Temkin<br>Temkin | Pseudo-second order<br>Pseudo-second order | 256 mg/g<br>271 mg/g | N.R<br>N.R | [100] |
| Sunflower seed hulls | MB<br>AB-15 | Langmuir<br>Langmuir | Pseudo-second order<br>Pseudo-second order | 473.44 mg/g<br>430.37 mg/g | N.R<br>N.R | [31] |
| Tamarind seeds ($H_2SO_4$-modified) | Pb(II) | Langmuir | Pseudo-first order | 18.34 mg/g | 99.52 | [32] |

**Table 3.** *Cont.*

| Biowaste | Contaminant | Isothermal Model Adjusted | Kinetic Model Adjusted | $q_m/q_e$ | Removal Efficiency (%) | Ref. |
|---|---|---|---|---|---|---|
| Tamarind seeds (Ultrasound-modified) | Pb(II) | Langmuir | Pseudo-first order | 18.86 mg/g | 99.52 | [32] |
| Walnut shell | Mn(II) | Langmuir | Pseudo-second order | 28.6 mg/g | 96.5 | [89] |
| | Zn | Langmuir | Pseudo-second order | 33.3 mg/g | N.R | |
| | Fe | Langmuir | Pseudo-second order | 62.6 mg/g | N.R | |
| | Cd | Langmuir | Pseudo-second order | 76.9 mg/g | N.R | |
| | Cu | Langmuir | Pseudo-second order | 38.8 mg/g | N.R | |
| | Ni | Langmuir | Pseudo-second order | 29.4 mg/g | N.R | |
| Walnut shell (NaOH-modified) | Ni | Langmuir | Pseudo-second order | 38.9 mg/g | N.R | [89] |
| Waste coffee grounds ($CO_2$-activated carbon) | MB | Langmuir | Pseudo-second order | 678 mg/g | N.R | [101] |
| | MO | Langmuir | Pseudo-second order | 612 mg/g | N.R | |

N.R, not reported; $q_m$, Maximum sorption capacity; $q_e$, Adsorption capacity at equilibrium; DFC, Diclofenac; CBZ, Carbamazepine; SMX, Sulfamethoxazole; SD, Sodium diclofenac; α-Lac, α-Lactalbumin; BSA, Bovine serum albumin; NFX; Norfloxacin; RR-120, Reactive red 120; AY-36, Metanil yellow; AO7, Acid orange 7; MB, Methylene blue; NPXLK, Naproxen lake water; DCFLK, Diclofenac lake water; IBULK, Ibuprofen lake water; NPXWW, Naproxen wastewater; DCFWW, Diclofenac wastewater; IBUWW, Ibuprofen wastewater; BPA, Bisphenol A; AB-15, Acid blue 15; MO, Methyl orange; $Fe_3O_4$ NPs, $Fe_3O_4$ nanoparticles.

### 5.1.2. Industrial Contaminants

Although a wide variety of industrial contaminants exist in water sources, the main substances that have been removed by employing adsorbents derived from grains and seeds are dyes, bisphenol A, and industrial proteins such as BSA and α-Lac (Table 3). Garg et al. [42] used peanut shell AC to adsorb the azo dye metanil yellow in a spontaneous and endothermic process at 35 °C by adding biochar at an adsorbent/contaminant ratio of 20:1 at pH 2. The chemical association of the dye molecules with the nitrogen and sulfur groups present in the adsorbent material contributed to the highest removal efficiency (98.55%). The Langmuir, Freundlich, and pseudo-second kinetics models were chosen to represent the adsorption dynamics. In the same year, Çelekli et al. [57] reported interaction between moringa residues (amino, carbonyl, and amide groups) and reactive red 120 dye, reaching the maximum adsorption equilibrium of 413.32 mg/g (Langmuir isotherm) in 30 min at 50 °C and pH 1. Thermodynamic analysis revealed that the adsorption process was spontaneous. Other organic dyes such as MB and methyl orange have been retained with high adsorption capacity ($q_e$ = 678 mg/g and 612 mg/g, respectively) in biochar from waste coffee grounds activated by $CO_2$ [101] or pyrolysis to remove acid orange 7 ($q_m$ = 665.9 mg/g) [44]. Interestingly, this last material can be recycled and used for at least seven cycles at a temperature of 20 °C. Foo and Hameed [31] modeled acid blue 15 and MB adsorption behavior in a sunflower seed hull-based bioadsorbent, obtaining the best result by implementing the Langmuir model ($q_m$ = 430.3 mg/g and 473.4 mg/g respectively), and suggested that the adsorbed layer on the AC is only one molecule thick. It is worth mentioning the advantage of the capacity to simultaneously retain anionic (acid blue 15) and cationic (MB) dyes at 30 °C.

The high toxicity of the chemical substances used in the synthesis of plastics and resins (e.g., bisphenol A) has drawn the attention of researchers seeking to eliminate them from water matrices; one alternative is an adsorbent based on coffee waste. Although no removal efficiencies have been reported, sulfonated coffee waste showed a biosorption capacity of 271 mg/g for bisphenol A at room temperature [100]. Pereira et al. [45] studied the capture of BSA and α-Lac by employing AC based on siriguela seeds and cacao shells. Although the adsorption of both whey proteins was evaluated under the same conditions (200 mg/L protein solution, 5 mg AC, room temperature, pH 7) the retention capacity remained constant over different times (16 h and 3 h, respectively), presenting removal efficiencies that varied between 70.3–96.6% for α-Lac and 21.2–92.2% for BSA. The Temkin, Freundlich, Toth, and Langmuir models most efficiently represented the adsorption data for coffee waste, siriguela seeds, and cacao shells.

#### 5.1.3. Pharmaceuticals

The removal of the antibiotic norfloxacin in real water samples was tested using rice and coffee husks with similar conditions and particle sizes [71]. The results showed that the maximum removal was achieved in distilled water (96.9 and 99.6%) after 1 h at pH 6.2, rather than in municipal samples (81.4 and 95.6%). Particularly, the highest $q_m$ value was observed for coffee husk (33.56 vs. 20.12 mg/g). Another pharmaceutical recovered from the water with an alternative biosorbent was sulfamethoxazole, which is commonly used to treat bacterial infections. The biological adsorbent consisted of carbonized and sulfonated coffee wastes, and although no removal efficiency was reported it exhibited higher sulfamethoxazole biosorption capacity (256 mg/g) at room temperature [100]. Spent coffee biochar (NaOH-activated) displayed efficient interactions with naproxen ($q_m$ = 263.3 and 269 μmol/g), diclofenac ($q_m$ = 97.1 μmol/g), and ibuprofen ($q_m$ = 74.0 and 76.1 μmol/g) in lake water and wastewater effluent at pH 7 and 25 °C [38]. The pristine activation of this biomaterial improved the adsorptive removal of naproxen (344.4 μmol/g) and diclofenac (202.9 μmol/g) in wastewater, as well as ibuprofen in lake water (86.2 μmol/g) and wastewater (124.1 μmol/g). The complete recovery (100%) of diclofenac through a $TiO_2$ composite obtained from *Argania Spinosa* tree nut shells was found to follow the Langmuir and pseudo-first order models for 50 mg/L analgesic and 0.1 g/L adsorbent at 25 °C [52]. In this study, carbamazepine and sulfamethoxazole were taken up using the same biomaterial with lower removal efficiencies (85 and 67% respectively) and adsorption capacities (105.3 and 125 mg/g respectively) under the same experimental conditions. Sodium diclofenac was retained using AC from cacao pod husk, with an efficiency of 76%; this value increased to 93.6% when decreasing biochar dosage (0.25 g) and increasing adsorbate concentration at neutral pH [51]. As shown in Table 3, the adsorption kinetics for these pharmaceuticals followed the pseudo-second order reactions and other isotherm models, as well as Langmuir.

### *5.2. Bioadsorbents Based on Fruits and Vegetables*

#### 5.2.1. Heavy Metals

Biosorbents have proven to be a cheap and effective alternative for the removal of heavy metals in polluted water thanks to their high capacity and selectivity for metal ions [102]. Removal efficiencies for Cu(II) and other heavy metals employing adsorbents derived from fruits and vegetables are presented in Table 4, including the most suitable kinetic and isothermal models. Bhattacharjee et al. [3] reviewed watermelon rinds as a copper adsorbent, reporting removal efficiencies between 58.4% and 88%. An increase in adsorption capacity was observed as the temperature approached the optimal drying temperature (120 °C), which increased the presence of active sites. Another fruit-based adsorbent that showed a high removal percentage of copper ions (78.8%) was pomegranate peel [103]; four models were considered for the analysis of its adsorption behavior, Langmuir, Freundlich, Dubinin–Radushkevic, and Temkin. These indicated the maximum pomegranate peel adsorption capacity at 30.12 mg/g; the heavy metal was indirectly adsorbed as a mono- or multilayer on a heterogeneous and porous surface [103]. AC derived from coconut buttons efficiently retained Cu(II), Pb(III), and Hg(II), with adsorption capacities (Q°) from 76.3 to 97.7 mg/g under batch conditions at two pH values (6 and 7) [37]. Kumar et al. [104] obtained removal percentages of 99.7% for Cu(II) from industrial wastewater by implementing raw corn cob at pH 5 and 40 °C with a contact time of 60 min. The maximum adsorption capacity was 6.24 mg/g (Langmuir isotherm) following pseudo-second order kinetics.

**Table 4.** Adsorption parameters and removal efficiencies of different biosorbents derived from fruit and vegetables for the removal of pollutants in aqueous matrices.

| Biowaste | Contaminant | Isothermal Model Adjusted | Kinetic Model Adjusted | $q_m/Q^\circ$ | Removal Efficiency (%) | Ref. |
|---|---|---|---|---|---|---|
| *Aegle marmelos* shell | IBU | Freundlich | Pseudo-second order | 5 mg/g | 90 | [53] |
| *Aegle marmelos* shell (Steam-activated) | IBU | Langmuir | Pseudo-second order | 12.65 mg/g | 95 | [53] |
| Banana peel | MO<br>MB<br>RB<br>CR<br>MV<br>AB-10B<br>Fluoride | Freundlich<br>Freundlich<br>Freundlich<br>Freundlich<br>Freundlich<br>Freundlich<br>N.R | Pseudo-first order<br>Pseudo-first order<br>Pseudo-first order<br>Pseudo-first order<br>Pseudo-first order<br>Pseudo-first order<br>N.R | 17.2 mg/g<br>15.9 mg/g<br>13.2 mg/g<br>11.2 mg/g<br>7.9 mg/g<br>7.9 mg/g<br>8.15 mg/g | N.R<br>N.R<br>N.R<br>N.R<br>N.R<br>N.R<br>86.5 | [10,67,68] |
| Banana peel (MeOH/HCl-treated) | Cd(II)<br>Cr(VI) | Langmuir<br>Langmuir and Dubinin-Radushkevic | Pseudo-first order<br>Pseudo-first order | 35.52 mg/g<br>131.56 mg/g | 95<br>98 | [60,61] |
| Beet Pulp (NaOH-treated) | Mn(II) | N.R | Pseudo-second order | N.R | 86.36 | [88] |
| Coconut buttons ($H_2SO_4$-treated) | Pb(II)<br>Hg(II)<br>Cu(II) | Freundlich<br>Freundlich<br>Freundlich | Pseudo-second order<br>Pseudo-second order<br>Pseudo-second order | 97.72 mg/g<br>78.84 mg/g<br>73.6 mg/g | 98.7<br>95.8<br>90.6 | [37] |
| Coconut coir pith | Co(II)<br>Cr(III)<br>Ni(II)<br>RB<br>AV<br>CR | Langmuir<br>Langmuir<br>Langmuir<br>Langmuir and Freundlich<br>Langmuir and Freundlich<br>Langmuir | Pseudo-second order<br>Pseudo-second order<br>Pseudo-second order<br>Pseudo-first order<br>Pseudo-first order<br>Pseudo-second order | 12.82 mg/g<br>11.56 mg/g<br>15.95 mg/g<br>2.56 mg/g<br>8.06 mg/g<br>6.72 mg/g | 49.64–85.4<br>46.08–98.2<br>62.88–98<br>43.6–79.4<br>47.0–78.7<br>30.5–66.5 | [10,46,47,59] |
| Coconut coir | MB | Langmuir and Temkin | Pseudo-second order | 29.5 mg/g | 99.5 | [72] |
| Corn cob | Mn(II)<br>Cu(II)<br>Cu<br>Cd<br>Ni<br>Pb | Langmuir<br>Langmuir<br>N.R<br>N.R<br>N.R<br>N.R | Pseudo-second order<br>Pseudo-second order<br>N.R<br>N.R<br>N.R<br>N.R | 6.24 mg/g<br>6.24 mg/g<br>0.034 mmol/g<br>0.096 mmol/g<br>0.097 mmol/g<br>0.019 mmol/g | 99.8<br>99.7<br>22<br>58<br>60<br>12 | [73,104] |
| Durian peel (HCl-treated) | AG25 | Langmuir | Pseudo-second order | 63.29 mg/g | N.R | [74] |
| Grapefruit peel | CV | Langmuir | Pseudo-second order | 254.16 mg/g | 96 | [75] |
| Jackfruit peel | MB | Type 2 Langmuir | Pseudo-second order | 285.71 mg/g | 58.2–89.8 | [76] |
| Kiwi peel (NaOH-treated) | Cd(II)<br>Cr(III)<br>Zn(II) | Langmuir<br>Langmuir<br>Langmuir | Pseudo-second order<br>Pseudo-second order<br>Pseudo-second order | 15.87 mg/g<br>41.66 mg/g<br>37.03 mg/g | 78<br>98<br>57 | [77,94] |
| Mango Peel | Cd(II)<br>Pb(II) | Langmuir<br>Langmuir | Pseudo-second order<br>Pseudo-second order | 68.92 mg/g<br>99.05 mg/g | 90.56<br>92.5 | [10,78] |
| Mangosteen pericarps (ABR) | Iodide ($I^-$) | Langmuir | Pseudo-second order | 79.4 mg/g | 100 | [79] |
| Olive waste ($H_3PO_4$-activated) | IBU<br>NPX<br>KTP<br>DCF | Langmuir<br>Langmuir<br>Langmuir<br>Langmuir | Pseudo-second order<br>Pseudo-second order<br>Pseudo-second order<br>Pseudo-second order | 12.6 mg/g<br>39.5 mg/g<br>24.7 mg/g<br>56.2 mg/g | 79<br>95<br>90<br>96 | [105] |
| Orange peel | V-17<br>CR<br>PO<br>RB<br>DR23<br>DR80 | Langmuir and Freundlich<br>Langmuir and Freundlich<br>Langmuir and Freundlich<br>Langmuir and Freundlich<br>Langmuir<br>Langmuir | N.R<br>Pseudo-first order<br>Pseudo-first order<br>Pseudo-first order<br>Pseudo-second order<br>Pseudo-second order | 19.88 mg/g<br>22.44 mg/g<br>1.3 mg/g<br>3.23 mg/g<br>10.72 mg/g<br>21.05 mg/g | 87<br>76.6<br>49<br>59.0–67.5<br>92<br>91 | [10,80,81,95] |
| Papaya Peel ($H_3PO_4$-activated) | Pb(II) | Langmuir | Pseudo-second order | 38.31 mg/g | 93.22 | [39] |
| Passion fruit rinds (ABR) | Iodide ($I^-$) | Langmuir | Pseudo-second order | 6.67 mg/g | 87.5 | [79] |
| Peach stones ($K_2CO_3$-activated) | ACE | Langmuir | Pseudo-second order | 204 mg/g | 82 | [54] |

**Table 4.** *Cont.*

| Biowaste | Contaminant | Isothermal Model Adjusted | Kinetic Model Adjusted | $q_m/Q^{\circ}$ | Removal Efficiency (%) | Ref. |
|---|---|---|---|---|---|---|
| Pineapple waste ($ZnCl_2$-activated) | MB | Langmuir | N.R | 288.34 mg/g | 67–76 | [48] |
| Pomegranate peel | Cu(II) | Langmuir, Freundlich, Dubinin Radushkevich and Temkin | Pseudo-second order | 30.12 mg/g | 78.85 | [103] |
| Pomelo peel | MB<br>RB-114 | Langmuir and Temkin<br>Langmuir | Pseudo-second order<br>Pseudo-second order | 133 mg/g<br>16.3 mg/g | 83<br>89 | [1,83,96] |
| Red dragon fruit peels (ABR) | Iodide ($I^-$) | Freundlich | Pseudo-second order | 68.6 mg/g | 68.4 | [79] |
| Red onion peels (ABR) | Iodide ($I^-$) | Langmuir | Pseudo-second order | 75.8 mg/g | 92 | [79] |
| Rice husk and fruit juice residue | Phosphate | Freundlich | Pseudo-first order | 13.89 mg/g | 95.85 | [49] |
| Tangerine peel (NaOH-treated) | Cd(II)<br>Cr(III)<br>Zn(II) | Langmuir<br>Langmuir<br>Langmuir | Pseudo-second order<br>Pseudo-second order<br>Pseudo-second order | 17.54 mg/g<br>47.61 mg/g<br>38.41 mg/g | 73<br>96<br>52 | [77,94] |
| Watermelon rinds | Cu(II)<br>Ni(II)<br>Zn(II)<br>Pb(II)<br>Cd(II)<br>Cr(III)<br>Tl(I)<br>As(III)<br>As(V)<br>Fe(II)<br>Mn(II)<br>Co(II)<br>MB<br>BG<br>RBBR<br>CR<br>BR2<br>OG | Langmuir<br>N.R<br>N.R<br>Langmuir, Thomas and Yoon–Nelson<br>Langmuir<br>Langmuir<br>Langmuir and Temkin<br>Langmuir<br>Langmuir<br>N.R<br>N.R<br>N.R<br>Langmuir<br>Langmuir<br>Freundlich<br>Langmuir<br>Extended Langmuir<br>Extended Langmuir | Pseudo-second order<br>N.R<br>N.R<br>Pseudo-second order<br>Pseudo-second order<br>Pseudo-second order<br>Pseudo-second order<br>Pseudo-second order<br>Pseudo-second order<br>N.R<br>N.R<br>N.R<br>Pseudo-second order<br>Pseudo-second order<br>Pseudo-second order<br>Pseudo-second order<br>Pseudo-first order<br>Pseudo-first order | 5.7–111.1 mg/g<br>18.4–38.9 mg/g<br>22.5 mg/g<br>19.33–116.2 mg/g<br>40.16–63.29 mg/g<br>172.6 mg/g<br>178.4–1123 mg/g<br>NR<br>NR<br>4.98 mg/g<br>1.37 mg/g<br>23.3 mg/g<br>188.6–489.8 mg/g<br>188.6 mg/g<br>333.33 mg/g<br>17 mg/g<br>125.79 mg/g<br>27.24 mg/g | 58.4–88<br>69–70<br>52.4–90.3<br>72–99.9<br>80<br>91<br>98–98.5<br>99<br>98<br>98.3<br>98.9<br>57<br>83–99<br>98<br>92–97<br>101.45<br>75<br>85 | [3] |

N.R, not reported; IBU, Ibuprofen; MO, Methyl orange; MB, Methylene blue; RB, Rhodamine B; CR, Congo red; MV, Methyl violet; AB-10B, Amido black 10B; AV, Acid violet; AG25, Acid green 25; CV, Crystal violet; NPX, Naproxen; KTP, Ketoprofen; DCF, Diclofenac; V-17, Violet 17; PO, Procion orange; DR23, Direct Red 23; DR80, Direct red 80; ACE, Acetaminophen; RB-114; Reactive blue 114; BG, Brilliant green; RBBR, Remazol brilliant blue reactive; BR2, Basic red 2; OG, Orange G; ABR, Anthocyanin based residues.

Cr(III) and Pb(II) are other metals that commonly appear in water matrix analyses. Cr(III) was removed by watermelon rinds and coconut residues with an efficiency and maximum adsorption capacity up to 98.2% and 172.6 mg/g, respectively, under acidic conditions (pH 3) [3,59]. Kiwi and tangerine peels exhibited notable adsorbing efficiencies (98% and 96% respectively) for Cr(III), reaching equilibrium in 60 min at pH 6.0 [77,94]. As for Cr(VI), banana peels adsorbed it following the Langmuir isotherm with a $q_m$ of 131.56 mg/g after 30 min of contact time [61]. Watermelon rinds sequestered Pb(II), reaching removal percentages between 72% and 99.9% within a temperature range of 20–30 °C [3]. When an acidic and alkaline treatment was carried out on the watermelon rinds it was possible to capture from 52.4 to 90.3% of Zn(II); although the lowest removal percentages were observed in the presence of other metals in solution, this demonstrated that the simultaneous bioadsorption of multiple pollutants is possible. Similarly, tangerine and kiwi peels achieved a Zn(II) adsorption efficiency of approximately 52% and 57%, respectively, at pH 6.0, respective with $q_m$ values of 38.4 and 37.0 mg/g [77]. After identifying the optimal adsorption conditions, papaya peels displayed a removal efficiency of 93.2% for Pb(II), and according to the Langmuir and kinetic models the adsorptive capacity ranged from 38.31 to 42.5 mg/g [39]. Another fruit peel that presented good removal efficiency of lead after chemical modification was mango peel waste, retaining 76.2% of this metal rapidly (30–60 min) with a $q_m$ of 99.05 mg/g based on the Langmuir isotherm model [10].

Many fruit residues have been reported to have a certain affinity for other heavy metals that may represent a hazard for human beings. The ability of watermelon rinds, mango peel waste, and banana, kiwi, and tangerine peel residues to retain Cd(II) from water sources was studied; the respective removal efficiencies obtained were 80%, 90.5%, 95%, 78%, and 73% [3,10,60,61,94]. The Langmuir isotherm and pseudo-first and second-order kinetic models were best suited for describing Cd uptake by these biomaterials. Šćiban, et al. [73] reported similar adsorption capacities for cadmium and nickel (0.096 and 0.097 mmol/g respectively) using raw corn cob as biosorbent, while coconut waste presented a maximum Ni(II) adsorption of 15.95 mg/g at pH 5.3 [59]. Raw corn cob was studied by Kumar et al. [104] for the almost complete reduction of Mn(II) (99.8%) in industrial wastewater after 60 min at pH 5.0 and 40 °C. The maximum adsorption capacity was 6.24 mg/g following the pseudo-second order kinetics. This removal efficiency was the highest when compared to that obtained with watermelon rinds (98.9%) or beet pulp (86.3%) at an adsorbent concentration of 1 g/L and pH 6 [3,88].

Coconut waste represented an adequate option for the removal of Co(II) from water, showing a maximum adsorption capacity of 12.82 mg/g at pH 4.3 [59]. The kinetic and equilibrium parameters were well-explained by the Langmuir model, which demonstrated the presence of stable chemical bonds between the molecules across the surface area. Recently, Bhattacharjee et al. [3] used watermelon rinds as an adsorbent material in water sources and reported removal efficiencies for Co(II) (57%), Tl(I) (98–98.5%), As(III) (99%), As(V) (98%), Ni(II) (69–70%), and Fe(II) (98.3%).

#### 5.2.2. Industrial Contaminants

Removing dyes from aqueous solutions can be a difficult task with conventional treatment methods. However, the bioadsorbent application of fruit biowastes as AC or raw material has been studied for dye removal (Table 4). For instance, watermelon rind-based AC adsorbed brilliant green, remazol brilliant blue reactive, basic red 2, and orange G, showing removal efficiencies of 98, 94.5 ± 2.5, 75 and 85%, respectively, at 30 °C and $q_m$ values in the range of 27.4–333.3 mg/g [3]. It is important to mention that the dye adsorption was both endothermic (remazol brilliant blue reactive and basic red 2) and exothermic (brilliant green and orange G). In the same study a dominance of ion-exchange sorption towards MB was present, with maximum capacity values of 188.6–489.8 mg/g and removal efficiencies from 83% to 98%. The models that best described the kinetic and equilibrium data were pseudo-second order and Langmuir isotherm. Additionally, coconut coir powder without chemical treatment exhibited a monolayer adsorption capacity of 29.50 mg/g for MB under optimal conditions (30 °C and pH 6.0), achieving maximum removal efficiency when the adsorbent dosage increased from 0.05 to 0.20 g [72]. In this case, the authors determined that the Temkin isotherm model was adequate to describe the adsorption behavior, as it considers that the adsorption heat of the molecules decreases linearly along the surface by adsorbent–adsorbate interactions. Depending on the initial adsorbent concentration, after 180 min of contact time the removal efficiency for MB using jackfruit peels without pre-treatment varied from 58.2 to 89.8% with a maximum adsorption capacity of 285.71 mg/g [74]. AC from pineapple waste was identified as another efficient bioadsorbent for MB recovery, presenting an outstanding adsorption capacity (288.34 mg/g) and removal efficiencies from 67 to 76%, according to the Langmuir isotherm [48]. Other biowastes of interest include pomelo and banana peels; the former exhibited a maximum MB removal efficiency of 83% with a $q_m$ of 133 mg/g, while the latter reached a maximum adsorption capacity of 15.9 mg/g at the same temperature (30 °C) and different pH (8 and 6, respectively) [67,96]. Pomelo peels without any chemical treatment were studied by Argun et al. [83], retaining antraquinonic dye reactive blue 114 from aqueous solutions; adsorption reached equilibrium (16.3 mg/g) within a 90 min reaction time, with a maximum removal efficiency of 96%.

Congo red is another widely-used dye with carcinogenic properties that needs to be monitored and removed from water sources. Bhattacharjee et al. [3] reported that AC from watermelon rinds adsorbed it at a capacity of up to 17 mg/g, with efficiency above 100% when using sonication and modification with $TiO_2$ nanoparticles. When the adsorption was assisted by the ultrasound method, the acoustic waves caused the water to create gas bubbles or cavities, increasing the surface area and facilitating contact between the dye and the biosorbent [106]. Coir pith presented notable removal percentages of congo red under acidic conditions, varying from 30.5 to 66.5% depending on the dye concentration [46]. This bioresidue is interesting for violet acid recovery, because after thermal treatment it exhibited up to 78.7% dye removal and attained equilibrium after the first 10 min of adsorption [46]. Congo red was removed using orange peel with a maximum removal efficiency reaching up to 76.6% at pH 5 and decaying to 49% at pH 12 [80]. Banana peel was evaluated for the adsorption of congo red, methyl orange, methyl violet, and amido black 10B at initial concentrations of 100 mg/g, achieving maximum adsorption capacities of 11.2, 17.2, and 7.9 mg/g, respectively [67]. The Langmuir and Freundlich isotherms were used to describe the adsorption behaviors, with the latter a better fit for the data.

Orange peels were tested as an effective bioadsorbent (87% removal) for violet 17 at pH 2 and showed the ability to recover more than 60% of the adsorbate with increasing pH [95]. Procion orange, direct red 23, and direct red 80 were effectively adsorbed by non-treated orange peels with a maximum removal efficiency of 49% [80], 92%, and 91% respectively, at initial dye concentrations of 50 mg/L [81]. Langmuir showed a better fit in most cases, indicating monolayer adsorption [10]. Rhodamine-B was adsorbed by orange peels as well, displaying a removal efficiency and $q_m$ of 63.5% and 3.2 mg/g, respectively, while calcined coir pith wastes removed 43.6–79.4% of this dye [47]. Additionally, a maximum adsorption capacity of 13.2 mg/g for rhodamine-B was achieved with banana peels at pH 7, following a Freundlich adsorption behavior [67].

Grapefruit peels were an excellent adsorbent for crystal violet, with higher desorption efficiency ($\leq$98%) after several cycles of regeneration with NaOH. Sorption equilibrium was reached within 60 min, resulting in 96% removal of the dye [75]. Durian peels showed an important adsorption capacity of 63.2 mg/g for acid green 25 at 30 °C; the good adjustment of the Langmuir model indicated that this biomaterial had equal adsorption activation energy for each dye molecule [74].

#### 5.2.3. Nutrients

Nutrients such as phosphate ions have been successfully recovered from aqueous solutions using an adsorbent derived from rice husk and lemon juice residues [49]. It displayed a removal efficiency of 95.8% at 298K and pH 6; the initial adsorbate concentration, adsorbent dose (5 and 3 g/L respectively), and contact time (three hours) were important to achieve the highest efficiency in the process. Banana peels were used as an environmentally friendly option for uptake of fluoride from groundwater, reaching a removal efficiency of 86.5% ($q_m$ = 8.15 mg/g) under optimal conditions for the sorbent [68]. Anthocyanin-based bioadsorbents were subsequently implemented to remove ions of radioactive iodine ($I^-$) from an aqueous environment. Mangosteen pericarps were the most efficient adsorbent, showing total removal of iodine (100%), followed by red onion peels (92%), passion fruit rinds (87.5%), and red dragon fruit peels (68.4%) after a contact period of 192 h [79]. The Freundlich, Langmuir, and pseudo-first and second order were the isothermal and kinetic models employed for the analysis of nutrients adsorption data (Table 4).

#### 5.2.4. Pharmaceuticals

Chakraborty et al. [53], among others (Table 4), evaluated fruit residues for the removal of common non-prescription drugs from water. AC from *Aegle marmelos* shells was subjected to steaming activation, achieving the highest removal efficiency of 95% for ibuprofen, while its raw form reached 90% under different pH and temperature conditions (15 °C and pH 3 vs. 20 °C and pH 2, respectively). In addition, AC produced from olive waste cakes followed the pseudo-second order kinetic model and Langmuir isotherm for the adsorption of ibuprofen, ketoprofen, naproxen, and diclofenac from surface water samples. The maximum removal of the four drugs was observed at pH 2.01, although this decreased to 90.4, 88.4, and 70.0% for naproxen, ketoprofen, and ibuprofen, respectively, when they were mixed in solution at 25 °C [105]. AC derived from peach stones was proposed by Cabrita et al. [54] for the adsorption of the analgesic acetaminophen from water, with an efficiency of 82% and $q_m$ of 204 mg/g at 30 °C while using only 10 mg of the adsorbent.

### 5.3. Bioadsorbents Based on Herbage and Forage

#### 5.3.1. Heavy Metals

Heavy metals have been removed from aqueous environments using adsorbents obtained from herbs and forage (Table 5). Ahmed et al. [88] investigated the potential of SCB for treating Mn(II) present in ground water, considering as optimum conditions an adsorbent dosage of 1.5 g and an initial heavy metal concentration of 2 mg/L at pH 6. Mn(II) was successfully removed with an efficiency of 62.5%, explained under a pseudo-second order kinetic model. Using a ten-fold quantity of raw SCB, it was possible to eliminate a higher concentration of this metal (12 mg/L) with similar efficiency [87]. Additional chemical pretreatment with hydrochloric acid had a significant positive effect on this adsorbent, increasing the removal to 99% due to the increase of carboxyl and hydroxyl groups on its surface. Tea and sunflower wastes were implemented for Mn(II) uptake, obtaining removal efficiencies of 95.5 and 81.6%, respectively, at pH 8 [89,92]. Particularly, the sunflower residue presented a higher maximum adsorption capacity (47.6 mg/g) than tea waste (0.15 mg/g); the Langmuir isotherm model was employed.

**Table 5.** Adsorption parameters and removal efficiencies of different biosorbents derived from herbage and forage for the removal of pollutants in aqueous matrices.

| Biowaste | Contaminant | Isothermal Model Adjusted | Kinetic Model Adjusted | $q_m/Q_e$ | Removal Efficiency (%) | Ref. |
|---|---|---|---|---|---|---|
| Barley straw | BBY<br>CV<br>MB<br>SO | Langmuir<br>Langmuir<br>Langmuir<br>Langmuir | Pseudo-second order<br>Pseudo-second order<br>Pseudo-second order<br>Pseudo-second order | 124.3 mg/g<br>95.8 mg/g<br>86.5 mg/g<br>99.7 mg/g | N.R<br>N.R<br>N.R<br>N.R | [69] |
| Barley straw (CA/NaOH-modified) | BBY<br>CV<br>MB<br>SO | Langmuir<br>Langmuir<br>Langmuir<br>Langmuir | Pseudo-second order<br>Pseudo-second order<br>Pseudo-second order<br>Pseudo-second order | 524.3 mg/g<br>473.2 mg/g<br>498.1 mg/g<br>296.6 mg/g | N.R<br>N.R<br>N.R<br>N.R | [69] |
| Barley straw (Magnetic) | BBY<br>CV<br>MB<br>SO | Langmuir<br>Langmuir<br>Langmuir<br>Langmuir | Pseudo-second order<br>Pseudo-second order<br>Pseudo-second order<br>Pseudo-second order | 137.6 mg/g<br>96.1 mg/g<br>94.1 mg/g<br>102 mg/g | N.R<br>N.R<br>N.R<br>N.R | [69] |
| Barley straw (Magnetic CA/NaOH-modified) | BBY<br>CV<br>MB<br>SO | Langmuir<br>Langmuir<br>Langmuir<br>Langmuir | Pseudo-second order<br>Pseudo-second order<br>Pseudo-second order<br>Pseudo-second order | 520.3 mg/g<br>410.8 mg/g<br>455.8 mg/g<br>460.7 mg/g | N.R<br>N.R<br>N.R<br>N.R | [69] |
| Black tea | Cu<br>Pb | Langmuir<br>Freundlich | Pseudo-second order<br>Pseudo-second order | 48 mg/g<br>65 mg/g | 77<br>94 | [90] |
| Cauliflower leaf powder | MB | Freundlich | Pseudo-second order | 149.22 mg/g | 88.1 | [70] |
| Corn stalks | Cu<br>Cd<br>Ni<br>Pb | N.R<br>N.R<br>N.R<br>N.R | N.R<br>N.R<br>N.R<br>N.R | 0.059 mmol/g<br>0.046 mmol/g<br>0.009 mmol/g<br>0.029 mmol/g | 35<br>30<br>8<br>20 | [73] |
| Date palm wastes (Surface fibres) | Phosphates | N.R | N.R | N.R | 85 | [107] |
| Date palm wastes (Date stones) | Phosphates | N.R | N.R | N.R | 87 | [107] |
| Date palm tree leaves ($H_2SO_4$-activated) | Pb | Langmuir | Pseudo-second order | 88.61 mg/g | 98.6 | [41] |
| Mixed tea waste | Cr(VI) | Freundlich | Pseudo-second order | 94.34 mg/g | ~97 | [93] |

**Table 5.** *Cont.*

| Biowaste | Contaminant | Isothermal Model Adjusted | Kinetic Model Adjusted | $q_m/Q_e$ | Removal Efficiency (%) | Ref. |
|---|---|---|---|---|---|---|
| Potato Stem powder | MB | Langmuir and Freundlich | Pseudo-second order | 41.6 mg/g | 82 | [84] |
| | MG | Langmuir and Freundlich | Pseudo-second order | 27.0 mg/g | 67 | |
| Potato Leaves powder | MB | Langmuir and Freundlich | Pseudo-second order | 52.6 mg/g | 87 | [84] |
| | MG | Langmuir and Freundlich | Pseudo-second order | 33.3 mg/g | 75 | |
| Raw wheat straw | Cu | N.R | N.R | 0.070 mmol/g | 42 | [73] |
| | Cd | N.R | N.R | 0.089 mmol/g | 55 | |
| | Ni | N.R | N.R | 0.051 mmol/g | 30 | |
| | Pb | N.R | N.R | 0.015 mmol/g | 10 | |
| Rice straw | Pb(II) | Langmuir | N.R | 42.55 mg/g | 94 | [16,85] |
| | CA | Freundlich | Pseudo-second order | 126.3 mg/g | 42.5 | |
| | CBZ | Freundlich | Pseudo-second order | 40.0 mg/g | 75.3 | |
| Soybean straws | Cu | N.R | N.R | 0.085 mmol/g | 60 | [73] |
| | Cd | N.R | N.R | 0.018 mmol/g | 10 | |
| | Ni | N.R | N.R | 0.007 mmol/g | 5 | |
| | Pb | N.R | N.R | 0.033 mmol/g | 25 | |
| Sugarcane bagasse | Mn(II) | N.R | Pseudo-second order | N.R | 62.5 | [87,88] |
| | Mn(II) | Langmuir | Pseudo-second order | 0.676 mg/g | 63 | |
| Sugarcane bagasse (HCl-treated) | Mn(II) | Freundlich | Pseudo-second order | 1.897 mg/g | 99 | [87] |
| Sugarcane bagasse (NaOH-treated) | Fe(III) | Langmuir | Pseudo-second order | 331.1 μmol/g | >95.0 | [86] |
| | Co(II) | N.R | N.R | 15.5 μmol/g | N.R | |
| | Cu(II) | N.R | N.R | 86 μmol/g | N.R | |
| | Cd(II) | N.R | N.R | 70 μmol/g | N.R | |
| | Pb(II) | N.R | N.R | 87 μmol/g | N.R | |
| | Zn(II) | N.R | N.R | 81 μmol/g | N.R | |
| Sulfonated tea waste | Cr(VI) | Langmuir | Pseudo-second order | 438.18 mg/g | 96 | [91] |
| | MB | Langmuir | Pseudo-second order | 1007.61 mg/g | >99 | |
| | Tet | Langmuir | Pseudo-second order | 380.97 mg/g | 97 | |
| Sunflower | Mn(II) | Langmuir | Pseudo-second order | 47.6 mg/g | 81.6 | [89] |
| | Cd | Langmuir | Pseudo-second order | 83.3 mg/g | N.R | |
| | Cu | Langmuir | Pseudo-second order | 30.3 mg/g | N.R | |
| | Zn | Langmuir | Pseudo-second order | 45.4 mg/g | N.R | |
| | Fe | Langmuir | Pseudo-second order | 71.4 mg/g | N.R | |
| | Ni | Langmuir | Pseudo-second order | 27 mg/g | N.R | |
| Sunflower (NaOH-modified) | Ni | Langmuir | Pseudo-second order | 41.7 mg/g | N.R | [89] |
| Tea waste | Mn(II) | Langmuir | Pseudo-second order | 0.157 mg/g | 95.5 | [92] |
| | Zn(II) | Langmuir | Pseudo-second order | 0.278 mg/g | 99.5 | |

N.R, not reported; BBY, Bismarck brown Y; CV, Crystal violet; MB, Methylene blue; SO, Safranin O; MG, Malachite green; CA, Clofibric acid; CBZ, Carbamazepine; Tet, Tetracycline; CA/NaOH, Citric acid and NaOH.

Tea waste was evaluated as a low-cost adsorbent for the effective recovery of Zn(II) from synthetic wastewater, following the pseudo-second order kinetic model and reaching 99.5% removal with three grams of tea waste per 100 mL of metal standard solution [92]. The residues of sunflower (specifically stalks and leaves) displayed good activity as a bioadsorbent, removing Zn, Ni, Cu, Cd, and Fe from aqueous solutions. The maximum sorption capacity observed for the five metals was about 50.8 ± 22.3 mg/g on average; the adsorption behavior for these metals was used the pseudo-second order model, showing that it is a competitive system [89]. Soliman et al. [86] employed SCB to remove Fe(III) in different water samples through the batch equilibrium technique at an optimum pH of 3. The average recovery efficiencies for this metal were 95–97.4% in Nile River water, 95–98.4% in groundwater, 95–98% in drinking tap water, 95% ± 0.2% in natural drinking water, and 97.2–98.2% in distilled water. The adsorption reaction was adjusted to pH 5 for the maximum uptake of Co(II) (15.5 μmol/g), pH 6 for Cd(II) (70.0 μmol/g) and Cu(II) (86.0 μmol/g), and pH 7 for Zn(II) (81.0 μmol/g). Amarasinghe and Williams [90] reported the sequestration of Cu using black tea waste in a pH range of 5–6, for a removal efficiency and maximum adsorption capacity of 77% and 48 mg/g. The adsorption kinetics showed a rapid initial adsorption rate followed by a slower rate (15–20 min contact time), fitting the Langmuir and pseudo-second order kinetic models. In the same year, Šćiban et al. [73] reported high adsorption efficiency for copper ions (0.085 mmol/g) by employing soybean straw, with low values for cadmium and nickel (0.018 and 0.007 mmol/g, respectively). The authors found that copper adsorption increased to over 80% when the adsorbent material was chemically modified with 1% NaOH. Other agricultural byproducts (raw wheat straw and corn stalks) were evaluated for the adsorption of these heavy metals, obtaining higher $q_m$ values for cadmium (0.089 mmol/g) and nickel (0.051 mmol/g) using raw wheat straw [73].

The adsorption of chromium and lead from synthetic water by applying mixed tea waste obtained from a coffee shop showed a maximum Cr(VI) removal of approximately 97% after 180 min of contact time at pH 2 [93]. This synthetic water consisted of tap water with trace concentrations of suspended organic and inorganic solids added to resemble municipal wastewater. Sulfonate-treated tea waste was an effective adsorbent for Cr(VI), exhibiting a removal efficiency of 96% and reaching equilibrium in the first 30 min [91]. Arris et al. [40] identified the optimal conditions for calcined cereal byproducts to achieve maximum chromium removal efficiency (90.3%) at pH values ranging from 6 to 8 and a high initial metal concentration (132 mg/L). In a similar vein, Amer et al. [16] studied the optimal conditions for the retention of lead employing rice straw, concluding that when increasing the pH from 3.5 to 6 the removal of the metal exceeded 90%. Date palm leaves, SCB, and black tea were effective herbage adsorbents for capturing Pb(II) in aqueous solutions, with the former retaining more than 98% of the metal at a pH of 5.5–6.0 [41,86] and the black tea showing a 94% removal and $q_m$ of 65 mg/g [90].

#### 5.3.2. Industrial Contaminants

Barley straw has been chemically and magnetically modified for the removal of four water-soluble dyes: bismarck brown Y, crystal violet, safranin O, and MB. The biomaterial modified with citric acid and NaOH presented the highest $q_m$ for bismarck brown Y (524.3 mg/g), followed by crystal violet (473.2 mg/g) and MB (498.1 mg/g), whereas its magnetic version attained the maximal adsorption for safranin O (460.7 mg/g). In contrast, the native and magnetic barley straw exhibited $q_m$ values below 124.3 and 137.6 mg/g, respectively [69]. Ansari et al. [70] reported the removal of MB using cauliflower leaves in a synthetic aqueous solution. The adsorption capacity of 149.22 mg/g fit the Freundlich model well and removed ~88% of the dye, increasing the adsorbent concentration to 0.1 g. Furthermore, Gupta et al. [84] found an alternative way to capture MB and malachite green using potato waste powder; potato leaves showed higher efficiency (87 and 75%, respectively) than potato stems (82 and 67%, respectively) for both dyes. Both adsorbents presented a porous, uneven, and rough surface structure, although the potato leaves evidenced a higher fixed carbon percentage (2.15% vs. 1.725%), which may explain its higher removing capacity. In addition to the dye removal efficiencies and adsorption capacities (Table 5), the equilibrium data fit the Langmuir, Freundlich, and the pseudo-second order kinetics models for the different adsorbents described.

#### 5.3.3. Nutrients and Pharmaceuticals

To resolve the imbalance caused by excess nutrients in water sources, Ismail [107] explored date palm wastes as an effective adsorbent for phosphate elimination. The removal percentages obtained at the initial $PO_4^{3-}$ concentration of 50 mg were 87% for granular date stones and 85% for palm surface fibers. Regarding removal of pharmaceuticals, Liu et al. [85] developed a sustainable method to adsorb clofibric acid and carbamazepine from aqueous solutions by employing rice straw at different pH conditions and biomaterial concentrations. The adsorption of both drugs depended on the adsorbent dosage and the pH. In the case of clofibric acid, 30 and 60 g/L of straw at pH 3.1 yielded the highest removal percentages (42.5 and 75.3%, respectively). The adsorption isotherms confirmed that the bioadsorbents had a heterogeneous surface structure, as both pharmaceuticals better fit the Freundlich isotherm model (Table 5).

## 6. Challenges and Future Perspectives

Agricultural waste has been demonstrated to be one of the most interesting materials for water remediation due to its ability to adsorb heavy metals, dyes, nutrients, pharmaceuticals, and other contaminants. Nevertheless, several variables that represent a challenging perspective in the near future need to be considered during the recovery process from water sources. One of these is the lifetime and final disposal of the biomaterial, which could negatively impact both the environment and the cost of the procedures. The most common disposal method, due to its low cost and reliability, is chemical neutralization of spent bioadsorbents, which are later placed in a landfill. Despite being a simple mechanism, this entails the possibility of secondary contamination of the soil, groundwater, surface water, and air, especially if the landfill lacks appropriate leachate and gas collection systems [108]. Another popular strategy to discard used biomaterials is incineration, which results in high reduction of the biomass at the cost of possible secondary pollution through the emission of gases and fly ash that may contain trace amounts of the adsorbed contaminants [109]. However, there are other heat treatments used to extend the lifecycle of adsorbents or regenerate biochars, such as hydrothermal modification, gasification, and pyrolysis; regardless of the benefits of recycling byproducts, high costs represent a considerable limitation for the industry [109].

Microbial degradation has been studied as a recovery method for exhausted bioadsorbents. Unfortunately, it has several disadvantages that make it an unsuitable alternative at industrial scales, for example, its long degradation time, high sediment production, and the high sensitivity of the processes [110]. On the other hand, the chemical extraction of contaminants with sulfuric acid or EDTA, among others, displays satisfactory results in recovering and recycling both heavy metals and extractants [111,112]. A fairly new strategy with promising outcomes has recently been studied, although it requires further research; this is the biosynthesis of nanomaterials using bioadsorbents as raw material [113].

Other challenges related to the use of bioadsorbents include the wide range of contaminants and the hydrological dynamics existing in different water sources. In this review, only four contaminant categories have been taken into consideration; however, it is necessary to explore other bioadsorbents that have been used to remove personal care and other lifestyle products such as galaxolide, tonalide [114–116], saccharin, sucralose, and caffeine [117–120]. Other known groups that form the growing list of emerging contaminants are chemical synthesizers and intermediates, corrosion inhibitors, flame retardants, plasticizers, per- and polyfluorinated compounds, biocides, hormones, resistance genes, etc. [121–125]. These emerging contaminants make more crucial the development of future research on the adsorption of these contaminants through the implementation of agricultural residues. Finally, the behavior of bioadsorbents in the treatment of other water sources, including groundwater, surface water, marine water, and different environmental matrices such as landfill leachates and industrial gas emissions, is another matter requiring further investigation. The current methodologies and protocols for the manufacturing of biowastes should be improved, with the aim of enhancing the adsorption process and favoring the implementation of bioadsorbents by further reducing costs.

## 7. Conclusions

In this review, the most notable outcomes were obtained for fruit-based bioadsorbents, with removal efficiencies of 100% or higher (predicted efficiency using response surface methodology) for adsorption of fluoride ions and congo red dye, respectively. Regarding adsorbents derived from seeds, AC from *Acacia erioloba* seed pods and *Argania Spinosa* tree nutshells–$TiO_2$ achieved the removal of 100% of MB and diclofenac. Forage and herbage adsorbents presented lower removal efficiencies; however, tea waste and SCB were suitable alternatives for the adsorption of Zn, Mn, and MB, with efficiencies around 99%. In most of these treatments, the effects of the initial adsorbate concentration, adsorbent dosage, contact time, and pH were critical for evaluation of biomaterial efficiency, in most cases involving a change in the ionic charges that govern the interaction between contaminants and adsorbents. However, it is necessary to continue studying these conditions in order to optimize the use and yield of bioadsorbents that could help solve the most challenging environmental problems involved in water pollution. Another aspect to consider is the use of recycled AC activators to reduce both environmental impacts and manufacturing costs. Moreover, in some cases the application of chemical activators is not required (e.g., in moringa biochar), avoiding the cost of chemical pretreatment.

In conclusion, this review highlights the wide availability of agricultural materials, reporting more than 60 bioadsorbents and the conditions under which they can most effectively adsorb heavy metals, dyes, pharmaceuticals, nutrients, and other contaminants that alter water sources. Further multidisciplinary studies are needed in order to apply the reviewed bioadsorbents on industrial scales while abiding by the relevant environmental policies of each sector. Lastly, it is imperative to enforce the regulations concerning industrial discharge, water safety, and waste control in order to reduce environmental impacts of anthropogenic origin.

**Author Contributions:** Writing—original draft preparation, J.A.-R., M.E.U.-L., B.E.R.-G. and S.X.L.-V.; writing—review and editing, I.E.L.-C. and D.L.C.-C.; supervision, D.L.C.-C. All authors have read and agreed to the published version of the manuscript.

**Funding:** This research received no external funding.

**Institutional Review Board Statement:** Not applicable.

**Informed Consent Statement:** Not applicable.

**Acknowledgments:** This work was supported by Consejo Nacional de Ciencia y Tecnología of Mexico (Scholarship No. 1103630).

**Conflicts of Interest:** The authors declare no conflict of interest.

## References

1. Bhatnagar, A.; Sillanpää, M.; Witek-Krowiak, A. Agricultural waste peels as versatile biomass for water purification—A review. *Chem. Eng. J.* **2015**, *270*, 244–271. [CrossRef]
2. Food and Agriculture Organization of the United Nations. Water Pollution from Agriculture: A Global Review. Executive Summary. Available online: https://www.fao.org/3/i7754e/i7754e.pdf (accessed on 21 August 2021).
3. Bhattacharjee, C.; Dutta, S.; Saxena, V.K. A review on biosorptive removal of dyes and heavy metals from wastewater using watermelon rind as biosorbent. *Environ. Adv.* **2020**, *2*, 100007. [CrossRef]
4. Karim, M.A.H.; Aziz, K.H.H.; Omer, K.M.; Salih, Y.M.; Mustafa, F.; Rahman, K.O.; Mohammad, Y. Degradation of aqueous organic dyes pollutants by heterogeneous photo-assisted Fenton-like process using natural mineral activator: Parameter optimization and degradation kinetics. *IOP Conf. Ser. Earth Environ. Sci.* **2021**, *958*, 012011. [CrossRef]
5. Abdulla, S.M.; Jamil, D.M.; Aziz, K.H.H. Investigation in heavy metal contents of drinking water and fish from Darbandikhan and Dokan Lakes in Sulaimaniyah Province—Iraqi Kurdistan Region. *IOP Conf. Ser. Earth Environ. Sci.* **2020**, *612*, 012023. [CrossRef]
6. Pesqueria, J.F.J.R.; Pereira, M.F.R.; Silva, A.M.T. Environmental impact assessment of advanced urban wastewater treatment technologies for the removal of priority substances and contaminants of emerging concern: A review. *J. Clean. Prod.* **2020**, *261*, 121078. [CrossRef]
7. Ramírez-Malule, H.; Quiñones-Murillo, D.H.; Manotas-Duque, D. Emerging contaminants as global environmental hazards. A bibliometric analysis. *Emerg. Contam.* **2020**, *6*, 179–193. [CrossRef]
8. Intisar, A.; Ramzan, A.; Sawaira, T.; Kareem, A.T.; Hussain, N.; Din, M.I.; Bilal, M.; Iqbal, H.M.N. Occurrence, toxic effects, and mitigation of pesticides as emerging environmental pollutants using robust nanomaterials—A review. *Chemosphere* **2022**, *293*, 133538. [CrossRef]
9. World Water Assessment Programme (UNESCO WWAP). The United Nations World Water Development Report 2015: Water for a Sustainable World. Available online: http://www.unesco.org/new/en/natural-sciences/environment/water/wwap/wwdr/2015-water-for-a-sustainable-world (accessed on 8 November 2021).
10. Bhatnagar, A.; Sillanpää, M. Utilization of agro-industrial and municipal waste materials as potential adsorbents for water treatment—A review. *Chem. Eng. J.* **2010**, *157*, 277–296. [CrossRef]
11. Sanchez-Silva, J.M.; González-Estrada, R.R.; Blancas-Benitez, F.J.; Fonseca-Cantabrana, N. Utilización de subproductos agroindustriales para la bioadsorción de metales pesados. *TIP* **2020**, *23*, 1–18. [CrossRef]

12. Saravanan, A.; Senthil Kumar, P.; Jeevanantham, S.; Karishma, S.; Tajsabreen, B.; Yaashikaa, P.; Reshma, B. Effective water/wastewater treatment methodologies for toxic pollutants removal: Processes and applications towards sustainable development. *Chemosphere* **2021**, *280*, 130595. [CrossRef]
13. Volesky, B. Biosorption and me. *Water Res.* **2007**, *41*, 4017–4029. [CrossRef] [PubMed]
14. De Gisi, S.; Lofrano, G.; Grassi, M.; Notarnicola, M. Characteristics and adsorption capacities of low-cost sorbents for wastewater treatment: A review. *Sustain. Mater. Technol.* **2016**, *9*, 10–40. [CrossRef]
15. Rudi, N.N.; Muhamad, M.S.; te Chuan, L.; Alipal, J.; Omar, S.; Hamidon, N.; Abdul Hamid, N.H.; Mohamed Sunar, N.; Ali, R.; Harun, H. Evolution of adsorption process for manganese removal in water via agricultural waste adsorbents. *Heliyon* **2020**, *6*, e05049. [CrossRef] [PubMed]
16. Amer, H.; El-Gendy, A.; El-Haggar, S. Removal of lead (II) from aqueous solutions using rice straw. *Water Sci. Technol.* **2017**, *76*, 1011–1021. [CrossRef] [PubMed]
17. Shafiq, M.; Alazba, A.A.; Amin, M.T. Removal of Heavy Metals from Wastewater using Date Palm as a Biosorbent: A Comparative Review. *Sains Malays.* **2018**, *47*, 35–49. [CrossRef]
18. De Andrade, J.R.; Oliveira, M.F.; da Silva, M.G.C.; Vieira, M.G.A. Adsorption of Pharmaceuticals from Water and Wastewater Using Nonconventional Low-Cost Materials: A Review. *Ind. Eng. Chem. Res.* **2018**, *57*, 3103–3127. [CrossRef]
19. Salman, A.; Ibrahim, I.; Tarek, M.; Abbas, S. Biosorption of heavy metals: A review. *J. Chem. Sci. Technol.* **2014**, *3*, 74–102.
20. Wang, J.; Guo, X. Adsorption isotherm models: Classification, physical meaning, application and solving method. *Chemosphere* **2020**, *258*, 127279. [CrossRef]
21. Kaur, S.; Rani, S.; Mahajan, R.K.; Asif, M.; Gupta, V.K. Synthesis and adsorption properties of mesoporous material for the removal of dye safranin: Kinetics, equilibrium, and thermodynamics. *J. Ind. Eng. Chem.* **2015**, *22*, 19–27. [CrossRef]
22. Piccin, J.S.; Dotto, G.L.; Pinto, A.A. Adsorption isotherms and thermochemical data of FD&C red n° 40 binding by chitosan. *Braz. J. Chem. Eng.* **2011**, *28*, 295–304. [CrossRef]
23. Ray, S.S.; Gusain, R.; Kumar, N. Adsorption equilibrium isotherms, kinetics and thermodynamics. In *Carbon Nanomaterial-Based Adsorbents for Water Purification Fundamentals and Applications*, 1st ed.; Ray, S.S., Gusain, R., Neeraj, K., Eds.; Elsevier: Amsterdam, The Netherlands, 2020; pp. 101–118. [CrossRef]
24. Largitte, L.; Pasquier, R. A review of the kinetics adsorption models and their application to the adsorption of lead by an activated carbon. *Chem. Eng. Res. Des.* **2016**, *109*, 495–504. [CrossRef]
25. Jones, A.N.; Bridgeman, J. A fluorescence-based assessment of the fate of organic matter in water treated using crude/purified Hibiscus seeds as coagulant in drinking water treatment. *Sci. Total Environ.* **2019**, *646*, 1–10. [CrossRef] [PubMed]
26. Ueda Yamaguchi, N.; Cusioli, L.F.; Quesada, H.B.; Camargo Ferreira, M.E.; Fagundes-Klen, M.R.; Salcedo Vieira, A.M.; Bergamasco, R. A review of *Moringa oleifera* seeds in water treatment: Trends and future challenges. *Process. Saf. Environ. Prot.* **2021**, *147*, 405–420. [CrossRef]
27. Jjagwe, J.; Olupot, P.W.; Menya, E.; Kalibbala, H.M. Synthesis and application of Granular activated carbon from biomass waste materials for water treatment: A review. *J. Bioresour. Bioprod.* **2021**, *4*, 292–322. [CrossRef]
28. Edathil, A.A.; Shittu, I.; Hisham Zain, J.; Banat, F.; Haija, M.A. Novel magnetic coffee waste nanocomposite as effective bioadsorbent for Pb(II) removal from aqueous solutions. *J. Environ. Chem. Eng.* **2018**, *6*, 2390–2400. [CrossRef]
29. Sciban, M.; Antov, M.G.; Klasnja, M. Extraction and partial purification of coagulation active components from common bean seed. *Acta Period. Technol.* **2006**, *37*, 37–43. [CrossRef]
30. Bodlund, I.; Pavankumar, A.R.; Chelliah, R.; Kasi, S.; Sankaran, K.; Rajarao, G.K. Coagulant proteins identified in mustard: A potential water treatment agent. *Int. J. Environ. Sci. Technol.* **2014**, *11*, 873–880. [CrossRef]
31. Foo, K.Y.; Hameed, B.H. Preparation and characterization of activated carbon from sunflower seed oil residue via microwave assisted $K_2CO_3$ activation. *Bioresour. Technol.* **2011**, *102*, 9794–9799. [CrossRef]
32. Jayasree, R.; Kumar, P.S.; Saravanan, A.; Hemavathy, R.V.; Yaashikaa, P.R.; Arthi, P.; Choi, K.C. Sequestration of toxic Pb(II) ions using ultrasonic modified agro waste: Adsorption mechanism and modelling study. *Chemosphere* **2021**, *285*, 131502. [CrossRef]
33. Aguayo-Villarreal, I.A.; Bonilla-Petriciolet, A.; Muñiz-Valencia, R. Preparation of activated carbons from pecan nutshell and their application in the antagonistic adsorption of heavy metal ions. *J. Mol. Liq.* **2017**, *230*, 686–695. [CrossRef]
34. Williams, P.T.; Reed, A.R. Development of activated carbon pore structure via physical and chemical activation of biomass fibre waste. *Biomass Bioenergy* **2006**, *30*, 144–152. [CrossRef]
35. Marsh, H.; Rodríguez-Reinoso, F. Activation Processes (Chemical). In *Activated Carbon*, 1st ed.; Marsh, H., Rodríguez-Reinoso, F., Eds.; Elsevier Science Ltd.: Amsterdam, The Netherlands, 2006; pp. 322–365. [CrossRef]
36. Yi, H.; Nakabayashi, K.; Yoon, S.-H.; Miyawaki, J. Pressurized physical activation: A simple production method for activated carbon with a highly developed pore structure. *Carbon* **2021**, *183*, 735–742. [CrossRef]
37. Anirudhan, T.; Sreekumari, S. Adsorptive removal of heavy metal ions from industrial effluents using activated carbon derived from waste coconut buttons. *J. Environ. Sci.* **2011**, *23*, 1989–1998. [CrossRef]
38. Shin, J.; Kwak, J.; Lee, Y.G.; Kim, S.; Choi, M.; Bae, S.; Chon, K. Competitive adsorption of pharmaceuticals in lake water and wastewater effluent by pristine and NaOH-activated biochars from spent coffee wastes: Contribution of hydrophobic and π-π interactions. *Environ. Pollut.* **2021**, *270*, 116244. [CrossRef]
39. Abbaszadeh, S.; Alwi, S.R.W.; Webb, C.; Ghasemi, N.; Muhamad, I.I. Treatment of lead-contaminated water using activated carbon adsorbent from locally available papaya peel biowaste. *J. Clean. Prod.* **2016**, *118*, 210–222. [CrossRef]

40. Arris, S.; Bencheikh Lehocine, M.; Meniai, A.H. Sorption study of chromium sorption from wastewater using cereal by-products. *Int. J. Hydrogen Energy* **2014**, *41*, 10299–10310. [CrossRef]
41. Soliman, A.M.; Elwy, H.M.; Thiemann, T.; Majedi, Y.; Labata, F.T.; Al-Rawashdeh, N.A. Removal of Pb(II) ions from aqueous solutions by sulphuric acid-treated palm tree leaves. *J. Taiwan Inst. Chem. Eng.* **2016**, *58*, 264–273. [CrossRef]
42. Garg, D.; Kumar, S.; Sharma, K.; Majumder, C.B. Application of waste peanut shells to form activated carbon and its utilization for the removal of Acid Yellow 36 from wastewater. *Groundw. Sustain. Dev.* **2019**, *8*, 512–519. [CrossRef]
43. Rahman, A.; Hango, H.J.; Daniel, L.S.; Uahengo, V.; Jaime, S.J.; Bhaskaruni, S.V.H.S.; Jonnalagadda, S.B. Chemical preparation of activated carbon from *Acacia erioloba* seed pods using $H_2SO_4$ as impregnating agent for water treatment: An environmentally benevolent approach. *J. Clean. Prod.* **2019**, *237*, 117689. [CrossRef]
44. Jung, K.W.; Choi, B.H.; Hwang, M.J.; Jeong, T.U.; Ahn, K.H. Fabrication of granular activated carbons derived from spent coffee grounds by entrapment in calcium alginate beads for adsorption of acid orange 7 and methylene blue. *Bioresour. Technol.* **2016**, *219*, 185–195. [CrossRef]
45. Pereira, R.G.; Veloso, C.M.; da Silva, N.M.; de Sousa, L.F.; Bonomo, R.C.F.; de Souza, A.O.; da Guarda Souza, M.O.; da Costa Ilhéu Fontan, R. Preparation of activated carbons from cocoa shells and siriguela seeds using $H_3PO_4$ and $ZnCl_2$ as activating agents for BSA and α-lactalbumin adsorption. *Fuel Process. Technol.* **2014**, *126*, 476–486. [CrossRef]
46. Namasivayam, C.; Kavitha, D. Removal of Congo red from water by adsorption onto activated carbon prepared from coir pith, an agricultural solid waste. *Dye. Pigm.* **2002**, *54*, 47–58. [CrossRef]
47. Namasivayam, C.; Radhika, R.; Suba, S. Uptake of dyes by a promising locally available agricultural solid waste: Coir pith. *J. Waste Manag.* **2001**, *21*, 381–387. [CrossRef]
48. Mahamad, M.N.; Zaini, M.A.A.; Zakaria, Z.A. Preparation and characterization of activated carbon from pineapple waste biomass for dye removal. *Int. Biodeterior. Biodegrad.* **2015**, *102*, 274–280. [CrossRef]
49. Yadav, D.; Kapur, M.; Kumar, P.; Mondal, M.K. Adsorptive removal of phosphate from aqueous solution using rice husk and fruit juice residue. *Process. Saf. Environ. Prot.* **2015**, *94*, 402–409. [CrossRef]
50. Isiuku, B.O.; Enyoh, C.E.; Duru, C.E.; Ibe, F.C. Phosphate ions removal from aqueous phase by batch adsorption on activated (activation before carbonization) biochar derived from rubber pod husk. *Curr. Res. Green Sustain. Chem.* **2021**, *4*, 100136. [CrossRef]
51. De Luna, M.D.G.; Murniati; Budianta, W.; Rivera, K.K.; Arazo, R.O. Removal of sodium diclofenac from aqueous solution by adsorbents derived from cocoa pod husks. *J. Environ. Chem. Eng.* **2017**, *5*, 1465–1474. [CrossRef]
52. El Mouchtari, E.M.; Daou, C.; Rafqah, S.; Najjar, F.; Anane, H.; Piram, A.; Hamade, A.; Briche, S.; Wong-Wah-Chung, P. $TiO_2$ and activated carbon of *Argania Spinosa* tree nutshells composites for the adsorption photocatalysis removal of pharmaceuticals from aqueous solution. *J. Photochem. Photobiol. A Chem.* **2020**, *388*, 112183. [CrossRef]
53. Chakraborty, P.; Banerjee, S.; Kumar, S.; Sadhukhan, S.; Halder, G. Elucidation of ibuprofen uptake capability of raw and steam activated biochar of *Aegle marmelos* shell: Isotherm, kinetics, thermodynamics and cost estimation. *Process. Saf. Environ. Prot.* **2018**, *118*, 10–23. [CrossRef]
54. Cabrita, I.; Ruiz, B.; Mestre, A.; Fonseca, I.; Carvalho, A.; Ania, C. Removal of an analgesic using activated carbons prepared from urban and industrial residues. *Chem. Eng. J.* **2010**, *163*, 249–255. [CrossRef]
55. Mora Alvarez, N.M.; Pastrana, J.M.; Lagos, Y.; Lozada, J.J. Evaluation of mercury ($Hg^{2+}$) adsorption capacity using exhausted coffee waste. *Sustain. Chem. Pharm.* **2018**, *10*, 60–70. [CrossRef]
56. Araújo, C.; Melo, E.; Alves, V.; Coelho, N.M.M. *Moringa oleifera* Lam. Seeds as a Natural Solid Adsorbent for Removal of AgI in Aqueous Solutions. *J. Braz. Chem. Soc.* **2010**, *21*, 1727–1732. [CrossRef]
57. Çelekli, A.; Al-Nuaimi, A.I.; Bozkurt, H. Adsorption kinetic and isotherms of Reactive Red 120 on *Moringa oleifera* seed as an eco-friendly process. *J. Mol. Struct.* **2019**, *1195*, 168–178. [CrossRef]
58. Obeng, G.Y.; Amoah, D.Y.; Opoku, R.; Sekyere, C.K.; Adjei, E.A.; Mensah, E. Coconut wastes as bioresource for sustainable energy: Quantifying wastes, calorific values and emissions in Ghana. *Energies* **2020**, *13*, 2178. [CrossRef]
59. Parab, H.; Joshi, S.; Shenoy, N.; Lali, A.; Sarma, U.S.; Sudersanan, M. Determination of kinetic and equilibrium parameters of the batch adsorption of Co(II), Cr(III) and Ni(II) onto coir pith. *Process. Biochem.* **2006**, *41*, 609–615. [CrossRef]
60. Memon, J.R.; Memon, S.Q.; Bhanger, M.I.; Memon, G.Z.; El-Turki, A.; Allen, G.C. Characterization of banana peel by scanning electron microscopy and FT-IR spectroscopy and its use for cadmium removal. *Colloids Surf. B Biointerfaces* **2008**, *66*, 260–265. [CrossRef]
61. Memon, J.R.; Memon, S.Q.; Bhanger, M.I.; El-Turki, A.; Hallam, K.R.; Allen, G.C. Banana peel: A green and economical sorbent for the selective removal of Cr(VI) from industrial wastewater. *Colloids Surf. B Biointerfaces* **2009**, *70*, 232–237. [CrossRef]
62. Liu, C.; Ngo, H.H.; Guo, W. Watermelon Rind: Agro-waste or Superior Biosorbent? *Appl. Biochem. Biotechnol.* **2012**, *167*, 1699–1715. [CrossRef]
63. Liu, C.; Ngo, H.H.; Guo, W.; Tung, K.L. Optimal conditions for preparation of banana peels, sugarcane bagasse and watermelon rind in removing copper from water. *Bioresour. Technol.* **2012**, *119*, 349–354. [CrossRef]
64. Banerjee, K.; Ramesh, S.T.; Gandhimathi, R.; Nidheesh, P.V.; Bharathi, K.S. A novel agricultural waste adsorbent, watermelon shell for the removal of copper from aqueous solutions. *Iran. J. Energy Environ.* **2012**, *3*, 143–156. [CrossRef]
65. Gupta, H.; Gogate, P.R. Intensified removal of copper from waste water using activated watermelon based biosorbent in the presence of ultrasound. *Ultrason. Sonochem.* **2016**, *30*, 113–122. [CrossRef] [PubMed]
66. Reddy, N.A.; Lakshmipathy, R.; Sarada, N.C. Application of *Citrullus lanatus* rind as biosorbent for removal of trivalent chromium from aqueous solution. *Alex. Eng. J.* **2014**, *53*, 969–975. [CrossRef]

67. Annadurai, G.; Juang, R.S.; Lee, D.J. Use of cellulose-based wastes for adsorption of dyes from aqueous solutions. *J. Hazard. Mater.* **2002**, *92*, 263–274. [CrossRef]
68. Mondal, N.K.; Roy, A. Potentiality of a fruit peel (banana peel) toward abatement of fluoride from synthetic and underground water samples collected from fluoride affected villages of Birbhum district. *Appl. Water Sci.* **2018**, *8*, 90. [CrossRef]
69. Baldikova, E.; Politi, D.; Maderova, Z.; Pospiskova, K.; Sidiras, D.; Safarikova, M.; Safarik, I. Utilization of magnetically responsive cereal by-product for organic dye removal. *J. Sci. Food Agric.* **2015**, *96*, 2204–2214. [CrossRef] [PubMed]
70. Ansari, S.A.; Khan, F.; Ahmad, A. Cauliflower Leave, an Agricultural Waste Biomass Adsorbent, and Its Application for the Removal of MB Dye from Aqueous Solution: Equilibrium, Kinetics, and Thermodynamic Studies. *Int. J. Anal. Chem.* **2016**, 1–10. [CrossRef]
71. Paredes-Laverde, M.; Silva-Agredo, J.; Torres-Palma, R.A. Removal of norfloxacin in deionized, municipal water and urine using rice (*Oryza sativa*) and coffee (*Coffea arabica*) husk wastes as natural adsorbents. *J. Environ. Manag.* **2018**, *213*, 98–108. [CrossRef]
72. Etim, U.; Umoren, S.; Eduok, U. Coconut coir dust as a low cost adsorbent for the removal of cationic dye from aqueous solution. *J. Saudi Chem. Soc.* **2016**, *20*, S67–S76. [CrossRef]
73. Šćlban, M.; Klašnja, M.; ŠKrbić, B. Adsorption of copper ions from water by modified agricultural by-products. *Desalination* **2008**, *229*, 170–180. [CrossRef]
74. Hameed, B.H.; Hakimi, H. Utilization of durian (*Durio zibethinus* Murray) peel as low cost sorbent for the removal of acid dye from aqueous solutions. *Biochem. Eng. J.* **2008**, *39*, 338–343. [CrossRef]
75. Saeed, A.; Sharif, M.; Iqbal, M. Application potential of grapefruit peel as dye sorbent: Kinetics, equilibrium and mechanism of crystal violet adsorption. *J. Hazard. Mater.* **2010**, *179*, 564–572. [CrossRef] [PubMed]
76. Hameed, B.H. Removal of cationic dye from aqueous solution using jackfruit peel as non-conventional low-cost adsorbent. *J. Hazard. Mater.* **2009**, *162*, 344–350. [CrossRef]
77. Al-Qahtani, K.M. Water purification using different waste fruit cortexes for the removal of heavy metals. *J. Taibah Univ. Sci.* **2016**, *10*, 700–770. [CrossRef]
78. Iqbal, M.; Saeed, A.; Zafar, S.I. FTIR spectrophotometry, kinetics and adsorption isotherms modeling, ion exchange, and EDX analysis for understanding the mechanism of $Cd^{2+}$ and $Pb^{2+}$ removal by mango peel waste. *J. Hazard. Mater.* **2009**, *164*, 161–171. [CrossRef]
79. Phanthuwongpakdee, J.; Babel, S.; Laohhasurayotin, K.; Sattayaporn, S.; Kaneko, T. Anthocyanin based agricultural wastes as bio-adsorbents for scavenging radioactive iodide from aqueous environment. *J. Environ. Chem. Eng.* **2020**, *8*, 104147. [CrossRef]
80. Namasivayam, C.; Muniasamy, N.; Gayatri, K.; Rani, M.; Ranganathan, K. Removal of dyes from aqueous solutions by cellulosic waste orange peel. *Bioresour. Technol.* **1996**, *57*, 37–43. [CrossRef]
81. Arami, M.; Limaee, N.Y.; Mahmoodi, N.M.; Tabrizi, N.S. Removal of dyes from colored textile wastewater by orange peel adsorbent: Equilibrium and kinetic studies. *J. Colloid Interface. Sci.* **2005**, *288*, 371–376. [CrossRef] [PubMed]
82. Aryee, A.A.; Dovi, E.; Shi, X.; Han, R.; Li, Z.; Qu, L. Zirconium and iminodiacetic acid modified magnetic peanut husk as a novel adsorbent for the sequestration of phosphates from solution: Characterization, equilibrium and kinetic study. *Colloids Surf. A Physicochem. Eng. Asp.* **2021**, *615*, 126260. [CrossRef]
83. Argun, M.E.; Güclü, D.; Karatas, M. Adsorption of Reactive Blue 114 dye by using a new adsorbent: Pomelo peel. *J. Ind. Eng. Chem.* **2014**, *20*, 1079–1084. [CrossRef]
84. Gupta, N.; Kushwaha, A.K.; Chattopadhyaya, M. Application of potato (*Solanum tuberosum*) plant wastes for the removal of methylene blue and malachite green dye from aqueous solution. *Arab. J. Chem.* **2016**, *9*, S707–S716. [CrossRef]
85. Liu, Z.; Zhou, X.; Chen, X.; Dai, C.; Zhang, J.; Zhang, Y. Biosorption of clofibric acid and carbamazepine in aqueous solution by agricultural waste rice straw. *J. Environ. Sci.* **2013**, *25*, 2384–2395. [CrossRef]
86. Soliman, E.M.; Ahmed, S.A.; Fadl, A.A. Reactivity of sugar cane bagasse as a natural solid phase extractor for selective removal of Fe(III) and heavy-metal ions from natural water samples. *Arab. J. Chem.* **2011**, *4*, 63–70. [CrossRef]
87. Esfandiar, N.; Nasernejad, B.; Ebadi, T. Removal of Mn(II) from groundwater by sugarcane bagasse and activated carbon (a comparative study): Application of response surface methodology (RSM). *J. Ind. Eng. Chem.* **2014**, *20*, 3726–3736. [CrossRef]
88. Ahmed, S.; El-Roudi, A.M.; Salem, A.A. Removal of Mn(II) from Ground Water by Solid Wastes of Sugar Industry. *J. Environ. Sci. Technol.* **2015**, *8*, 338–351. [CrossRef]
89. Feizi, M.; Jalali, M. Removal of heavy metals from aqueous solutions using sunflower, potato, canola and walnut shell residues. *J. Taiwan Inst. Chem. Eng.* **2015**, *54*, 125–136. [CrossRef]
90. Amarasinghe, B.; Williams, R. Tea waste as a low cost adsorbent for the removal of Cu and Pb from wastewater. *Chem. Eng. J.* **2007**, *132*, 299–309. [CrossRef]
91. Ahsan, M.A.; Katla, S.K.; Islam, M.T.; Hernandez-Viezcas, J.A.; Martinez, L.M.; Díaz-Moreno, C.A.; Lopez, J.; Singamaneni, S.R.; Banuelos, J.; Gardea-Torresdey, J.; et al. Adsorptive removal of methylene blue, tetracycline and Cr(VI) from water using sulfonated tea waste. *Environ. Technol. Innov.* **2018**, *11*, 23–40. [CrossRef]
92. Badrealam, S.; Darrell, V.C.; Dollah, Z.; Mohamed-Latiff, M.F.P.; Handan, R. Adsorption of manganese and zinc in synthetic wastewater by tea waste (TW) as a low cost adsorbent. *J. Phys. Conf. Ser.* **2019**, *1349*, 012061. [CrossRef]
93. Cherdchoo, W.; Nithettham, S.; Charoenpanich, J. Removal of Cr(VI) from synthetic wastewater by adsorption onto coffee ground and mixed waste tea. *Chemosphere* **2019**, *221*, 758–767. [CrossRef]
94. Solangi, N.H.; Kumar, J.; Mazari, S.A.; Ahmed, S.; Fatima, N.; Mujawar, N.M. Development of fruit waste derived bio-adsorbents for wastewater treatment: A review. *J. Hazard. Mater.* **2021**, *416*, 125848. [CrossRef]

95. Sivaraj, R.; Namasivayam, C.; Kadirvelu, K. Orange peel as an adsorbent in the removal of acid violet 17 (acid dye) from aqueous solutions. *Waste Manag.* **2001**, *21*, 105–110. [CrossRef]
96. Hou, S.X. Adsorption properties of pomelo peels against methylene blue in dye wastewater. *Adv. Mater. Res.* **2013**, *634*, 178–181. [CrossRef]
97. Nasrollahzadeh, M.; Atarod, M.; Sajjadi, M.; Sajadi, S.M.; Issaabadi, Z. Plant-Mediated Green Synthesis of Nanostructures: Mechanisms, Characterization, and Applications. *Interface Sci. Technol.* **2019**, *28*, 199–322. [CrossRef]
98. Bardestani, R.; Patience, G.S.; Kaliaguine, S. Experimental methods in chemical engineering: Specific surface area and pore size distribution measurements—BET, BJH, and DFT. *Can. J. Chem. Eng.* **2019**, *97*, 2781–2791. [CrossRef]
99. Yan, Y. Tribology and tribo-corrosion testing and analysis of metallic biomaterials. In *Metals for Biomedical Devices: Woodhead Publishing Series in Biomaterials*, 1st ed.; Niinomi, M., Ed.; Woodhead Publishing Ltd.: Sawston, UK, 2010; pp. 178–201. [CrossRef]
100. Ahsan, M.A.; Islam, M.T.; Imam, M.A.; Hyder, A.G.; Jabbari, V.; Dominguez, N.; Noveron, J.C. Biosorption of bisphenol A and sulfamethoxazole from water using sulfonated coffee waste: Isotherm, kinetic and thermodynamic studies. *J. Environ. Chem. Eng.* **2018**, *6*, 6602–6611. [CrossRef]
101. Chiang, C.H.; Chen, J.; Lin, J.H. Preparation of pore-size tunable activated carbon derived from waste coffee grounds for high adsorption capacities of organic dyes. *J. Environ. Chem. Eng.* **2020**, *8*, 103929. [CrossRef]
102. Wang, J.; Chen, C. Biosorbents for heavy metals removal and their future. *Biotechnol. Adv.* **2009**, *27*, 195–226. [CrossRef]
103. Ben-Ali, S.; Jaouali, I.; Souissi-Najar, S.; Ouederni, A. Characterization and adsorption capacity of raw pomegranate peel biosorbent for copper removal. *J. Clean. Prod.* **2017**, *142*, 3809–3821. [CrossRef]
104. Kumar, G.V.S.R.; Rao, K.S.; Yadav, A.; Kumar, M.L.; Sarathi, T.V.N. Biosorption of copper(II) and manganese(II) from waste water using low cost bio adsorbents. *J. Indian Chem. Soc.* **2018**, *95*, 1–8.
105. Baccar, R.; Sarrà, M.; Bouzid, J.; Feki, M.; Blánquez, P. Removal of pharmaceutical compounds by activated carbon prepared from agricultural by-product. *Chem. Eng. J.* **2012**, *211*, 310–317. [CrossRef]
106. Masoudian, N.; Rajabi, M.; Ghaedi, M. Titanium oxide nanoparticles loaded onto activated carbon prepared from bio-waste watermelon rind for the efficient ultrasonic-assisted adsorption of congo red and phenol red dyes from wastewaters. *Polyhedron* **2019**, *173*, 114105. [CrossRef]
107. Ismail, Z.Z. Kinetic study for phosphate removal from water by recycled date-palm wastes as agricultural by-products. *Int. J. Environ. Stud.* **2012**, *69*, 135–149. [CrossRef]
108. Ye, M.; Sun, M.; Chen, X.; Feng, Y.; Wan, J.; Liu, K.; Tian, D.; Liu, M.; Wu, J.; Schwab, A.P.; et al. Feasibility of sulfate-calcined eggshells for removing pathogenic bacteria and antibiotic resistance genes from landfill leachates. *Waste Manag.* **2017**, *63*, 275–283. [CrossRef]
109. Liu, Z.; Tran, K.-Q. A review on disposal and utilization of phytoremediation plants containing heavy metals. *Ecotoxicol. Environ. Saf.* **2021**, *226*, 112821. [CrossRef]
110. El Gamal, M.; Mousa, H.A.; El-Naas, M.H.; Zacharia, R.; Judd, S. Bio-regeneration of activated carbon: A comprehensive review. *Sep. Purif. Technol.* **2018**, *197*, 345–359. [CrossRef]
111. Reza, M.S.; Hasan, A.B.M.; Afroze, S.; Abu Bakar, M.; Taweekun, J.; Azad, A. Analysis on Preparation, Application, and Recycling of Activated Carbon to Aid in COVID-19 Protection. *Int. J. Integr. Eng.* **2020**, *12*, 233–244. [CrossRef]
112. Wu, F.; Zhao, T.; Yao, Y.; Jiang, T.; Wang, B.; Wang, M. Recycling supercapacitor activated carbons for adsorption of silver (I) and chromium (VI) ions from aqueous solutions. *Chemosphere* **2020**, *238*, 124638. [CrossRef]
113. Hassan, M.; Naidu, R.; Du, J.; Liu, Y.; Qi, F. Critical review of magnetic biosorbents: Their preparation, application, and regeneration for wastewater treatment. *Sci. Total Environ.* **2020**, *702*, 134893. [CrossRef]
114. Gao, Q.; Blum, K.M.; Gago-Ferrero, P.; Wiberg, K.; Ahrens, L.; Andersson, P.L. Impact of on-site wastewater infiltration systems on organic contaminants in groundwater and recipient waters. *Sci. Total Environ.* **2019**, *651*, 1670–1679. [CrossRef]
115. Moreau, M.; Hadfield, J.; Hughey, J.; Sanders, F.; Lapworth, D.J.; White, D.; Civil, W. A baseline assessment of emerging organic contaminants in New Zealand groundwater. *Sci. Total Environ.* **2019**, *686*, 425–439. [CrossRef] [PubMed]
116. Llamas-Dios, M.I.; Vadillo, I.; Jiménez-Gavilán, P.; Candela, L.; Corada-Fernández, C. Assessment of a wide array of contaminants of emerging concern in a Mediterranean water basin (Guadalhorce river, Spain): Motivations for an improvement of water management and pollutants surveillance. *Sci. Total Environ.* **2021**, *788*, 147822. [CrossRef] [PubMed]
117. Riva, F.; Castiglioni, S.; Fattore, E.; Manenti, A.; Davoli, E.; Zuccato, E. Monitoring emerging contaminants in the drinking water of Milan and assessment of the human risk. *Int. J. Hyg. Environ. Health* **2018**, *221*, 451–457. [CrossRef] [PubMed]
118. Yang, Y.Y.; Zhao, J.L.; Liu, Y.S.; Liu, W.R.; Zhang, Q.Q.; Yao, L.; Hu, L.X.; Zhang, J.N.; Jiang, Y.X.; Ying, G.G. Pharmaceuticals and personal care products (PPCPs) and artificial sweeteners (ASs) in surface and ground waters and their application as indication of wastewater contamination. *Sci. Total Environ.* **2018**, *616*, 816–823. [CrossRef] [PubMed]
119. Lee, H.-L.; Kim, K.Y.; Hamm, S.-Y.; Kim, M.S.; Kim, H.K.; Oh, J.-E. Occurrence and distribution of pharmaceutical and personal care products, artificial sweeteners, and pesticides in groundwater from an agricultural area in Korea. *Sci. Total Environ.* **2019**, *659*, 168–176. [CrossRef]
120. Sharma, B.J.; Bečanová, J.; Scheringer, M.; Sharma, A.; Bharat, G.K.; Whitehead, P.G.; Klánová, J.; Nizzetto, L. Health and ecological risk assessment of emerging contaminants (pharmaceuticals, personal care products, and artificial sweeteners) in surface and groundwater (drinking water) in the Ganges River Basin, India. *Sci. Total Environ.* **2019**, *646*, 1459–1467. [CrossRef]

121. Alquwaizany, A.S.; Alfadul, S.M.; Khan, M.A.; Alabdulaaly, A.I. Occurrence of organic compounds in groundwater of Saudi Arabia. *Environ. Monit. Assess.* **2019**, *191*, 601. [CrossRef]
122. Hou, L.; Jiang, J.; Gan, Z.; Dai, Y.Y.; Yang, P.; Yan, Y.; Ding, S.; Su, S.; Bao, X. Spatial Distribution of Organophosphorus and Brominated Flame Retardants in Surface Water, Sediment, Groundwater, and Wild Fish in Chengdu, China. *Arch. Environ. Contam. Toxicol.* **2019**, *77*, 279–290. [CrossRef]
123. Close, M.E.; Humphries, B.; Northcott, G. Outcomes of the first combined national survey of pesticides and emerging organic contaminants (EOCs) in groundwater in New Zealand 2018. *Sci. Total Environ.* **2021**, *754*, 142005. [CrossRef]
124. Huang, H.; Liu, H.; Xiong, S.; Zeng, F.; Bu, J.; Zhang, B.; Liu, W.; Zhou, H.; Qi, S.; Xu, L.; et al. Rapid transport of organochlorine pesticides (OCPs) in multimedia environment from karst area. *Sci. Total Environ.* **2021**, *775*, 145698. [CrossRef]
125. Thompson, T.J.; Briggs, M.A.; Phillips, P.J.; Blazer, V.S.; Smalling, K.L.; Kolpin, D.W.; Wagner, T. Groundwater discharges as a source of phytoestrogens and other agriculturally derived contaminants to streams. *Sci. Total Environ.* **2021**, *755*, 142873. [CrossRef]

applied sciences

MDPI

Review

# Biodegradable Solvents: A Promising Tool to Recover Proteins from Microalgae

**David Moldes [1,2], Elena M. Rojo [1,3], Silvia Bolado [1,3], Pedro A. García-Encina [1,3,*] and Bibiana Comesaña-Gándara [1,3]**

1 Institute of Sustainable Processes, University of Valladolid, Dr. Mergelina s/n, 47011 Valladolid, Spain; david.moldes@uva.es (D.M.); elenamaria.rojo@uva.es (E.M.R.); silvia.bolado@uva.es (S.B.); bibiana.comesana@uva.es (B.C.-G.)

2 Department of Analytical Chemistry, Faculty of Sciences, Campus Miguel Delibes, University of Valladolid, Paseo de Belén 7, 47011 Valladolid, Spain

3 Department of Chemical Engineering and Environmental Technology, School of Industrial Engineering, University of Valladolid, Dr. Mergelina s/n, 47011 Valladolid, Spain

* Correspondence: pedroantonio.garcia@uva.es

**Abstract:** The world will face a significant protein demand in the next few decades, and due to the environmental concerns linked to animal protein, new sustainable protein sources must be found. In this regard, microalgae stand as an outstanding high-quality protein source. However, different steps are needed to separate the proteins from the microalgae biomass and other biocompounds. The protein recovery from the disrupted biomass is usually the bottleneck of the process, and it typically employs organic solvents or harsh conditions, which are both detrimental to protein stability and planet health. Different techniques and methods are applied for protein recovery from various matrices, such as precipitation, filtration, chromatography, electrophoresis, and solvent extraction. Those methods will be reviewed in this work, discussing their advantages, drawbacks, and applicability to the microalgae biorefinery process. Special attention will be paid to solvent extraction performed with ionic liquids (ILs) and deep eutectic solvents (DESs), which stand as promising solvents to perform efficient protein separations with reduced environmental costs compared to classical alternatives. Finally, several solvent recovery options will be analyzed to reuse the solvent employed and isolate the proteins from the solvent phase.

**Keywords:** biodegradable solvents; cell disruption; deep eutectic solvents; extraction with solvents; ionic liquids; microalgae biorefinery; protein recovery; protein release; solvent recovery

**Citation:** Moldes, D.; Rojo, E.M.; Bolado, S.; García-Encina, P.A.; Comesaña-Gándara, B. Biodegradable Solvents: A Promising Tool to Recover Proteins from Microalgae. *Appl. Sci.* **2022**, *12*, 2391. https://doi.org/10.3390/app12052391

Academic Editor: Carlos Rico de la Hera

Received: 28 January 2022
Accepted: 23 February 2022
Published: 25 February 2022

**Publisher's Note:** MDPI stays neutral with regard to jurisdictional claims in published maps and institutional affiliations.



## 1. Introduction

Proteins are one of the three essential macronutrients in the diet and the most important to prevent malnutrition and hunger [1]. According to the United Nations, the global population will increase to 9.5 billion people by 2050 [2]; as a result, the demand for food, and therefore also for protein, will rise [3]. There are two primary protein sources of similar nutritional quality: animal (which includes meat, fish, eggs, and dairy products) and plant (legumes, and to a lesser extent, seeds, and nuts). However, animal protein is known for its high environmental and climate impact, not only for the emissions related to their production but also due to the arable land used for the intensive crop production for feed and a large amount of water required (for both feed and animals) [4]. Thus, it is necessary to find sustainable plant-based sources and production processes to meet future protein requirements without damaging the planet and human health [5]. One of the most promising alternative feedstocks are microalgae, which are photosynthetic organisms mainly composed of proteins, carbohydrates, and lipids [6,7]. Microalgae use solar energy to store carbon from $CO_2$ conversion, and they also uptake nutrients from the culture medium for their growth [8]. This type of biomass has several advantages over others, including superior photosynthetic efficiencies and biomass production potentials than terrestrial crops, growth over a wide range of pH and temperature, capacity to uptake nutrients from different types of sources, such as wastewater, and most of them have a short generation cycle [9].

The content of proteins, carbohydrates, lipids, and other microalgae biocompounds can be utilized as feedstock for various bioproducts applying biorefinery processes [10]. Nevertheless, the current approaches are still not profitable and have some challenges to be addressed [11]. The price of microalgae cultivation is a notable drawback of microalgae-based biorefineries, but the use of low-cost culture medium as wastewater could contribute to achieve economically sustainable processes. As a result, the microalgae biorefinery approach can be integrated within the circular bioeconomy, whose main goals are to reduce, reuse, and recycle raw materials [11]. Additionally, biomass's integral and fractional valorization could maximize the process benefits by taking advantage of waste streams, promoting this circular bioeconomy [7].

The characteristics and composition of microalgae depend on the species and the environmental and operational conditions during cultivation [12]. However, in most cases, the main constituents of microalgae biomass are proteins, ranging from 30 to 70% mass percentage in dry weight [13], achieving the higher protein concentrations in microalgae grown in stressful environments such as wastewater. Due to the proteins' high added value and lability, it is reasonable to first recover the protein fraction in the process of fractional valorization of this biomass. The major issue with protein recovery is related to where the proteins are found in the cell. Typically, these biomolecules are inside a very recalcitrant cell wall that is difficult to break and whose thickness and chemical composition depend on various factors such as growth stage, cellular form, and environmental conditions [14]. Hence, the first step of a biorefinery process to recover biomolecules from microalgae is the cell wall disruption by physical, chemical, or biological methods to extract intracellular molecules by different solvents [15]. The use of organic solvents is a well-known technique for the extraction and separation of microalgae biomolecules, but some of them are toxic and not aligned with environmental and human health concerns [16]. Thus, alternative solvents have recently begun to be studied, such as ionic liquids (ILs) and deep eutectic solvents (DESs), with lower environmental impact than conventional solvents and promising applications in microalgae processing [17].

## 2. Microalgae Biorefinery: Biomass Treatment

According to Chew et al., the biorefinery is a "process to obtain biofuels, energy and high-value products through biomass transformation and process equipment" [18]. In the case of microalgal biorefinery, microalgae biomass is used as feedstock. Different steps are involved in the process, including cultivation, harvesting, cell disruption, component recovery, and final product purification [19]. The development of these new industries is faced with critical challenges; as an example, industrial photobioreactors for cultivation in synthetic media have high costs and reach low biomass concentrations. At the same time, current downstream processes are designed for a single main product, and the rest of the microalgae biomass is frequently "waste" that is not useful [11]. Because of all this, it is necessary to develop more efficient technologies to achieve a sustainable process for microalgal cultivation, but also to profit and valorize all the components of this microalgal biomass, establishing a fractional biorefinery process. This sequential valorization main concern is the release of different fractions without damaging the other fractions to attain an integral recovery of the biomass components [18].

The first step for the microalgae valorization is the cell disruption (Figure 1), which is a critical process for obtaining the desired biocomponents from intracellular contents [20]. Microalgae cells present a high resistance to intracellular extraction, usually attributed to their rigid cell walls [21]. This cell wall consists of a microfibrillar network composed of diverse polysaccharides, proteins (mainly glycoproteins), and a biopolymer called algaenan, which is very difficult to degrade [14]. This cell disruption can be achieved by several processes, which can be classified into physical (e.g., bead milling, high-pressure homogenization, microwave irradiation, ultrasonication, pulse electric field), chemical (e.g., acid, alkali, and thermal hydrolysis) and biological methods (e.g., enzymatic hydrolysis) [16]. However, most of these techniques require harsh conditions (such as chemical methods), which allow obtaining high release efficiencies but at the expense of degrading the released bioproducts, resulting in low recovery yields. Thus, the use of mild cell disruption will maintain the biomolecule's activity and functionality to provide high-quality products [20]. Moreover, in order to take advantage of all the algal biomass and achieve a sustainable process, cell wall components (including some types of proteins) may themselves also be target molecules within the biorefinery process [15].

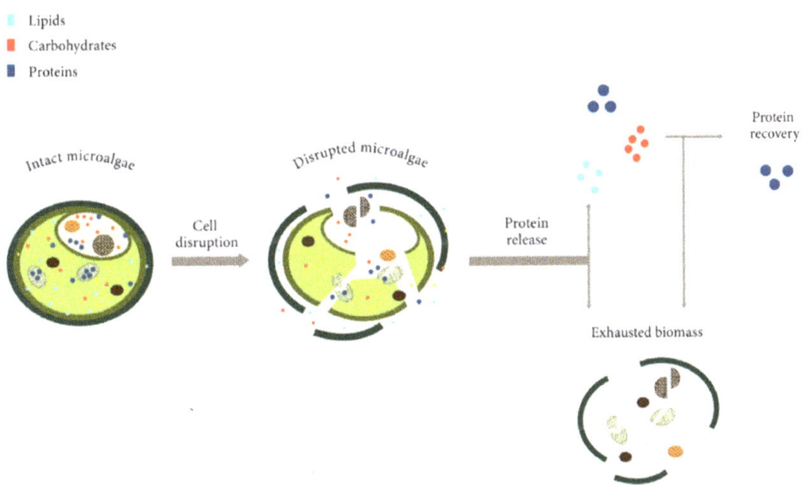


**Figure 1.** Scheme of proteins obtaining from biomass. Adapted from [22].

Once the cellular disruption occurs (Figure 1), the biomolecules contained inside the cell are released (proteins, carbohydrates, or lipids), usually employing an extraction solvent. Those soluble biomolecules will be in a liquid phase, whose composition will depend on the cell disruption method and extraction conditions. Therefore, a recovery step is needed to isolate the required product from the other released compounds [16].

In the case of protein recovery, there are many possible techniques, including chemical precipitation, electrophoresis, membrane filtration, chromatography, or solvent extraction [23]. Although some of these methods provide good separation results, they have essential disadvantages such as low resolution, use of toxic solvents, presence of plenty of impurities, time-consuming process, and the use of harsh conditions such as high temperatures or energy consumption, which may lead to protein denaturation [23,24]. Furthermore, a single methodology usually cannot achieve the desired purification effect. Among all these alternatives, solvent-extraction stands as an attractive technique because it is well known, efficient, and easily scalable. Nonetheless, the use of traditional toxic organic solvents can lead to changes in the stereochemistry and degradation of the released proteins [12]. Thus, there is a high demand for suitable non-toxic solvents that do not damage proteins. Two new types of solvents have been developed recently to substitute conventional organic solvents, which are in the category of biodegradable solvents: ionic liquids (ILs) and deep eutectic solvents (DESs) [25]. Biodegradable solvents are an interesting replacement for classical solvents, as they have lower toxicity and are a more environmentally friendly alternative. In this regard, DESs are considered excellent substitutes to address these bottlenecks [26], since they are instrumental as a suitable solvent for the release and purification of several proteins without altering their native structure [27]. In the following sections, different approaches for protein recovery will be considered, particularly from microalgae biomass, emphasizing biodegradable solvents.

## 3. Protein Recovery

Once the microalgal biomass has been adequately treated, obtaining proteins selectively from the liquid phase with all the released compounds requires the use of recovery approaches. The recovery stands as one of the bottlenecks of the process [28,29], as the purity levels of the proteins obtained will determine their subsequent applications [23]. Generally, using a single recovery step is not enough to isolate proteins of enough purity. Thus, the current trend is to follow a multiple-step approach [19]. The process becomes complicated because of the hydrophilic nature of proteins, which makes them prone to denaturation, precipitation, and degradation. Proteins are fragile molecules, so mild operation conditions are required to preserve their native structure and biological activity. A brief description of the most used conventional separation techniques, along with new emerging alternatives such as biodegradable solvents, which are up-and-coming tools to perform efficient protein purification [22], will be related in this section, with a focus on their physicochemical basis, pros, and cons. Other techniques such as molecular imprinting [30], magnetic separation [31], or electrostatic separation [32] are attracting researchers' attention, but they are still in a very incipient state.

### 3.1. Precipitation Methods

Precipitation is a process in which a solid is obtained from a solution, in the case of proteins, by adding a precipitation agent. Precipitation is a suitable method for crude protein recovery, and it preserves the structural activity of the proteins [33,34], but it is inefficient for further purification stages as it does not provide high recovery yields and efficiencies [23]. Depending on the precipitation agent used, three distinct approaches have been tried for microalgae protein recovery: (i) isoelectric point precipitation, which deals with pH changes; the compounds usually employed are NaOH and HCl [35]; (ii) salting-out, consisting of using a high concentration of salt solutions such as NaCl or $K_2HPO_4$ [36], and (iii) organic solvent precipitation, for which classical organic solvents such trichloroacetic acid or mixtures of ethanol and acetone, are commonly used [37]. This latter approach affords a larger protein recovery yield than isoelectric point precipitation or salting-out; however, classical organic solvents are not advisable due to their environmental concerns and the complexity of solvent recovery.

### 3.2. Membrane Separations

Membranes are boundaries between two phases. They have been widely used in the separation field, achieving high selectivity. Non-solvent or harsh conditions are involved, and the purification can be carried out at low temperatures, preventing protein degradation [38,39]. However, classical filtration usually is a slow process. To solve this issue, tangential and cross-flow processes of microfiltration (MF) and ultrafiltration (UF) have been developed. MF membranes are good at separating components ranging from 10.0 to 0.1 μm [40], even though modest recovery yields are usually achieved after filtration [41]. On the other hand, UF employs membranes with a pore size of 0.1–0.01 μm, requiring slightly higher pressures than MF (0.5–5 bar versus 0.1–2 bar required for MF) or the use of centrifugal force [40]. There are some handicaps for the implementation of membranes to industry, and fouling is probably the most troublesome point, as it reduces membrane permeability, decreasing the filtration performance. That is one of the reasons why tangential filtration of cross-flow filtration has gained importance in recent years [42,43]. Only a couple of studies have been found applying membrane filtration for microalgae protein recovery [44,45].

Despite MF and UF consisting of semipermeable membranes, the driving force in both cases is the external pressure applied. In opposition, dialysis is a separation method based on the concentration gradient established between the phases [23]. Dialysis can aid in removing ionic species from the protein fractions obtained [46] as part of a combination of protein purification methods.

### 3.3. Electrophoresis and Chromatographic Techniques

Electrophoresis separates analytes according to their mobility through an aqueous buffer in response to an electric field. The most common mode for protein purification is capillary electrophoresis [16]. Electrophoresis is an excellent technique to identify and quantify amino acids [47], but it is scarcely used for industrial protein purification because of its long operational times and low sample intake.

Chromatographic methods, in turn, allow the separation of closely related molecules in complex mixtures [48]. Chromatographic techniques used for protein purification are gel permeation, ion-exchange, affinity, and hydrophobic interaction chromatography [49,50]. Sometimes, protein purification might require the combined use of several chromatographic techniques. For example, the use of a first stage in which ion-exchange chromatography separates protein fractions according to their charge, and then a gel permeation stage to obtain proteins with similar charge and size, resulting in the so-called two-dimensional approach, used for proteomic separation. Although chromatography is often oriented to quantitative analysis, some operational modes can be applied to purify molecule fractions, known as preparative chromatography [51]. Nevertheless, chromatography is an expensive technique and not easy to scale up [52]. In addition, it demands high amounts of organic mobile phases, which is not environmentally sustainable [53]. Therefore, chromatographic techniques are not a good alternative for microalgae proteins at the industrial level [23].

### 3.4. Solvent Extraction

As mentioned above, solvents can be used to release the proteins from inside the biomass cell, but they can also be employed to selectively obtain proteins from the released compounds through conventional liquid–liquid extraction (LLE). This separation technique consists of the analyte's migration to the extracting phase, relying on the different affinity for the medium (measured by the distribution constant, $K_D$) and the concentration gradient [54]. This section will discuss classical and new emerging solvents for protein LLE, along with innovative solution combination strategies.

Commonly employed solvents for protein recovery are hydrocarbons, alkalis, acetone, alcohols, and chloroalkanes [55]. Even though these solvents are low-cost and have good solubility capacities [56], they also present important drawbacks such as long operational times, high solvents and energy consumption, and the high odds of separated proteins damage [57]. Furthermore, most of them are volatile organic compounds (VOCs) derived from non-renewable sources such as petroleum, being both detrimental to human health (as the product obtained might introduce pollutants) and the earth's well-being [56,58]. The use exclusively of classical solvents to obtain proteins from microalgae is not common. However, several studies reported their use in combination with greener options such as aqueous two-phase systems, in which the organic solvent volume is significantly reduced [59].

Due to the drawbacks of classical solvents, alternative strategies are needed to achieve more sustainable processes, with reduced energy consumption and lower presence of hazardous substances [60]. One approach to accomplish these goals consists of replacing classical for green solvents, which must be cheap, non-toxic, safe, stable, biodegradable, renewable, and atom economic in their synthesis; however, telling which solvents are green is still a controversial issue [61]. Moreover, most typical green solvents are not preferred for protein recovery because proteins are not very soluble in water [56] nor supercritical fluids such as $CO_2$ [62]. Thus, the aim is to find greener solvent options for LLE, and one class of solvents that have shown promising results in the recovery of bioactive compounds are the so-called biodegradable solvents, which are not necessarily fully green (although every green solvent must be biodegradable). The advantage is that defining what a biodegradable solvent is can be quantified in a rigorous manner [63,64]. These solvents can be combined with innovative assisted technologies such as microwave, or ultrasound, to achieve an optimum purification process [55,62]. This is the case for ionic liquids (ILs) and deep eutectic solvents (DESs), whose advantages and applications will be examined in Section 4.

Another possibility is to combine two different solutions to carry an LLE process, as applied in an aqueous two-phase system (ATPS), which can also be found in literature as aqueous biphasic system (ABS) [65]. As its name indicates, ATPS consists of two aqueous phases that are immiscible due to the repulsive behavior of their components [66,67]. Compared to classical single solvent extraction, ATPSs allow obtaining in one step proteins in high concentration, saving time and costs, as the equilibrium is rapidly reached [68]. One phase is usually formed by a water-soluble polymer solution or a hydrophilic organic solvent (such as alcohols or organic acids), while the other is generally a salt solution [69]. ATPSs allow mild operation conditions, reducing the chance of denaturation or loss of structural activity of the proteins [69]. Moreover, ATPSs enable continuous mode operation conditions, so they are easy to scale up, unlike other techniques, restricted to batch mode [65]. ATPS containing conventional organic solvents has proved potential for microalgae protein purification; for example, Phong et al. [59] prepared an ATPS formed by aqueous solutions of ammonium sulfate and 2-propanol; or Baskaran et al. [70], who, instead of the 2-propanol, used polyethylene glycol 600. Nevertheless, the interest is focused on ATPS based on green or biodegradable solvents, as they show the excellent properties of a typical ATPS using non-toxic components (or at least, less detrimental). The main drawback of these methodologies is the insufficient knowledge about the recovery from the phase in which the proteins are concentrated, and, in fact, this is still the subject of extensive research [23].

In summary, solvent extraction can be performed through single-solvent extraction and with an ATPS system. Substituting classical solvents with biodegradable solvents will ensure a greener process compared to the traditional approaches. In this regard, ILs and DESs emerged as suitable alternatives, as they enable more efficient and selective purifications, not damaging the structural activity of the proteins extracted and being more environmentally friendly than classical solvents.

## 4. Biodegradable Solvents for Protein Extraction

Biodegradable solvents are compounds able to be degraded in smaller chemical entities by biological organisms' actions and other factors such as temperature or light. The two types of biodegradable solvents more studied for biocompounds recovery, and in particular for proteins, are ILs and DESs, which are also known as "designer solvents" because their physicochemical properties can be tuned if needed for further improvements on the separation performance, achieving high protein recovery yields [71]. Most authors do not grant them the category of "green" because, despite their good biodegradability and environmental advantages, other requisites needed are not fulfilled, such as high stability under high temperature and low pressure, flammability, and potential toxicity [61]. In addition, other aspects related to solvent preparation must be examined, for example, the synthesis process of ILs employs organic solvents and demands high temperature and reaction times, not to mention the final purification step always required, so overall, the process could not be considered green [72]. Hence, it is advisable to refer to them as biodegradable solvents.

In this section, ILs and DESs for solvent protein extractions will be reviewed, detailing their characteristics, differences, strengths, weaknesses, application range, together with a revision of the current greenness state of both types of solvents. Finally, the solvent recovery options available will be evaluated.

### 4.1. Ionic Liquids (ILs)

ILs are liquid salt mixtures at room temperature [73]. This low melt temperature is a result of their low-charge density and low symmetry ions of the organic salts, along with ionic interactions, hydrogen bonds, and dipole–dipole intermolecular forces [74]. Generally, to obtain salts in the liquid state, very high temperatures are required to disrupt the strong ionic interactions. However, the strong electrostatic interactions of ILs lead to a lowering of the melting point below 100 °C [75]. Although there are many options for organic salts, cations are usually exclusively organic derivatives, while anions might have either organic or inorganic nature [76]. Because of their molecular structure, ILs possess outstanding physicochemical properties compared to organic solvents: low vapor pressure (<1 Pa), high conductivity, moderate thermal stability, and great extraction capacities for many organic and inorganic compounds [77]. They also have a wide electrochemical window, and valuable polarity properties, depending on the ions used [78]. Moreover, thanks to their physicochemical characteristics, some ILs (such as 1-tetradecyl-3-methylimidazolium bromide, or 1-ethoxyethanol-3-methylimidazolium chloride, etc.) have demonstrated promising results in guaranteeing the stability and activity of proteins, as shown in the review published by Patel et al. [79]. Nevertheless, details of the mechanism of action of how the IL interacts with proteins to extract them still need to be elucidated [72].

The first IL was synthesized in 1914 by Paul Walden, yet until 2007, they were not explicitly applied to extract biocompounds [76]. To synthesize an IL, diluted solutions of each ion must be prepared and mixed correctly. When the reaction ends (which usually takes up to 48 h and needs 25–100 °C), the solid residue is removed and washed with an organic solvent. As stated by Xiao et al., ILs have been used extensively to extract natural products [80]. However, attention must be paid to the ion components chosen to maximize the analyte–solvent interactions, leading to a more efficient process [72].

Extraction of proteins from microalgae biomass can be carried out by single IL solvent extraction or through an ATPS system. Single IL extraction is usually aided with a complementary technique such as ultrasonication or microwave. Rodrigues et al. evaluated different ILs and their mixture for obtaining phycobiliproteins from *Spirulina platensis* [81]. The authors extracted the highest protein concentration using an equimolar mixture of 2-hydroxyethylammonium acetate and 2-hydroxyethylammonium formate using an ultrasonication bath (25 °C and 25 kHz) and a solvent:biomass ratio of 7.93 mL $g^{-1}$. The pH was fixed at 6.50, and the extraction time was set to 30 min. The protein recovered in a more significant amount was allophycocyanin, attaining values of 6.34 mg $g^{-1}$, but also 5.95 mg $g^{-1}$ of phycocyanin and 2.62 mg $g^{-1}$ of phycoerythrin were recovered. As another example, Desai et al. also extracted proteins and astaxanthin by treating the liquid microalgae extract from *Haematococcus pluvialis* with the commercial trihexyl(tetradecyl)phosphonium bistriflamide ionic liquid and a microgel-stabilized emulsion. They reached an average extraction efficiency greater than 90% using a solvent:biomass ratio of 3:7, a microgel particle concentration of 1 g $L^{-1}$, and mixing time of 120 min; the proteins recovered preserved their native state after separation [82]. Similarly, Lee et al. employed an IL-buffer (cholinium 2-hydroxy-3-morpholinopropanesulfonate) at pH of 7.00 combined with ultrasonication (400 W; 24 kHz; 5 s on/5 s off pulse mode) to obtain proteins from *Chlorella vulgaris*. Proteins were recovered in a single step operation (extraction from the cell and purification occurred simultaneously) with an almost complete recovery yield of 95.0 $\pm$ 1.9% [83]. Instead of ultrasonication, Desai et al. applied bead milling as a cell disruption technique and managed to separate lipids of carbohydrates and proteins from *Neochloris oleoabundans*. They used a 40% solution of tributylmethylphosphonium methyl sulfate, a biomass concentration of 6.67 g $L^{-1}$ at 45 °C after 30 min. After centrifugation, the lipids were obtained in the solvent phase (68% recovery yield of total lipids). The proteins and carbohydrates remained in the microalgae biomass, which was subsequently mechanically disrupted, leading to an aqueous phase rich in those components, achieving 80% recovery yield for proteins and 77% for carbohydrates [84].

Concerning IL-based ATPS, there is more research when compared to single ILs extraction, as it combines the advantages of using ionic liquids with the benefits of following an ATPS approach: short extraction time, low viscosity, and options of recyclability [85]. The most common IL-based ATPS is composed of one ionic liquid and an aqueous polymer or salt solution [86–88]. For instance, Suarez Ruiz et al. studied the fractionation of several microalgal biomolecules: proteins, carbohydrates, lipids, and pigments, from *Neochloris oleoabundans*, grown in saline freshwater conditions [86]. After cell disruption, it was treated with three different ILs-ATPSs with a polymer-rich phase. A multistep fractionation process was implemented, and the system formed by PPG400-Ch DHp provided the best results, attaining an 82% of protein recovery in the IL phase, along with pigments (98%) and carbohydrates (93%), while lipids and starch were isolated from the interface, with recovery yields of 79 and 71%, respectively. Then, they proposed a further purification step applying ultrafiltration, selectively obtaining 80% of proteins from sugars and IL, allowing the recycling of the solvent. Rodrigues et al. also concluded that the electrostatic interaction between proteins and the IL might be the driving force of the separation process. Using salts instead of polymers ($[C8_{mim}]Cl$—$K_3PO_4$) to *Isochrysis galbana*, Santos et al. achieved a selective and efficient carbohydrate-protein separation, with 100% of proteins in the top phase, and 71% of carbohydrates in the bottom phase [87]. García et al. also employed an IL-salt ATPS, consisted of Iolilyte 221PG as IL, and potassium citrate tribasic monohydrate as salt, obtaining protein extracts with industrial relevance from crude microalgae *Neochloris oleoabundans* and *Tetraselmis suecica* applying mild conditions (pH = 7.0 and room temperature) [88].

The number of published articles concerning ILs grew until 2016, when concerns about their toxicity surpassed their assumed benefits. Even though ILs do not pollute the air due to their low vapor pressure at room temperature, some of the ions used have displayed ecotoxicological effects, flammability, and have proved to be poorly biodegradable [89]. However, not all ILs are indeed equally harmful. Some choline-based ILs derived from biomaterials (Bio-ILs) have shown low toxicity and high stability [90], substituting the cation (typically pyridinium or imidazolium) by choline. Nonetheless, the appearance of Bio-ILs has not prevailed among scientist investigations [91,92]. Furthermore, synthesizing ILs is expensive compared to classical solvents [93], mainly due to the starting products, which are obtained from fossil fuels sources, making the process less green, as waste is created indirectly [94]. Other important aspects of any green extraction are the recovery of the value-added compounds (hindered by their low vapor pressures) and the capacity of the solvent to be recovered and reused. These issues remain a challenge, and there is a need for further research, as generally, studies focused only on measuring protein content on the IL and are not concerned about these aspects (more information in Section 4.3) [76]. These drawbacks prevent the widespread ILs in the industry and engender the need for alternative greener options.

### 4.2. Deep Eutectic Solvents (DESs)

To avoid the problems associated with ILs, scientists turned their interest into a newly emerging type of solvents, DESs, which was first described in 2003 by Abbott et al. [95]. DESs are liquid mixtures at room temperature, formed by the combination of two solids (or a solid and a liquid) in a specific molar ratio, for which the eutectic point temperature is below that of an ideal mixture, much lower than the compounds separately [96]. The liquid obtained shows exceptional physicochemical properties similar to ILs [97]. Generally, the eutectic point is found in a significant depression, around 150 °C lower than if the mixture had an ideal behavior [98], and it is thought to be caused by the charge delocalization produced between the molecules mixed [99].

Some authors consider DESs as a subclass of ILs. However, most agree that DESs are a different kind of solvent because they are not formed exclusively by ionic species (existing DESs with non-ionic compounds), and no chemical reaction is involved in their preparation [17,62,98,100]. DESs have two components: a hydrogen bond donor (HBD) or a metal halide and a hydrogen bond acceptor (HBA). The hydrogen bond interaction between those components forms the eutectic mixture [98]. The variety of those two groups is enormous, so it is estimated there could be around $10^6$–$10^8$ possible combinations to form DESs [101]. Depending on the chemical nature of the components, DESs are classified into five distinct types (Table 1) [102,103]. However, most of the research has focused on Type III: DESs consisting of a quaternary ammonium salt as HBA (usually choline chloride) and a suitable hydrogen bond donor such as alcohols, carboxylic acids, or amides [104]. This is because type III shows a stronger hydrogen bond interaction, which directly affects the physicochemical properties of DESs [105].

**Table 1.** General classification of DESs [102,103].

| Type | General Formula | Terms |
|---|---|---|
| I—HBA and metal chloride | $Cat^+ X^- z MCl_n$ | M = Zn, Sn, Fe, Al, Ga, In |
| II—HBA and metal chloride hydrate | $Cat^+ X^- z MCl_n y H_2O$ | M = Cr, Co, Cu, Ni, Fe |
| III—HBA and HBD | $Cat^+ X^- z RZ$ | Z = OH, COOH, $CONH_2$ |
| IV—Metal chloride and HBD | $MCl_n + RZ = MCl_{n-1}{}^+{\cdot}RZ + MCl_{n+1}$ | M = Al, Zn; Z = OH, $CONH_2$, |
| V—Molecular compounds | Non-ionic DESs | Molecular substances |

$Cat^+$: ammonium, phosphonium, or sulfonium cation. $X^-$: Lewis base (generally, a halide). HBA: hydrogen bond acceptor. HBD: hydrogen bond donor. R: alkyl rest. z: stoichiometric coefficient.

As mentioned, DESs have similar physicochemical properties compared to ILs, since they present low volatility and vapor pressure (slightly larger than ILs), thermal stability, and tunable polarity, which allow solubilization of various molecules, including proteins [104]. The separation mechanism is meant to imply several forces such as hydrophilic-hydrophobic or electrostatic interactions, although it is thought that their excellent protein extraction capabilities are driven by the formation of protein-DES aggregates [24,106]. Contrary to ILs, DESs starting materials are cheap, easy to produce, and do not entail associated pollution (most are allowed for human consumption). DESs are also more biodegradable than ILs (>69.3% in 28 days, according to Zhao et al. [107]), but the nature of the ions highly determines their biodegradability [72]. Moreover, DESs are considered mixtures rather than pure compounds [96], so there is no need for by-product removal as no chemical reaction has occurred (process with 100% of the atom economy, as opposed to ILs). Due to their outstanding properties, such as great biodegradability, comparable physicochemical properties of ILs, and small or no toxicity, DESs are applied in many fields such as electrochemistry, organic and inorganic chemistry. In addition, DESs are being used in separation strategies replacing the role of ILs, as they are considered a greener alternative [108].

However, as reported by Chen et al. [61], not only the greenness of ILs must be questioned. Some DESs should also be critically reevaluated, as they encompass some downsides such as hygroscopicity, toxicity, regenerability and renewability, and decomposition at high temperatures. To circumvent some of these problems, Choi et al. found that a DES-like liquid could be formed when mixing natural primary cellular metabolites such as choline, sugars, betaine, alcohols, organic acids, and amino acids [109]. Thus, using natural products to prepare DESs (natural deep eutectic solvents or NADESs) will ensure low or negligible environmental impact and toxicity, as these new solvents are biodegraded >70% in 7 days [110], along with ubiquity and low cost [62]. A rigorous comparison of the key properties of ILs, DESs, and NADESs, has been recently performed by Mehariya et al., highlighting that NADESs are more eco-friendly and exhibit lower toxicity than DESs, which in turn are better in these and other aspects such as biodegradability, cost, and synthesis facility than ILs [111]. NADESs have proved to be an ideal extraction media since many macromolecules such as DNA, proteins, carbohydrates, and other small molecules are highly soluble in some of these solvents such as 1,2-propanediol–choline chloride–water, glucose–choline chloride–water, or lactic acid–glucose–water, as demonstrated by Dai et al. [112]. Analogous to ILs, the complete understanding of the underlying extraction mechanism is yet to be elucidated. Even though NADESs show numerous advantages and are considered capable of dissolving all sorts of biomolecules in living things [109], they also present some limitations, such as their high viscosity, which may prevent further industrial application. Nonetheless, this issue can be solved by increasing the operation temperature, or by adding water, which is preferred to maintain the structural activity of the proteins extracted [111].

Many works have proved the good performance of DESs to selectively extract proteins for several matrices, such as animal tissues or plant-derived sources like oilseeds [24,113–117]. For instance, Bai et al. achieved >90% of extraction efficiency for collagen proteins from cod skin, with the corresponding purity of 93%, using the single-DES formed by choline chloride and oxalic acid in optimum conditions [115]. Turning the attention to the research focused on matrices similar to microalgae biomass, it is worth mentioning Grudniewska and co-workers research, in which protein-rich extracts from rapeseed cake and evening primrose cake, residues derived from the oil industry were obtained using a choline chloride:glycerol DES, in 1:2 molar ratio. Once the proteins are in the DES phase, they can be precipitated using water as an anti-solvent, increasing the percentage of protein in the final extract up to 20% compared to the starting biomass [114]. However, most of the research is focused on DES-based ATPS, as it is usually the most efficient technique to recover and purify proteins due to the low interfacial tension of these systems. For instance, Li et al. developed a biphasic system based on a NADES containing betaine as HBA, urea as HBD, and water in a 1:2:1 molar ratio, reaching an extraction efficiency of 99.82% for BSA, and therefore showing ATPS are valuable tools for protein obtaining [117].

Regarding protein recovery from microalgae using DESs, only a couple of studies were found [118,119]. Firstly, Cicci et al. carried out three sets of experiments applying different pretreatments to study the extraction of proteins, lipids, carbohydrates, carotenoids and chlorophylls from *Scenedesmus dimorphus* (UTEX 1237) [118]. The NADES was prepared by combining 1,2-propanediol, choline chloride, and water in a 1:1:1 molar ratio. The experimental set-up consisted of putting in contact the solvent with the biomass and simultaneously applying different cell disruption methods (ball mill/ultrasonication/none). The highest protein extraction yield was achieved with the ultrasound-assisted experiments, attaining a 27%. No further purification was performed to isolate the extracted proteins. Subsequently, Sed et al. conducted a study with the same microalgae, but using a new NADES (based on a mixture of octanoic and dodecanoic fatty acids) and a sequential two-stage extraction, consisting of a first extraction step using the solvent in its hydrophilic form (recovering the proteins), and then a second phase with the solvent in its hydrophobic mode, providing the solvent and other hydrophobic compounds. The protein extraction efficiency achieved was 100% [119].

Hence, solvent extraction with DESs or ATPS-DES is a promising sustainable technique for protein recovery. However, there is still much scope to further investigate these methodologies for microalgae protein extraction and also for the purification of the proteins once they are extracted.

Table 2 compiles the protein recovery studies reviewed in this article, summarizing the IL or DES employed, the salt or polymer used (in case of following ATPS approach), the specific process conditions, and finally the protein yield addressed.

**Table 2.** Protein recovery from different microalgae with ILs, DESs, and ATPS shown in this review.

| Microalgae Strain | DES/IL Used (Molar Ratio) | Salt/Polymer Used (in Case of ATPS) | Extraction Method and Conditions | Yield | Reference |
|---|---|---|---|---|---|
| *Spirulina platensis* | 2-hydroxy ethylammonium acetate and 2-hydroxy ethylammonium formate (1:1) | None | Single IL extraction<br>US bath: 25 °C and 25 kHz<br>Solvent:Biomass concentration: 7.93 mL $g^{-1}$<br>pH: 6.50<br>Extraction time: 30 min | Allophycocyanin: 6.34 mg $g^{-1}$<br>Phycocyanin: 5.95 mg $g^{-1}$<br>Phycoerythrin: 2.62 mg $g^{-1}$ | [81] |

**Table 2.** *Cont.*

| Microalgae Strain | DES/IL Used (Molar Ratio) | Salt/Polymer Used (in Case of ATPS) | Extraction Method and Conditions | Yield | Reference |
|---|---|---|---|---|---|
| *Haematococcus pluvialis* | Trihexyl(tetradecyl) phosphonium bis-325 triflamide | None | Single IL extraction, aided with microgel particle emulsion (concentration: 1 g $L^{-1}$)<br>Solvent:Biomass ratio 3:7<br>Mixing time: 120 min | Protein recovery: >90% | [82] |
| *Chlorella vulgaris* | Cholinium 2-hydroxy-3-morpho linopropanesulfonate | None | Single IL extraction<br>US: 400 W; 24 kHz; 5 s on/5 s off pulse mode<br>pH: 7.00 | Protein recovery: 95.0 ± 1.9% | [83] |
| *Neochloris oleabundans* | Tributylmethyl phosphonium methyl sulfate (40% solution) | None | Single IL extraction<br>Biomass:Solvent concentration: 6.67 g $L^{-1}$<br>Temperature: 45 °C<br>Extraction time: 30 min | Protein recovery: 80% | [84] |
| *Neochloris oleabundans* | Cholinium dihydrogen phosphate (14 wt%) | Polypropylene glycol 440 (40 wt%) | IL-ATPS extraction<br>Microalgae suspension: 2.5 mg (dry weight)/mL<br>Extraction conditions: 50 rpm, 1 h<br>Room Temperature (RT) | Protein recovery: 82% | [86] |
| *Isochrysis galbana* | 1-decyl-3-methylimidazolium chloride (15 wt%) | $K_3PO_4$ (20 wt%) | IL-ATPS extraction: 10,000 rpm, 10 min<br>Temperature: RT | Protein separation: 100% in the top phase | [87] |
| *Neochloris oleabundans* and *Tetraselmis suecica* | Iolilyte 221PG | Potassium citrate | IL-ATPS extraction: 14,000 rpm and 30 min<br>Temperature: RT<br>pH: 7.0 | Protein extraction efficiencies: 75–85% (depending on the tie lines) | [88] |
| *Scenedesmus dimorphus* (UTEX 1237) | 1,2-propanediol; chloine chloride; water (1:1:1) | None | Single DES extraction: 250 rpm and 24 h<br>Biomass:Solvent ratio: 10 g $L^{-1}$<br>Temperature: RT<br>US: 40 min with a frequency of 20 kHz and a range of 70% | Protein recovery yield: 27% | [118] |
| | Octanoic acid and dodecanoic acid (1:1), plus a dilute (5%) aqueous solution of Jeffamine D-230 | None | Single DES extraction, conditions not correctly described in article | Protein extraction efficiency: 100% | [119] |

### 4.3. Solvent Recovery Options

From an industrial point of view, recycling and reusing the solvents is essential to achieve a cost-efficient process [120]. However, it is also vital to obtain the product (proteins) without any impurity of solvent or by-product. Thus far, research has focused on transferring the proteins from the matrix to the biodegradable solvent. However, several issues such as solvent recycling, protein recovery from the biodegradable solvent, and potential solvent regeneration are not usually addressed and remain to be solved. Mainly because of the low vapor pressure of the solvents used, the isolation of the proteins from ILs or DESs by vaporization (the most extended method) is complicated, implying protein degradation [100]. New approaches have been attempted, such as the use of resins, ultrafiltration, adsorption [120,121]. Liquid–liquid re-extraction can be an efficient approach to isolate the biocompounds and recover the solvent [122], although organic solvents could be involved, limiting the greenness of the process.

The liquid–liquid re-extraction (or back-extraction) method is preferred for studies concerning protein recovery. Nonetheless, little research has been reported about the recovery problem of these new biodegradable solvents. For example, Zeng et al. used a sodium dodecyl sulfate solution as a stripping reagent to recover the DESs employed in a $K_2HPO_4$-ATPS system (choline chloride-urea, tetramethylammonium chloride-urea, tetrapropylammonium bromide-urea, and choline chloride-methylurea). However, they did not succeed, due to their hydrophilic character [24]. Li et al. managed to back-extract the proteins from a DES-ATPS ($K_3PO_4$ and betaine-urea) by forming a new ATPS with fresh $K_3PO_4$ aqueous phase mixed with ethanol (to improve the phase-forming ability of the DES). The ATPS created was shaken mechanically for 12 min, and then the DES phase was collected. Unfortunately, the highest back-extraction efficiency obtained was relatively low (32.66%) [117]. In turn, Xu et al. studied the extraction efficiency for BSA and trypsin with a DES-ATPS formed by $K_2HPO_4$ and choline chloride-glycerol, obtaining extraction yields of 98.71 and 94.36%, respectively. Then, to obtain the BSA from the DES-rich phase, 1.8 g of $K_2HPO_4$ were added to 0.27 mL to the DES-rich top phase, forming a new DES-ATPS, fixing the ion strength by adding 3% solution of NaCl, they obtain a highest back-extraction yield of 32.96% [116]. Similarly, Zhang et al. study BSA's extraction and back extraction efficiencies for two DES: tetramethylammonium chloride-urea (TMAC-U) and tetramethylammonium chloride-glycerol-urea ternary (TMAC-G-U). Once the extraction was performed and the DES phase-separated, they carried out the back extraction by adding a solution of $K_2HPO_4$ (0.70 g $mL^{-1}$) to create a new ATPS system. For TMAC-G-U, the back extraction efficiency achieved was 71.88%, while for the TMAC-U, only a modest 21.02% was reached. Thus, Zhang et al. confirmed that the main handicap for back-extraction was the high interfacial mass transfer resistance between the two phases, as the addition of glycerol reduced the mass transfer resistance. Furthermore, they also checked the extraction capability of the DES used, showing that it still can be used for further experiments (97.99% of extraction efficiency for new BSA added) [113].

Regarding ILs, there is even less investigation. The research discussed in Section 4.1 from Rodrigues et al. studied the recovery of the IL by using a precipitating agent (ammonium sulfate), concluding that it could be reused up to three times, although the extraction capabilities were reduced [81]. Alternatively, Orr et al. studied the recycling of the IL employed for selective lipid extraction through anti-solvent precipitation, adding methanol to the biomass-IL mixture. Then, the precipitated IL was separated from the biomass by centrifugation, and finally, the methanol was eliminated via evaporation. Although chlorophyll was also recovered along with the IL, they obtained an average recovery of 98.0%, showing a similar performance in further extraction cycles. Thus, more studies focusing on the recovery of other impurities should be performed [123].

Furthermore, in the context of a biorefinery, these solvent recovery processes have not been put into effect at an industrial level, although given the low toxicity of NADESs, they might not need to be recovered, which would diminish the overall cost of the process [124]. Therefore, despite the bright present and future of biodegradable solvents, there is an urgent need to develop green, reliable, and straightforward methods that will enable us to recover and purify the proteins and that allows correct recycling and regeneration of the solvent used so that the environmental impact is minimized or negligible. The current state of the art is in its infancy, and there are very few studies dealing with protein purification or solvent recovery issues.

## 5. Conclusions and Future Perspectives

To solve the upcoming protein shortage, new alternatives to animal protein must appear, reducing the environmental consequences of current protein production processes. Microalgae biorefinery is a bright answer to this problem because it is an excellent protein source (around half of the microalgae dry weight is protein, containing all nine essential amino acids). At the same time, these microorganisms can reduce atmospheric $CO_2$. However, the recovery of proteins from the microalgae matrix is complicated, and it is usually considered the bottleneck of the process. For protein recovery, multiple available techniques allow protein obtaining, and depending on their fundamentals, different purity levels will be attained. All the techniques discussed in this review have different advantages, but they also present some drawbacks. For example, despite precipitation being a straightforward approach, still low yields are obtained, while chromatography provides high efficiencies, but it is challenging to scale up the process. The solvent extraction method is one of the most employed, yet organic solvents go against the sustainability desirable for the process. In this context, ILs, and especially DESs, have been revealed as a greener alternative for protein extraction due to their excellent physicochemical properties that enable an efficient and environmentally friendly recovery.

There is no doubt that biodegradable solvents will be the key to new green processes for obtaining many bioactive compounds. Nevertheless, more efforts are needed to investigate ILs and DESs limitations, and consequently, the ATPS approach. In particular, from our view, forthcoming studies and developments should focus on: (I) preparing more efficient DESs or NADESs to extract protein from microalgae, as they stand as a more environmentally friendly protein source; (II) recovering the proteins sustainably from the solvent, paying attention to how to reuse the solvents employed in further extraction procedures; (III) fully understanding the nature and behavior of DESs and ATPS systems, in order to obtain more selective solvents for the analyte to be separated. Shedding light on these items would provide incredible advances not only for microalgae protein extraction but also in separation science.

**Author Contributions:** Conceptualization, S.B., P.A.G.-E. and B.C.-G.; analysis of the information, investigation, and writing of the paper D.M., E.M.R., S.B., P.A.G.-E. and B.C.-G.; writing of the paper, D.M. and E.M.R.; review and editing S.B., P.A.G.-E. and B.C.-G. All authors have read and agreed to the published version of the manuscript.

**Funding:** This research was funded by MICINN-FEDER (projects PID2020-113544RB-I00 and CTQ2017-84006-C3-1-R). The authors also thank the Regional Government of Castilla y León and the EU-FEDER (CLU 2017-09). Elena M. Rojo would like to thank the "Ministerio de Ciencia, Innovación y Universidades" for her doctorate scholarship (PRE2018-083845). David Moldes would like to thank the "Ministerio de Universidades" for his FPU grant (FPU20/02086).

**Institutional Review Board Statement:** Not applicable.

**Informed Consent Statement:** Not applicable.

**Data Availability Statement:** Not applicable.

**Acknowledgments:** The authors would also thank Laura García for Figure 1 illustration.

**Conflicts of Interest:** The authors declare no conflict of interest.

## References

1. Chou, C.J.; Affolter, M.; Kussmann, M. A Nutrigenomics View of Protein Intake. In *Progress in Molecular Biology and Translational Science*; Elsevier: Amsterdam, The Netherlands, 2012; Volume 108, pp. 51–74.
2. UN Department of Economics and Social Affairs World Population Prospects—Population Division—United Nations. Available online: https://esa.un.org/unpd/wpp/ (accessed on 27 January 2022).
3. Shahid, A.; Malik, S.; Zhu, H.; Xu, J.; Nawaz, M.Z.; Nawaz, S.; Asraful Alam, M.; Mehmood, M.A. Cultivating microalgae in wastewater for biomass production, pollutant removal, and atmospheric carbon mitigation; a review. *Sci. Total Environ.* **2020**, *704*, 135303. [CrossRef] [PubMed]

4. Röös, E.; Bajželj, B.; Smith, P.; Patel, M.; Little, D.; Garnett, T. Protein futures for Western Europe: Potential land use and climate impacts in 2050. *Reg. Environ. Chang.* **2017**, *17*, 367–377. [CrossRef]
5. Naseri, A.; Marinho, G.S.; Holdt, S.L.; Bartela, J.M.; Jacobsen, C. Enzyme-assisted extraction and characterization of protein from red seaweed Palmaria palmata. *Algal Res.* **2020**, *47*, 101849. [CrossRef]
6. Molino, A.; Iovine, A.; Casella, P.; Mehariya, S.; Chianese, S.; Cerbone, A.; Rimauro, J.; Musmarra, D. Microalgae characterization for consolidated and new application in human food, animal feed and nutraceuticals. *Int. J. Environ. Res. Public Health* **2018**, *15*, 2436. [CrossRef]
7. Rojo, E.M.; Piedra, I.; González, A.M.; Vega, M.; Bolado, S. Effect of process parameters on the valorization of components from microalgal and microalgal-bacteria biomass by enzymatic hydrolysis. *Bioresour. Technol.* **2021**, *335*, 125256. [CrossRef]
8. Venkata Mohan, S.; Hemalatha, M.; Chakraborty, D.; Chatterjee, S.; Ranadheer, P.; Kona, R. Algal biorefinery models with self-sustainable closed loop approach: Trends and prospective for blue-bioeconomy. *Bioresour. Technol.* **2020**, *295*, 122128. [CrossRef]
9. Karemore, A.; Sen, R. Downstream processing of microalgal feedstock for lipid and carbohydrate in a biorefinery concept: A holistic approach for biofuel applications†. *RSC Adv.* **2016**, *6*, 29486–29496. [CrossRef]
10. Bhattacharya, M.; Goswami, S. Microalgae—A green multi-product biorefinery for future industrial prospects. *Biocatal. Agric. Biotechnol.* **2020**, *25*, 101580. [CrossRef]
11. Gifuni, I.; Pollio, A.; Safi, C.; Marzocchella, A.; Olivieri, G. Current Bottlenecks and Challenges of the Microalgal Biorefinery. *Trends Biotechnol.* **2019**, *37*, 242–252. [CrossRef]
12. Shahid, A.; Khan, F.; Ahmad, N.; Farooq, M.; Mehmood, M.A. Microalgal carbohydrates and proteins: Synthesis, extraction, applications, and challenges. In *Microalgae Biotechnology for Food, Health and High Value Products*; Springer: Singapore, 2020; pp. 433–468.
13. Dolganyuk, V.; Belova, D.; Babich, O.; Prosekov, A.; Ivanova, S.; Katserov, D.; Patyukov, N.; Sukhikh, S. Microalgae: A promising source of valuable bioproducts. *Biomolecules* **2020**, *10*, 1153. [CrossRef]
14. Córdova, O.; Chamy, R. Microalgae to biogas: Microbiological communities involved. In *Microalgae Cultivation for Biofuels Production*; Academic Press: Cambridge, MA, USA, 2019; pp. 227–249.
15. Nitsos, C.; Filali, R.; Taidi, B.; Lemaire, J. Current and novel approaches to downstream processing of microalgae: A review. *Biotechnol. Adv.* **2020**, *45*, 107650. [CrossRef] [PubMed]
16. Corrêa, P.S.; Júnior, W.G.M.; Martins, A.A.; Caetano, N.S.; Mata, T.M. Microalgae Biomolecules: Extraction, Separation and Purification Methods. *Processes* **2021**, *9*, 10. [CrossRef]
17. Santana-Mayor, Á.; Rodríguez-Ramos, R.; Herrera-Herrera, A.V.; Socas-Rodríguez, B.; Rodríguez-Delgado, M.Á. Deep eutectic solvents. The new generation of green solvents in analytical chemistry. *TrAC Trends Anal. Chem.* **2020**, *134*, 116108. [CrossRef]
18. Chew, K.W.; Yap, J.Y.; Show, P.L.; Suan, N.H.; Juan, J.C.; Ling, T.C.; Lee, D.J.; Chang, J.S. Microalgae biorefinery: High value products perspectives. *Bioresour. Technol.* **2017**, *229*, 53–62. [CrossRef]
19. Grossmann, L.; Hinrichs, J.; Weiss, J. Cultivation and downstream processing of microalgae and cyanobacteria to generate protein-based technofunctional food ingredients. *Crit. Rev. Food Sci. Nutr.* **2020**, *60*, 2961–2989. [CrossRef]
20. Timira, V.; Meki, K.; Li, Z.; Lin, H.; Xu, M.; Pramod, S.N. A comprehensive review on the application of novel disruption techniques for proteins release from microalgae. *Crit. Rev. Food Sci. Nutr.* **2021**, 1–17. [CrossRef]
21. Papachristou, I.; Akaberi, S.; Silve, A.; Navarro-López, E.; Wüstner, R.; Leber, K.; Nazarova, N.; Müller, G.; Frey, W. Analysis of the lipid extraction performance in a cascade process for Scenedesmus almeriensis biorefinery. *Biotechnol. Biofuels* **2021**, *14*, 20. [CrossRef]
22. Eppink, M.H.M.; Ventura, S.P.M.; Coutinho, J.A.P.; Wijffels, R.H. Multiproduct Microalgae Biorefineries Mediated by Ionic Liquids. *Trends Biotechnol.* **2021**, *39*, 1131–1143. [CrossRef]
23. Liu, S.; Li, Z.; Yu, B.; Wang, S.; Shen, Y.; Cong, H. Recent advances on protein separation and purification methods. *Adv. Colloid Interface Sci.* **2020**, *284*, 102254. [CrossRef]
24. Zeng, Q.; Wang, Y.; Huang, Y.; Ding, X.; Chen, J.; Xu, K. Deep eutectic solvents as novel extraction media for protein partitioning. *Analyst* **2014**, *139*, 2565–2573. [CrossRef]
25. Zainal-Abidin, M.H.; Hayyan, M.; Hayyan, A.; Jayakumar, N.S. New horizons in the extraction of bioactive compounds using deep eutectic solvents: A review. *Anal. Chim. Acta* **2017**, *979*, 1–23. [CrossRef] [PubMed]
26. Meng, J.; Wang, Y.; Zhou, Y.; Chen, J.; Wei, X.; Ni, R.; Liu, Z.; Xu, F. Development of different deep eutectic solvent aqueous biphasic systems for the separation of proteins. *RSC Adv.* **2019**, *9*, 14116–14125. [CrossRef]
27. Sequeira, R.A.; Bhatt, J.; Prasad, K. Recent Trends in Processing of Proteins and DNA in Alternative Solvents: A Sustainable Approach. *Sustain. Chem.* **2020**, *1*, 116–137. [CrossRef]
28. Amorim, M.L.; Soares, J.; dos Reis Coimbra, J.S.; de Oliveira Leite, M.; Albino, L.F.T.; Martins, M.A. Microalgae proteins: Production, separation, isolation, quantification, and application in food and feed. *Crit. Rev. Food Sci. Nutr.* **2020**, *61*, 1976–2002. [CrossRef]
29. Skoog, D.A.; Holler, F.J.; Crouch, S.R. *Principles of Instrumental Analysis*; Cengage Learning: Boston, MA, USA, 2017.
30. Phan, N.V.H.; Sussitz, H.F.; Ladenhauf, E.; Pum, D.; Lieberzeit, P.A. Combined layer/particle approaches in surface molecular imprinting of proteins: Signal enhancement and competition. *Sensors* **2018**, *18*, 180. [CrossRef]

31. Farzi-Khajeh, H.; Safa, K.D.; Dastmalchi, S. Arsanilic acid modified superparamagnetic iron oxide nanoparticles for Purification of alkaline phosphatase from hen's egg yolk. *J. Chromatogr. B* **2017**, *1061*, 26–33. [CrossRef]
32. Zhu, H.-G.; Tang, H.-Q.; Cheng, Y.-Q.; Li, Z.-G.; Tong, L.-T. Electrostatic separation technology for obtaining plant protein concentrates: A review. *Trends Food Sci. Technol.* **2021**, *113*, 66–76. [CrossRef]
33. Bertsch, P.; Böcker, L.; Mathys, A.; Fischer, P. Proteins from microalgae for the stabilization of fluid interfaces, emulsions, and foams. *Trends Food Sci. Technol.* **2021**, *108*, 326–342. [CrossRef]
34. Böcker, L.; Bertsch, P.; Wenner, D.; Teixeira, S.; Bergfreund, J.; Eder, S.; Fischer, P.; Mathys, A. Effect of *Arthrospira platensis* microalgae protein purification on emulsification mechanism and efficiency. *J. Colloid Interface Sci.* **2021**, *584*, 344–353. [CrossRef]
35. Pereira, A.M.; Lisboa, C.R.; Costa, J.A.V. High protein ingredients of microalgal origin: Obtainment and functional properties. *Innov. Food Sci. Emerg. Technol.* **2018**, *47*, 187–194. [CrossRef]
36. Pylaeva, S.; Brehm, M.; Sebastiani, D. Salt bridge in aqueous solution: Strong structural motifs but weak enthalpic effect. *Sci. Rep.* **2018**, *8*, 13626. [CrossRef]
37. Barbarino, E.; Lourenço, S.O. An evaluation of methods for extraction and quantification of protein from marine macro-and microalgae. *J. Appl. Phycol.* **2005**, *17*, 447–460. [CrossRef]
38. Saxena, A.; Tripathi, B.P.; Kumar, M.; Shahi, V.K. Membrane-based techniques for the separation and purification of proteins: An overview. *Adv. Colloid Interface Sci.* **2009**, *145*, 1–22. [CrossRef] [PubMed]
39. Yang, F.; Yang, P. Protein-Based Separation Membranes: State of the Art and Future Trends. *Adv. Energy Sustain. Res.* **2021**, *2*, 2100008. [CrossRef]
40. Zeman, L.J.; Zydney, A.L. *Microfiltration and Ultrafiltration: Principles and Applications*; CRC Press: Boca Raton, FL, USA, 2017.
41. Liu, S.; Gifuni, I.; Mear, H.; Frappart, M.; Couallier, E. Recovery of soluble proteins from Chlorella vulgaris by bead-milling and microfiltration: Impact of the concentration and the physicochemical conditions during the cell disruption on the whole process. *Process Biochem.* **2021**, *108*, 34–47. [CrossRef]
42. Zhang, Y.; Fu, Q. Algal fouling of microfiltration and ultrafiltration membranes and control strategies: A review. *Sep. Purif. Technol.* **2018**, *203*, 193–208. [CrossRef]
43. Zhang, Z.; Li, L. Efficient synthesis of molecularly imprinted polymers with bio-recognition sites for the selective separation of bovine hemoglobin. *J. Sep. Sci.* **2018**, *41*, 2479–2487. [CrossRef]
44. Md Saleh, N.I.; Wan Ab Karim Ghani, W.A.; Mustapa Kamal, S.M.; Harun, R. Performance of Single and Two-Stage Cross-Flow Ultrafiltration Membrane in Fractionation of Peptide from Microalgae Protein Hydrolysate (*Nannochloropsis gaditana*). *Processes* **2021**, *9*, 610. [CrossRef]
45. Balti, R.; Zayoud, N.; Hubert, F.; Beaulieu, L.; Massé, A. Fractionation of *Arthrospira platensis* (Spirulina) water soluble proteins by membrane diafiltration. *Sep. Purif. Technol.* **2021**, *256*, 117756. [CrossRef]
46. Vidya, B.; Palaniswamy, M.; Angayarkanni, J.; Nawaz, K.A.; Thandeeswaran, M.; Chaithanya, K.K.; Tekluu, B.; Muthusamy, K.; Gopalakrishnan, V.K. Purification and characterization of β-galactosidase from newly isolated Aspergillus terreus (KUBCF1306) and evaluating its efficacy on breast cancer cell line (MCF-7). *Bioorg. Chem.* **2020**, *94*, 103442. [CrossRef]
47. Ta, H.Y.; Collin, F.; Perquis, L.; Poinsot, V.; Ong-Meang, V.; Couderc, F. Twenty years of amino acid determination using capillary electrophoresis: A review. *Anal. Chim. Acta* **2021**, *1174*, 338233. [CrossRef] [PubMed]
48. Lundanes, E.; Reubsaet, L.; Greibrokk, T. *Chromatography: Basic Principles, Sample Preparations and Related Methods*; John Wiley & Sons: Hoboken, NJ, USA, 2013.
49. Wang, A.; Islam, M.N.; Qin, X.; Wang, H.; Peng, Y.; Ma, C. Purification, identification, and characterization of d-galactose-6-sulfurylase from marine algae (*Betaphycus gelatinus*). *Carbohydr. Res.* **2014**, *388*, 94–99. [CrossRef] [PubMed]
50. Tang, Z.; Zhao, J.; Ju, B.; Li, W.; Wen, S.; Pu, Y.; Qin, S. One-step chromatographic procedure for purification of B-phycoerythrin from *Porphyridium cruentum*. *Protein Expr. Purif.* **2016**, *123*, 70–74. [CrossRef]
51. Staby, A.; Rathore, A.S.; Ahuja, S. *Preparative Chromatography for Separation of Proteins*; John Wiley & Sons: Hoboken, NJ, USA, 2017.
52. Przybycien, T.M.; Pujar, N.S.; Steele, L.M. Alternative bioseparation operations: Life beyond packed-bed chromatography. *Curr. Opin. Biotechnol.* **2004**, *15*, 469–478. [CrossRef]
53. Kress, C.; Sadowski, G.; Brandenbusch, C. Solubilization of proteins in aqueous two-phase extraction through combinations of phase-formers and displacement agents. *Eur. J. Pharm. Biopharm.* **2017**, *112*, 38–44. [CrossRef] [PubMed]
54. Rice, N.M.; Irving, H.; Leonard, M.A. Nomenclature for liquid-liquid distribution (solvent extraction) (IUPAC Recommendations 1993). *Pure Appl. Chem.* **1993**, *65*, 2373–2396. [CrossRef]
55. Kumar, M.; Tomar, M.; Potkule, J.; Verma, R.; Punia, S.; Mahapatra, A.; Belwal, T.; Dahuja, A.; Joshi, S.; Berwal, M.K.; et al. Advances in the plant protein extraction: Mechanism and recommendations. *Food Hydrocoll.* **2021**, 106595. [CrossRef]
56. Chemat, F.; Abert Vian, M.; Ravi, H.K.; Khadhraoui, B.; Hilali, S.; Perino, S.; Fabiano Tixier, A.-S. Review of alternative solvents for green extraction of food and natural products: Panorama, principles, applications and prospects. *Molecules* **2019**, *24*, 3007. [CrossRef]
57. Cravotto, G.; Boffa, L.; Mantegna, S.; Perego, P.; Avogadro, M.; Cintas, P. Improved extraction of vegetable oils under high-intensity ultrasound and/or microwaves. *Ultrason. Sonochem.* **2008**, *15*, 898–902. [CrossRef]
58. Chemat, F.; Vian, M.A.; Cravotto, G. Green extraction of natural products: Concept and principles. *Int. J. Mol. Sci.* **2012**, *13*, 8615–8627. [CrossRef]

59. Phong, W.N.; Show, P.L.; Teh, W.H.; Teh, T.X.; Lim, H.M.Y.; binti Nazri, N.S.; Tan, C.H.; Chang, J.-S.; Ling, T.C. Proteins recovery from wet microalgae using liquid biphasic flotation (LBF). *Bioresour. Technol.* **2017**, *244*, 1329–1336. [CrossRef]
60. Clarke, C.J.; Tu, W.-C.; Levers, O.; Brohl, A.; Hallett, J.P. Green and sustainable solvents in chemical processes. *Chem. Rev.* **2018**, *118*, 747–800. [CrossRef] [PubMed]
61. Chen, Y.; Mu, T. Revisiting greenness of ionic liquids and deep eutectic solvents. *Green Chem. Eng.* **2021**, *2*, 174–186. [CrossRef]
62. Bubalo, M.C.; Vidović, S.; Redovniković, I.R.; Jokić, S. New perspective in extraction of plant biologically active compounds by green solvents. *Food Bioprod. Process.* **2018**, *109*, 52–73. [CrossRef]
63. Chowdhury, S.; Rakshit, A.; Acharjee, A.; Saha, B. Biodegradability and biocompatibility: Advancements in synthetic surfactants. *J. Mol. Liq.* **2021**, *324*, 115105. [CrossRef]
64. Brown, D.M.; Lyon, D.; Saunders, D.M.V.; Hughes, C.B.; Wheeler, J.R.; Shen, H.; Whale, G. Biodegradability assessment of complex, hydrophobic substances: Insights from gas-to-liquid (GTL) fuel and solvent testing. *Sci. Total Environ.* **2020**, *727*, 138528. [CrossRef]
65. Asenjo, J.A.; Andrews, B.A. Aqueous two-phase systems for protein separation: Phase separation and applications. *J. Chromatogr. A* **2012**, *1238*, 1–10. [CrossRef]
66. Hatti-Kaul, R. *Aqueous Two-Phase Systems: Methods and Protocols*; Springer Science & Business Media: Berlin, Germany, 2000; Volume 11.
67. Pereira, J.F.B.; Freire, M.G.; Coutinho, J.A.P. Aqueous two-phase systems: Towards novel and more disruptive applications. *Fluid Phase Equilib.* **2020**, *505*, 112341. [CrossRef]
68. Xie, Y.; Xing, H.; Yang, Q.; Bao, Z.; Su, B.; Ren, Q. Aqueous Biphasic System Containing Long Chain Anion-Functionalized Ionic Liquids for High-Performance Extraction. *ACS Sustain. Chem. Eng.* **2015**, *3*, 3365–3372. [CrossRef]
69. Schuur, B.; Brouwer, T.; Smink, D.; Sprakel, L.M.J. Green solvents for sustainable separation processes. *Curr. Opin. Green Sustain. Chem.* **2019**, *18*, 57–65. [CrossRef]
70. Baskaran, D.; Chinnappan, K.; Manivasagan, R.; Mahadevan, D.K. Partitioning of crude protein from aqua waste using PEG 600-inorganic salt Aqueous Two-Phase Systems. *Chem. Data Collect.* **2018**, *15*, 143–152. [CrossRef]
71. Chen, X.; Liu, J.; Wang, J. Ionic liquids in the assay of proteins. *Anal. Methods* **2010**, *2*, 1222–1226. [CrossRef]
72. Benvenutti, L.; Zielinski, A.A.F.; Ferreira, S.R.S. Which is the best food emerging solvent: IL, DES or NADES? *Trends Food Sci. Technol.* **2019**, *90*, 133–146. [CrossRef]
73. Choi, Y.H.; Verpoorte, R. Green solvents for the extraction of bioactive compounds from natural products using ionic liquids and deep eutectic solvents. *Curr. Opin. Food Sci.* **2019**, *26*, 87–93. [CrossRef]
74. Seddon, K.R. Ionic liquids for clean technology. *J. Chem. Technol. Biotechnol.* **1997**, *68*, 351–356. [CrossRef]
75. Cvjetko Bubalo, M.; Vidović, S.; Radojčić Redovniković, I.; Jokić, S. Green solvents for green technologies. *J. Chem. Technol. Biotechnol.* **2015**, *90*, 1631–1639. [CrossRef]
76. Passos, H.; Freire, M.G.; Coutinho, J.A.P. Ionic liquid solutions as extractive solvents for value-added compounds from biomass. *Green Chem.* **2014**, *16*, 4786–4815. [CrossRef]
77. Nasirpour, N.; Mohammadpourfard, M.; Zeinali Heris, S. Ionic liquids: Promising compounds for sustainable chemical processes and applications. *Chem. Eng. Res. Des.* **2020**, *160*, 264–300. [CrossRef]
78. Bubalo, M.C.; Radošević, K.; Redovniković, I.R.; Halambek, J.; Srček, V.G. A brief overview of the potential environmental hazards of ionic liquids. *Ecotoxicol. Environ. Saf.* **2014**, *99*, 1–12. [CrossRef]
79. Patel, R.; Kumari, M.; Khan, A.B. Recent advances in the applications of ionic liquids in protein stability and activity: A review. *Appl. Biochem. Biotechnol.* **2014**, *172*, 3701–3720. [CrossRef]
80. Xiao, J.; Chen, G.; Li, N. Ionic liquid solutions as a green tool for the extraction and isolation of natural products. *Molecules* **2018**, *23*, 1765. [CrossRef] [PubMed]
81. Rodrigues, R.D.P.; de Castro, F.C.; de Santiago-Aguiar, R.S.; Rocha, M.V.P. Ultrasound-assisted extraction of phycobiliproteins from Spirulina (Arthrospira) platensis using protic ionic liquids as solvent. *Algal Res.* **2018**, *31*, 454–462. [CrossRef]
82. Desai, R.K.; Monteillet, H.; Li, X.; Schuur, B.; Kleijn, J.M.; Leermakers, F.A.M.; Wijffels, R.H.; Eppink, M.H.M. One-step mild biorefinery of functional biomolecules from microalgae extracts. *React. Chem. Eng.* **2018**, *3*, 182–187. [CrossRef]
83. Lee, S.Y.; Show, P.L.; Ling, T.C.; Chang, J.-S. Single-step disruption and protein recovery from Chlorella vulgaris using ultrasonication and ionic liquid buffer aqueous solutions as extractive solvents. *Biochem. Eng. J.* **2017**, *124*, 26–35. [CrossRef]
84. Desai, R.K.; Fernandez, M.S.; Wijffels, R.H.; Eppink, M.H.M. Mild fractionation of hydrophilic and hydrophobic components from Neochloris oleoabundans using ionic liquids. *Front. Bioeng. Biotechnol.* **2019**, *7*, 284. [CrossRef]
85. Zeng, Q.; Wang, Y.; Li, N.; Huang, X.; Ding, X.; Lin, X.; Huang, S.; Liu, X. Extraction of proteins with ionic liquid aqueous two-phase system based on guanidine ionic liquid. *Talanta* **2013**, *116*, 409–416. [CrossRef]
86. Suarez Ruiz, C.A.; Kwaijtaal, J.; Peinado, O.C.; Van Den Berg, C.; Wijffels, R.H.; Eppink, M.H.M. Multistep Fractionation of Microalgal Biomolecules Using Selective Aqueous Two-Phase Systems. *ACS Sustain. Chem. Eng.* **2020**, *8*, 2441–2452. [CrossRef]
87. Santos, J.H.P.M.; Trigo, J.P.; Maricato, E.; Nunes, C.; Coimbra, M.A.; Ventura, S.P.M. Fractionation of isochrysis galbana proteins, arabinans, and glucans using ionic-liquid-based aqueous biphasic systems. *ACS Sustain. Chem. Eng.* **2018**, *6*, 14042–14053. [CrossRef]

88. Suarez Garcia, E.; Suarez Ruiz, C.A.; Tilaye, T.; Eppink, M.H.M.; Wijffels, R.H.; van den Berg, C. Fractionation of proteins and carbohydrates from crude microalgae extracts using an ionic liquid based-aqueous two phase system. *Sep. Purif. Technol.* **2018**, *204*, 56–65. [CrossRef]
89. Durand, E.; Lecomte, J.; Villeneuve, P. Are emerging deep eutectic solvents (DES) relevant for lipase-catalyzed lipophilizations? *OCL* **2015**, *22*, D408. [CrossRef]
90. Zhao, W.; Chi, X.; Li, H.; He, J.; Long, J.; Xu, Y.; Yang, S. Eco-friendly acetylcholine-carboxylate bio-ionic liquids for controllable N-methylation and N-formylation using ambient $CO_2$ at low temperatures. *Green Chem.* **2019**, *21*, 567–577. [CrossRef]
91. Pham, T.P.T.; Cho, C.-W.; Yun, Y.-S. Environmental fate and toxicity of ionic liquids: A review. *Water Res.* **2010**, *44*, 352–372. [CrossRef] [PubMed]
92. Petkovic, M.; Seddon, K.R.; Rebelo, L.P.N.; Pereira, C.S. Ionic liquids: A pathway to environmental acceptability. *Chem. Soc. Rev.* **2011**, *40*, 1383–1403. [CrossRef] [PubMed]
93. Francisco, M.; van den Bruinhorst, A.; Kroon, M.C. New natural and renewable low transition temperature mixtures (LTTMs): Screening as solvents for lignocellulosic biomass processing. *Green Chem.* **2012**, *14*, 2153–2157. [CrossRef]
94. Ratti, R. Ionic liquids: Synthesis and applications in catalysis. *Adv. Chem.* **2014**, *2014*, 729842. [CrossRef]
95. Abbott, A.P.; Capper, G.; Davies, D.L.; Rasheed, R.K.; Tambyrajah, V. Novel solvent properties of choline chloride/urea mixtures. *Chem. Commun.* **2003**, 70–71. [CrossRef] [PubMed]
96. Martins, M.A.R.; Pinho, S.P.; Coutinho, J.A.P. Insights into the nature of eutectic and deep eutectic mixtures. *J. Solution Chem.* **2019**, *48*, 962–982. [CrossRef]
97. Tang, S.; Baker, G.A.; Zhao, H. Ether- and alcohol-functionalized task-specific ionic liquids: Attractive properties and applications. *Chem. Soc. Rev.* **2012**, *41*, 4030–4066. [CrossRef]
98. Zhang, Q.; Vigier, K.D.O.; Royer, S.; Jerome, F. Deep eutectic solvents: Syntheses, properties and applications. *Chem. Soc. Rev.* **2012**, *41*, 7108–7146. [CrossRef]
99. Abbott, A.P.; Boothby, D.; Capper, G.; Davies, D.L.; Rasheed, R.K. Deep eutectic solvents formed between choline chloride and carboxylic acids: Versatile alternatives to ionic liquids. *J. Am. Chem. Soc.* **2004**, *126*, 9142–9147. [CrossRef]
100. Huang, J.; Guo, X.; Xu, T.; Fan, L.; Zhou, X.; Wu, S. Ionic deep eutectic solvents for the extraction and separation of natural products. *J. Chromatogr. A* **2019**, *1598*, 1–19. [CrossRef] [PubMed]
101. Beyersdorff, T.; Schubert, T.J.S.; Welz-Biermann, U.; Pitner, W.; Abbott, A.P.; McKenzie, K.J.; Ryder, S. Chapter 2: Synthesis of Ionic Liquids. In *Electrodeposition from Ionic Liquids*; Wiley: Weinheim, Germany, 2017; pp. 17–53.
102. Smith, E.L.; Abbott, A.P.; Ryder, K.S. Deep eutectic solvents (DESs) and their applications. *Chem. Rev.* **2014**, *114*, 11060–11082. [CrossRef] [PubMed]
103. Abranches, D.O.; Martins, M.A.R.; Silva, L.P.; Schaeffer, N.; Pinho, S.P.; Coutinho, J.A.P. Phenolic hydrogen bond donors in the formation of non-ionic deep eutectic solvents: The quest for type v des. *Chem. Commun.* **2019**, *55*, 10253–10256. [CrossRef] [PubMed]
104. Hansen, B.B.; Spittle, S.; Chen, B.; Poe, D.; Zhang, Y.; Klein, J.M.; Horton, A.; Adhikari, L.; Zelovich, T.; Doherty, B.W.; et al. Deep eutectic solvents: A review of fundamentals and applications. *Chem. Rev.* **2020**, *121*, 1232–1285. [CrossRef] [PubMed]
105. Özel, N.; Elibol, M. A review on the potential uses of deep eutectic solvents in chitin and chitosan related processes. *Carbohydr. Polym.* **2021**, *262*, 117942. [CrossRef] [PubMed]
106. Farooq, M.Q.; Abbasi, N.M.; Anderson, J.L. Deep eutectic solvents in separations: Methods of preparation, polarity, and applications in extractions and capillary electrochromatography. *J. Chromatogr. A* **2020**, *1633*, 461613. [CrossRef]
107. Zhao, B.-Y.; Xu, P.; Yang, F.-X.; Wu, H.; Zong, M.-H.; Lou, W.-Y. Biocompatible deep eutectic solvents based on choline chloride: Characterization and application to the extraction of rutin from Sophora japonica. *ACS Sustain. Chem. Eng.* **2015**, *3*, 2746–2755. [CrossRef]
108. Florindo, C.; Lima, F.; Ribeiro, B.D.; Marrucho, I.M. Deep eutectic solvents: Overcoming 21st century challenges. *Curr. Opin. Green Sustain. Chem.* **2019**, *18*, 31–36. [CrossRef]
109. Choi, Y.H.; van Spronsen, J.; Dai, Y.; Verberne, M.; Hollmann, F.; Arends, I.W.C.E.; Witkamp, G.-J.; Verpoorte, R. Are natural deep eutectic solvents the missing link in understanding cellular metabolism and physiology? *Plant Physiol.* **2011**, *156*, 1701–1705. [CrossRef]
110. Huang, Y.; Feng, F.; Jiang, J.; Qiao, Y.; Wu, T.; Voglmeir, J.; Chen, Z.-G. Green and efficient extraction of rutin from tartary buckwheat hull by using natural deep eutectic solvents. *Food Chem.* **2017**, *221*, 1400–1405. [CrossRef]
111. Mehariya, S.; Fratini, F.; Lavecchia, R.; Zuorro, A. Green extraction of value-added compounds form microalgae: A short review on natural deep eutectic solvents (NaDES) and related pre-treatments. *J. Environ. Chem. Eng.* **2021**, *9*, 105989. [CrossRef]
112. Dai, Y.; van Spronsen, J.; Witkamp, G.-J.; Verpoorte, R.; Choi, Y.H. Natural deep eutectic solvents as new potential media for green technology. *Anal. Chim. Acta* **2013**, *766*, 61–68. [CrossRef]
113. Zhang, H.; Wang, Y.; Xu, K.; Li, N.; Wen, Q.; Yang, Q.; Zhou, Y. Ternary and binary deep eutectic solvents as a novel extraction medium for protein partitioning. *Anal. Methods* **2016**, *8*, 8196–8207. [CrossRef]
114. Grudniewska, A.; de Melo, E.M.; Chan, A.; Gniłka, R.; Boratynski, F.; Matharu, A.S. Enhanced protein extraction from oilseed cakes using glycerol–choline chloride deep eutectic solvents: A biorefinery approach. *ACS Sustain. Chem. Eng.* **2018**, *6*, 15791–15800. [CrossRef]

115. Bai, C.; Wei, Q.; Ren, X. Selective extraction of collagen peptides with high purity from cod skins by deep eutectic solvents. *ACS Sustain. Chem. Eng.* **2017**, *5*, 7220–7227. [CrossRef]
116. Xu, K.; Wang, Y.; Huang, Y.; Li, N.; Wen, Q. A green deep eutectic solvent-based aqueous two-phase system for protein extracting. *Anal. Chim. Acta* **2015**, *864*, 9–20. [CrossRef] [PubMed]
117. Li, N.; Wang, Y.; Xu, K.; Huang, Y.; Wen, Q.; Ding, X. Development of green betaine-based deep eutectic solvent aqueous two-phase system for the extraction of protein. *Talanta* **2016**, *152*, 23–32. [CrossRef] [PubMed]
118. Cicci, A.; Sed, G.; Bravi, M. Potential of choline chloride–based natural deep eutectic solvents (NaDES) in the extraction of microalgal metabolites. *Chem. Eng. Trans.* **2017**, *57*, 61–66.
119. Sed, G.; Cicci, A.; Jessop, P.G.; Bravi, M. A novel switchable-hydrophilicity, natural deep eutectic solvent (NaDES)-based system for bio-safe biorefinery. *RSC Adv.* **2018**, *8*, 37092–37097. [CrossRef]
120. Lee, S.Y.; Khoiroh, I.; Ling, T.C.; Show, P.L. Enhanced recovery of lipase derived from Burkholderia cepacia from fermentation broth using recyclable ionic liquid/polymer-based aqueous two-phase systems. *Sep. Purif. Technol.* **2017**, *179*, 152–160. [CrossRef]
121. Dai, Y.; Van Spronsen, J.; Witkamp, G.J.; Verpoorte, R.; Choi, Y.H. Ionic liquids and deep eutectic solvents in natural products research: Mixtures of solids as extraction solvents. *J. Nat. Prod.* **2013**, *76*, 2162–2173. [CrossRef] [PubMed]
122. Cláudio, A.F.M.; Ferreira, A.M.; Freire, M.G.; Coutinho, J.A.P. Enhanced extraction of caffeine from guarana seeds using aqueous solutions of ionic liquids. *Green Chem.* **2013**, *15*, 2002–2010. [CrossRef]
123. Orr, V.C.A.; Plechkova, N.V.; Seddon, K.R.; Rehmann, L. Disruption and wet extraction of the microalgae Chlorella vulgaris using room-temperature ionic liquids. *ACS Sustain. Chem. Eng.* **2016**, *4*, 591–600. [CrossRef]
124. Gullón, P.; Gullón, B.; Romaní, A.; Rocchetti, G.; Lorenzo, J.M. Smart advanced solvents for bioactive compounds recovery from agri-food by-products: A review. *Trends Food Sci. Technol.* **2020**, *101*, 182–197. [CrossRef]

MDPI
St. Alban-Anlage 66
4052 Basel
Switzerland
Tel. +41 61 683 77 34
Fax +41 61 302 89 18
www.mdpi.com

*Applied Sciences* Editorial Office
E-mail: applsci@mdpi.com
www.mdpi.com/journal/applsci

www.ingramcontent.com/pod-product-compliance
Lightning Source LLC
Chambersburg PA
CBHW040809250926
51850CB00007BA/1124
* 9 7 8 3 0 3 6 5 6 2 8 5 8 *